AF539274

Salt, Shale and Igneous Diapirs in and around Europe

Special Publication reviewing procedures

The Society makes every effort to ensure that the scientific and production quality of its books matches that of its journals. Since 1997, all book proposals have been refereed by specialist reviewers as well as by the Society's Publications Committee. If the referees identify weaknesses in the proposal, these must be addressed before the proposal is accepted.

Once the book is accepted, the Society has a team of series editors (listed above) who ensure that the volume editors follow strict guidelines on refereeing and quality control. We insist that individual papers can only be accepted after satisfactory review by two independent referees. The questions on the review forms are similar to those for *Journal of the Geological Society*. The referees' forms and comments must be available to the Society's series editors on request.

Although many of the books result from meetings, the editors are expected to commission papers that were not presented at the meeting to ensure that the book provides a balanced coverage of the subject. Being accepted for presentation at the meeting does not guarantee inclusion in the book.

Geological Society Special Publications are included in the ISI Science Citation Index, but they do not have an impact factor, the latter being applicable only to journals.

More information about submitting a proposal and producing a Special Publication can be found on the Society's web site: www.geolsoc.org.uk.

It is recommended that reference to all or part of this book should be made in one of the following ways.

VENDEVILLE, B., MART, Y. & VIGNERESSE, J.-L. (eds) 2000. *Salt, Shale and Igneous Diapirs in and around Europe*. Geological Society, London, Special Publications, **174**.

STEPHANESCU, M., DICEA, O. & TARI, G. 2000. Influence of extension and compression on salt diapirism in its type area, East Carphathian Bend area, Romania. *In*: VENDEVILLE, B., MART, Y. & VIGNERESSE, J.-L. (eds) *Salt, Shale and Igneous Diapirs in and around Europe*. Geological Society, London, Special Publications, **174**, 131–147.

GEOLOGICAL SOCIETY SPECIAL PUBLICATION NO. 174

Salt, Shale and Igneous Diapirs in and around Europe

EDITED BY

BRUNO C. VENDEVILLE
University of Texas, USA

YOSSI MART
University of Haifa, Israel

and

JEAN-LOUIS VIGNERESSE
Université Nancy, France

2000
Published by
The Geological Society
London

THE GEOLOGICAL SOCIETY

The Geological Society of London was founded in 1807 and is the oldest geological society in the world. It received its Royal Charter in 1825 for the purpose of 'investigating the mineral structure of the Earth' and is now Britain's national society for geology.

Both a learned society and a professional body, the Geological Society is recognized by the Department of Trade and Industry (DTI) as the chartering authority for geoscience, able to award Chartered Geologist status upon appropriately qualified Fellows. The Society has a membership of 8600, of whom about 1500 live outside the UK.

Fellowship of the Society is open to persons holding a recognized honours degree in geology or a cognate subject and who have at least two years' relevant postgraduate experience, or not less than six years' relevant experience in geology or a cognate subject. A Fellow with a minimum of five years' relevant postgraduate experience in the practice of geology may apply for chartered status. Successful applicants are entitled to use the designatory postnominal CGeol (Chartered Geologist). Fellows of the Society may use the letters FGS. Other grades of membership are available to members not yet qualifying for Fellowship.

The Society has its own Publishing House based in Bath, UK. It produces the Society's international journals, books and maps, and is the European distributor for publications of the American Association of Petroleum Geologists (AAPG), the Society for Sedimentary Geology (SEPM) and the Geological Society of America (GSA). Members of the Society can buy books at considerable discounts. The Publishing House has an online bookshop (*http://bookshop.geolsoc.org.uk*).

Further information on Society membership may be obtained from the Membership Services Manager, The Geological Society, Burlington House, Piccadilly, London W1V 0JU (Email: *enquiries@geolsoc.org.uk*; tel: +44 (0)171 434 9944).

The Society's Web Site can be found at *http://www.geolsoc.org.uk/*.The Society is a Registered Charity. number 210161.

Published by The Geological Society from:
The Geological Society Publishing House
Unit 7, Brassmill Enterprise Centre
Brassmill Lane
Bath BA1 3JN, UK
(*Orders*: Tel. +44 (0)1225 445046
Fax +44 (0)1225 442836)
Online bookshop: *http://bookshop.geolsoc.org.uk*

First published 1999

British Library Cataloguing in Publication Data
A catalogue record for this book is available from the British Library.

ISBN 1-86239-066-5
ISSN 0305-8719

Typeset by Wyvern 21, Bristol, UK

Printed by the Alden Group, Oxford, UK

Distributors

USA
AAPG Bookstore
PO Box 979
Tulsa
OK 74101-0979
USA
Orders: Tel. +1 918 584-2555
Fax +1 918 560-2652
Email *bookstore@aapg.org*

Australia
Australian Mineral Foundation Bookshop
63 Conyngham Street
Glenside
South Australia 5065
Australia
Orders: Tel. +61 88 379-0444
Fax +61 88 379-4634
Email *bookshop@amf.com.au*

India
Affiliated East-West Press PVT Ltd
G-1/16 Ansari Road, Daryaganj,
New Delhi 110 002
India
Orders: Tel. +91 11 327-9113
Fax +91 11 326-0538

Japan
Kanda Book Trading Co.
Cityhouse Tama 204
Tsurumaki 1-3-10
Tama-shi
Tokyo 206-0034
Japan
Orders: Tel. +81 (0)423 57-7650
Fax +81 (0)423 57-7651

Contents

Preface

Despite their differences in size and scale, shale, salt and igneous intrusions display many similarities. For example, at the time of emplacement, the material forming all three types of intrusions (viscous or not) is always much weaker than the country rocks. Units of halite, granitic magma, or overpressured shale cannot sustain large deviatoric stresses without deforming, which makes them highly mobile rocks, provided there is space available into which the intrusive material can flow. All intrusions pierce, or appear to pierce, surrounding country rocks having a wide variety of lithologies. All three types of intrusions have experienced much more internal strain and vertical movement than the country rocks. All have risen, at least partly, in the upper continental crust. Another similarity these three types of intrusion have is that their formation has traditionally been attributed to the same geological process, Rayleigh–Taylor instabilities. This is the process in which a weak, viscous, less-dense material (the intrusion) can spontaneously rise through and deform denser country rocks that are assumed to behave viscously and to have negligible yield strength. The latter assumption has been challenged by rock-mechanics data and geological observations. Apart from magmas, evaporites, and overpressured shales, most rocks in the upper continental crust exhibit a rheological behaviour controlled by elastic and frictional properties. The strength of the country rocks is typically considerably higher than (1) the strength of the intrusive rocks and (2) the deviatoric stress generated by small instabilities at the interface between the intrusion's source layer and the overlying rocks.

The three types of intrusions also differ. Evaporitic intrusions are fed typically by a laterally continuous layer, and the rheological properties of the intrusive material do not change drastically throughout the time of emplacement. In contrast, the material forming mud volcanoes and igneous intrusions does not originate from a continuous layer but is fed by more locally restricted sources, where thermal and mineralogical conditions make the source rocks weak and mobile. Both magma and overpressured shale tend to gain strength during their ascent, as the fluid pressure and temperature decrease.

Most recent literature suggests that researchers working in all three fields no longer support the traditional Rayleigh–Taylor instability mechanism and now favour a mechanism in which the weak, intrusive material rises through much stronger country rocks. Typically some amount of tectonic deformation is required to overcome the strength of the country rocks. For example, normal faulting caused by regional extension thins and weakens the country rocks and creates the accommodation space in which intrusions can rise. Regional contraction or transtension squeezes the weak, mobile intrusive material and thereby provides the additional pressure required for the intrusion to lift and deform the overlying country rocks.

Most of the articles herein illustrate how emplacement of intrusions is associated with regional or local tectonics. **Vigneresse & Clement** discuss arguments in favour of various mechanisms of granitic magma emplacement and emphasizes the role of regional tectonics in allowing igneous intrusions to rise. **Román-Berdiel *et al.*** illustrate, by means of experimental models and field observations in the Spanish Variscan belt, how transtensional tectonics can control the mode of emplacement of granitic plutons. **Merle & Donnadieu** address the combined influence of thin-skinned (gravity-driven) and thick-skinned (tectonic) deformation on the geometry and kinematics of faulting above magmatic intrusions. **Rossetti *et al.*** demonstrate how regional, post-orogenic extension has controlled the emplacement of monzogranitic intrusions in the northern Tyrrhenian region. **Rabinowitz & Mart** present seismic tomography data from the Dead Sea rift that suggest the presence of a magmatic intrusion at depth caused by transtension. **Talbot *et al.*** provide new data on the rates at which salt diapirs in the Zagros region rise and spread. **Gaullier *et al.*** illustrate the different styles of salt structures in and around the Nile deep-sea fan, where regional thick-skinned tectonics interact with thin-skinned, gravity-driven tectonics. **Stephanescu *et al.*** present seismic-reflection data from Romania, the area where the term *diapir* was first defined, and demonstrate how regional tectonics has triggered and controlled salt-diapir rise and evolution. **Miralles *et al.*** describe the internal geometry of evaporitic diapirs in the south Pyrenees. The last article, by **Kopf & Behrmann**, shows that mud diapirism in the Mediterranean Ridge was a rapid but episodic process related to tectonic-plate convergence.

We hope that the several examples provided in this publication will help future studies focus on the role of regional tectonics in controlling intrusion emplacement, a process that was previously thought to depend solely on the internal properties of source and country rocks. We also hope that the many similar characteristics of various types of intrusions illustrated herein will provide an impetus for further, multidisciplinary approaches in all three fields.

The content of this Special Publication was derived from a symposium entitled 'From the Arctic to the Mediterranean: Salt, Shale and Igneous Diapirs in and around Europe'. The symposium was convened in 1998 by Yossi Mart and Bruno Vendeville at the 23rd General Assembly of the European Geophysical Society in Nice, France.

We would like to thank Angharad Hills and Bob Holdsworth for their help in organizing this book and for their patience. We also thank A. W. Bally, K. Benn, B. C. Burchfield, J. P. Burg, A. Castro, A. Cruden, G. Eisenstadt, D. Grujic, M. P. A. Jackson, F. Kockel, R. Nelson, K. T. Nilsen, F. Odonne, M. G. Rowan, M. de Saint-Blanquat, H. Schmeling, D. D. Schultz-Ela, W. M. Schwerdtner, J. Simmons, R. Weijermars, and other anonymous reviewers for their careful reading of the manuscripts.

Bruno Vendeville
Yossi Mart
Jean-Louis Vigneresse

Granitic magma ascent and emplacement: neither diapirism nor neutral buoyancy

J. L. VIGNERESSE[1] & J. D. CLEMENS[2]

[1]*CREGU, UMR CNRS 7566 G2R, BP 23, 54501 Vandoeuvre Cédex, France (e-mail jean-louis.vigneresse@g2r.u-nancy.fr*

[2]*School of Geological Sciences, CEESR, Kingston University, Penrhyn Road, Kingston-upon-Thames, KT1 2EE, UK*

Abstract: It is probable that granitic magma ascent does not result from the intrinsic properties of the magmas. Within the uppermost crust, neither the reduced viscosity nor the density contrast between magma and surroundings are themselves sufficient to induce either low-inertia flow (diapirism) or fracture-induced magma propagation (dyking). Igneous diapirism is intrinsically restricted to the lower, ductile crust. Dyking is therefore the most probable ascent mechanism for granitic magmas that reach shallow crustal levels. A neutral buoyancy level in the crust, at which magma ascent should stall, is never observed. This is demonstrated by coeval emplacement of magmas with different compositions and densities, and the negative gravity anomalies measured over many granitic plutons. We suggest that deformation, through strain partitioning, is necessary to magma ascent. Pluton formation is controlled by local structures and rock types rather than by intrinsic magma properties. As a result of its intermittent character, deformation (both local and regional) induces magma pulses, and this may have important consequences for the chemical homogeneity of intruded magmas.

Ascent of granitic (leucogranitic to tonalitic) magma in the continental lithosphere is the most potent means of mass transfer and chemical segregation between the lower and upper crust. Granitic magmatism contributes to crustal recycling and, to a lesser extent, continental growth. In rapidly accreted crust, granitic rocks can occupy up to one-third of the volume of the middle crust (Meissner 1986). Together with shear zones, they are the most conspicuous effects of plate convergence and collision in orogenic zones. Consequently, they are a key element in understanding the evolution of Earth's mechanical workings.

As most felsic magmas are emplaced in a liquid, or near-liquid state, at temperatures over 800°C, a major heat source must be involved in their genesis. The existence of such a heat source also has consequences for the deformation regime of the crust. Commonly, a large temperature rise in the lithosphere will also increase the capacity of crustal material to respond to stress (by deformation). Such a thermal anomaly will contribute to the dissipation of energy and thus buffer the temperature effects. The stress regime will be imposed externally by the tectonic processes operating at the time. As a reaction to a significant increase in crustal heat flow, the rheological capacity of rocks to sustain stresses decreases, and this activates deformation. Therefore, most departures from a 'normal' heat distribution in the lithosphere will trigger deformation (e.g. formation of extensional rift zones, or shear zones in convergent situations).

From: VENDEVILLE, B., MART, Y. & VIGNERESSE, J.-L. (eds) *Salt, Shale and Igneous Diapirs in and around Europe*. Geological Society, London, Special Publications, **174**, 1–19. 1-86239-066-5/00/$15.00

Growing granitic intrusions are also sites at which chemical systems can be driven out of equilibrium. This occurs mainly because interactions between the incoming magma and upper-crustal rocks (commonly H_2O saturated) will drive local hydrothermal convection cells, generated to dissipate the excess heat from the cooling magma. The intense element circulation that this situation may trigger can lead to concentration and precipitation of economically interesting elements (e.g. Li, Be, Cu, Zn, Sn, W, Au, Pb, U).

For the above reasons, granitic plutons have been intensively surveyed for their internal structure, mineralogy and chemical composition, as well as to constrain the physical conditions of magma generation and emplacement. An important issue in such studies is the time-scale of the whole process of granite generation (thermal diffusion, melting, segregation, ascent and emplacement; Clemens *et al.* 1997). Granitic plutons represent only the terminal stages of the process, and studies of granitic bodies are particularly informative on the final emplacement conditions. However, they usually provide little or no insight into the magma ascent mechanism(s). At the other end of the process, migmatites are commonly considered to preserve the initial stages of granitic magma generation (Brown 1994). However, migmatites represent only a snapshot of the process during its development, and there is debate over the question of whether migmatites truly represent the outcomes of processes involved in the formation of most granitic magmas (the magma v. migma argument). Certain types of migmatites may typify the phase of initial magma segregation from its source. In some cases, migmatitic bodies are accompanied by small anatectic granites derived from the same source rocks. There is also evidence that some weakly mobile, restite-rich granitic magmas are formed as diatexitic migmatites (e.g. Finger & Clemens 1995). However, the emplacement of such bodies is far from being akin to the emplacement of large, high-level granitic batholiths displaying features that indicate high liquid contents on initial emplacement. Clemens & Droop (1998) have given a broad treatment of the theoretical outcomes of various partial melting scenarios (in terms of fluid presence or absence, the occurrence of segregation and the nature of the metamorphic P–T path).

Experiments on partial melting of common crustal rocks provide considerable insight into the nature and products of partial fusion. At present, most experiments are performed under constant pressure, temperature and compositional conditions, and do not address the role of stress in magma segregation. Although they have provided valuable structural insights, attempts at dynamic partial melting experiments (e.g. Rutter & Neumann 1995) have not produced the near-equilibrium melt compositions or rock textures observed in nature. This is largely due to the limitations on strain rates attainable on laboratory time-scales, and the consequent necessity of performing short experiments at very high temperatures. Additionally, because of slow diffusion rates, even fine-grained natural rocks do not reach chemical equilibrium on laboratory time-scales. Numerical and analogue experiments (Román-Berdiel *et al.* 1995; Benn *et al.* 1998; Barnichon *et al.* 1999) are useful, as their boundary conditions can be controlled and varied at will. Their limitation is essentially in their relative naivetJ and the relatively primitive state of our knowledge of the rheological properties of minerals and their aggregates under realistic geological conditions.

The paradigms developed from these approaches have been widely accepted as the best representations available to explain granitic magma generation, extraction and emplacement. This has led to the adoption of apparently attractive but unverified concepts such as metasomatic granitization (generally regarded as totally unverified) or diapiric upwelling of magmas. Features that most such models share are that they have been derived from single disciplines (e.g. petrology, fluid mechanics or tectonics), and they do not take account of knowledge gained through other disciplines. Here we attempt to formulate a convergent analysis of the processes that lead to granite generation, ascent and emplacement, using constraints from different fields of study. Granitic magma ascent has served as a model for many other forceful emplacement mechanisms (e.g. salt or mud) and this paper could provide useful constraints and concepts for understanding these phenomena.

In what follows, we consider ascent and emplacement as being intimately linked, even if the principal strain direction differs (dominantly vertical v. dominantly horizontal magma movement). We first briefly review what can be observed in outcrops of granitic intrusions (shapes, volumes and physical characteristics). We then examine the major proposed mechanisms of magma ascent (low-inertia flow or diapirism and fracture-driven flow or dyking). During diapirism, the response of the surrounding medium controls the ascent mechanism, and partly controls the volume emplaced in the case of dyking. We also examine the question of whether magma properties control ascent to a neutral buoyancy level, at which forces driving ascent will cease. We focus on the intrinsic limits of these two magma ascent models, and suggest that deformation (stress) actively drives granitic magma ascent. As stress is not temporally constant in the crust, granite emplacement is predicted to be episodic, which is reflected in the chemical evolution of the magmas.

Subsurface shapes of granitic bodies

The present shapes of granitic intrusions at depth reflect the emplacement modes of the magmas rather than their ascent. They do provide information on the geometrical factors that controlled emplacement. Seismic profiles reveal the 2D shapes of intrusions, although the internal structures remain transparent to this technique (Matthews 1987). On such profiles, the surrounding crust appears either laminated (corresponding to sedimentary or low-grade metasedimentary rocks), or unlaminated (corresponding to gneissic or higher-grade metamorphic rocks). No granitic intrusions have been encountered within highly reflective and laminated environments, such as those characteristic of the lower continental crust. Seismic data place limits on the depths to the floors of granitic intrusions at about 3–4 s two-way travel time; about 9–12 km in present-day depth. Indeed, numerous studies show shallow reflectors beneath granitic plutons. One interpretation of these features is that they represent the pluton floors. If this is correct, it implies overall tabular shapes, at least for the large bodies.

The gravity technique is particularly suited to determination of the bulk geometries of felsic intrusions (Améglio *et al.* 1997), provided there is sufficient density contrast between the plutons and their wall rocks. Modelled shapes fall broadly into two major categories: well-shaped and sheet-like bodies. Intrusion thickness appears to depend mainly on the chemistry of the magma, the more mafic being the least viscous and with the greatest tendency to invade planar discontinuities in the wall rocks. One general result of these studies is that the average volume of magma delivered by a single feeder ('root') is about 1500 km^3, which results in plutons around 5 ± 2 km thick.

The compilation of geological and geophysical data by McCaffrey & Petford (1997) suggests that granitic plutons are at least crudely scale invariant, as far as width–thickness ratios are concerned. The power-law relationship that they extracted is:

$$T = 0.12L^{0.88}$$

where T is the thickness and L is the largest horizontal dimension. This equation predicts that a batholith of 30 km diameter is likely to have a thickness of less than 2.5 km. The scatter in the dataset presented by McCaffrey & Petford leads to a sizeable error, but it appears that granitic intrusions are nowhere near as vertically extensive as has commonly been assumed.

Physical contrasts between the crust and felsic magmas

The materials of the continental crust react mechanically to stress rather independently of their composition and structure. The upper crust reacts elastically and fractures under high stress whereas the lower crust responds plastically, except at very high strain rates. This leads to discontinuous faults in the upper crust, and localized ductile shear zones in the lower crust. In the brittle crust, Hooke's law linearly relates strain to stress through a compliance tensor, and fracture occurs when the differential stress exceeds a value that relates essentially to the maximum frictional strength of the rocks. In ductile crust, a power law relates strain rate to stress. Accordingly, an equivalent viscosity can be computed for any strain rate.

In accordance with the physical conditions necessary for the generation of felsic magmas (>700°C and *c.* 500 MPa), an estimate of a typical source depth would be around 20 km, within amphibolite- to granulite-facies 'ductile' crust. In contrast, environmental conditions of emplacement are typically between 50 and 300°C, at depths of 2–10 km, within zeolite- to greenschist-facies 'brittle' crust. Nevertheless, granitic rocks are also encountered in higher-grade metamorphic envelopes, which could testify to deeper emplacement in some instances.

Given the chemical composition of a felsic magma (including its H_2O content) it is possible to compute its density (e.g. Bottinga *et al.* 1983; Lange & Carmichael 1987; Ochs & Lange 1997) and viscosity (e.g. Shaw 1972; Hess & Dingwell 1996). Corresponding properties of the solid wall rocks can be estimated from compilations of petrophysical data (e.g. Ahrens 1995). Thus, it is possible to estimate the magnitudes of the major physical factors that are assumed to control the ascent of felsic magmas (Table 1).

Density and viscosity contrasts are evidently amongst the major factors, but we have also computed the temperature and stress fields within and around magma bodies. Contrasts between these values are either simple differences between the magma properties and those of the surrounding rocks, or the ratios of such values, in cases where properties present large ranges of variation (e.g. viscosity and stress). For the stress values we indicate the vertical stress (lithostatic load) as well as the differential stress ($\sigma_1 - \sigma_3$), computed assuming either a vertical σ_1 (the extensional case) or a vertical σ_3 (the compressional case). These values are normalized with respect to the lithostatic load, and represent extremes of stress amplitude. The computed contrasts are given in Table 1.

Evidently, viscosity contrasts present the largest ratio, around 10^{13}. This results from the moderate magma viscosities (10^4–10^6 Pa s; see also Clemens & Petford

Table 1. *Differences (–) and/or ratios (/), indicated by the symbol (Δ), between physical properties of magma and its wall rocks*

Variable	Unit	Magma	Rock	Δ	– or /
Viscosity	log η Pa s	4–6	15–20	/	13
Density	g cm^{-3}	2.4–2.6	2.6–2.7	–	0.3–0.4
Temperature	°C	850–950	300–400	–	500
Stress	s_v MPa km^{-1}	4	28	/	7
	$\sigma_1 - \sigma_3$ MPa	4	220(ext)	/	55
	(10 km)	4	1030(cont)	/	257

Stress and viscosity values given are those from materials with the most contrasting properties. Stress magnitudes are given according to the lithostatic load (σ_v) and for the differential stress magnitudes, at a depth of 10 km in extensional and in contractional conditions.

1999) compared with those of the wall rocks (10^{15}–10^{20} Pa s). Values have been computed for magma compositions ranging from tonalite to peraluminous leucogranite. In comparison, density contrasts are rather weak (300–500 kg m^{-3}) though they represent about 14% of the value of the wall rock. Stress values are also important, as the buoyant stress developed by the magma is about 4 MPa km^{-1}, whereas the lithostatic load amounts to 28 MPa km^{-1}. Thus, a contrast of *c.* 14% exists, providing a driving force for magma ascent. Conversely, the maximum differential stress that a magma can generate, when emplaced at 10 km depth, is very restricted (4 MPa), whereas the crustal differential stress ranges from 220 MPa in extension to 1030 MPa in a compressional regime (Vigneresse *et al.* 1999). Clearly, the internal stress developed by magma cannot sustain the horizontal stress existing at that depth in the crust. The calculated stress ratios between the magma and its surrounding rocks are 55 for extension and 257 for compression. Finally, temperature contrasts are large, reaching values as high as 500°C at the moment of initial granite emplacement in the upper crust.

Ascent mechanisms

Two principal end-member mechanisms have been advanced to explain the ascent of granitic magmas. One considers low-inertia flow (derived from Stokes flow), as a result of a viscosity or occasionally a density contrast between magma and wall rocks. The other mechanism is through fracture propagation, driven by a pressure gradient (magma pressure in dykes and magma buoyancy). Magma upwelling is induced by the density contrast between the magma and its wall rocks, and sustained by a continuous volume increase of the decompressing magma. A fundamental difference between the two types of mechanisms is the respective importance of magma properties v. wall-rock properties. In low-inertia flow, the properties of the wall rocks govern ascent, whereas the intrinsic magma properties control ascent during fracture propagation.

Below, we briefly summarize the mechanics of the two types of model and their physical consequences. Numerical and experimental modelling are included under a same heading. As this paper is an attempt to synthesize published mechanisms, we typically cite only the most recent work of any particular group working on these problems. Thus, some pioneering papers have not been cited. Also, we only summarize final results and findings, and do not intend to produce an exhaustive list of

Table 2. *Models for magma ascent by low-inertia flow*

Low-inertia flow Ascent rate controlled by the host-rock viscosity	
Stokes flow	
Rigid sphere	Cruden (1988)
Soft sphere (strain)	Schmelling *et al.* (1988)
	Cruden (1990)
Viscous–viscous	Whitehead & Helfrich (1991)
Hot sphere	Mahon *et al.* (1988)
Rayleigh-Taylor	
$\Delta\rho$, no $\Delta\eta$	Ramberg (1981)
	Weijermars (1986)
$\Delta\rho$ and $\Delta\eta$	Weinberg (1996)
Brittle–ductile (numerical)	Barnichon *et al.* (1999)
Rayleigh Bénard	
Variable viscosity	Neugebauer & Reuther (1987)
	Kukowski & Neugebauer (1990)

Models are subdivided into Stokes flow, and Rayleigh–Taylor and Rayleigh–Bénard instabilities. In each case, major references are given for the important controlling factors. $\Delta\rho$ and $\Delta\eta$ are respectively density and viscosity contrasts.

published models and their historical development. For a recent review paper the reader is referred to Clemens *et al.* (1997).

Low-inertia flow

In this category of ascent model we include the hot sphere or Stokes flow model, together with Rayleigh–Taylor and Rayleigh–Bénard instabilities (Table 2).

In the hot sphere, or Stokes model, a body moves within its matrix as a result of its high density contrast with that matrix. Displacement of the surrounding material allows the body to move, so the simplest Stokes flow experiments use a rigid sphere falling through a viscous liquid (e.g. Dixon 1975; Cruden 1988). These experiments were later verified numerically (Schmelling *et al.* 1988). Depending on the value of the Reynolds number, the sphere internally reacts (or fails to react) to the fluid movement, and it reacts externally through its inertia (which may be very low). The surrounding medium is either uniformly viscous or shows a power-law-dependent viscosity. The case of a non-rigid sphere, in which strain may alter its shape, can also be considered (Cruden 1990), as can the case of a viscous sphere intrusive into a viscous medium (Whitehead 1986; Whitehead & Helfrich 1991). In such flows, the velocity of the sphere is a function of the viscosity (and density contrast) of the matrix.

Rayleigh–Taylor experiments are more sophisticated. These employ a layered medium with varying viscosity and density contrasts. As a result of these contrasts, the system soon becomes unstable and the lower, lighter and less viscous layer bulges upward and then ascends, forming the well-known diapiric structures. Experimental modelling of such systems was undertaken at Ramberg's laboratory at Uppsala, Sweden (Ramberg 1981). Weijermars (1986) used constant viscosity media, but with varying density contrasts, whereas Weinberg (1993, 1996) invoked variable viscosity or power-law-variable viscous media (Weinberg & Podladchikov 1994, 1995). Numerous experiments have been derived from the earlier Ramberg models (Anma & Sokoutis 1997). Recently, numerical modelling, incorporating a variable viscosity matrix overlain by a brittle layer (Barnichon *et al.* 1999) has illustrated the influence of the viscosity of the medium on ascent mechanisms.

Finally, using Rayleigh–Bénard instabilities, of which the Rayleigh–Taylor instability is a specific case, numerical and analogue models have also been developed. These models have failed to show any clear development of diapir-like structures when reasonable dimensional parameters have been used (Neugebauer & Reuther 1987; Kukowski & Neugebauer 1990, 1994; Weijermars 1986).

Arguments against granitic diapirism

One major conclusion from these models is that the characteristics of the external medium (wall rocks) severely constrain the ascent velocity of the unstable material (granitic magma in our case). This is partly inherent in the physics of the models themselves. By nature, the whole process is governed by the ability of the surrounding medium (wall rock) to deform, to accommodate magma ascent, and then to recover. The introduction, into the system, of a deformable body or of any strain-sustaining body does not fundamentally modify the outcome.

The introduction of a strong anisotropy into the surrounding medium, such as a décollement layer, will facilitate magma intrusion, as demonstrated in the experimental modelling of Román-Berdiel *et al.* (1995). This contrasts with previous models, in which layers of uniform viscosity allowed the mass to ascend through the entire system and to pierce it completely. The introduction of a power-law viscosity in the matrix also greatly reduces the ascent velocity and increases depth of intrusion (Fig. 1). The use of an increasing viscosity and a coefficient (n) of only two for the power law (Weinberg & Podladchikov 1994), though values around three are commonly applied to crustal materials (Kirby & Kronenberg 1987), drastically slows ascent at mid-crustal depths around 15 km. Similar values appeared in earlier modelling that used a power-law viscosity and took into account crustal thermal gradients (Mahon *et al.* 1988). In all presented models, the intrusions fail to reach the uppermost crust, with emplacement restricted to the region between 21 and 14 km depth (Fig. 3). Using a Rayleigh–Bénard model, the variable viscosity also constrains emplacement depth to >14 km (Neugebauer & Reuther 1987). Finally, with incorporation of an upper brittle layer, diapiric ascent terminates at this level (Barnichon *et al.* 1999).

In conclusion, all published models using low-inertia flow are in accord; such flow is essentially restricted to the lower and ductile crust. In any case, diapiric ascent of magma could proceed only to a depth more or less equivalent to the brittle–ductile transition. Diapirism cannot account for magma ascent up to depths of 2–10 km, which generally correspond to greenschist-facies rocks in which a large number of major granitic plutons are observed. Although stated explicitly in the results of numerical modelling (e.g. by Weinberg & Podladchikov 1995) it appears that these papers have commonly been improperly cited, by numerous workers, even one of us (Hanmer & Vigneresse 1983), as supporting granitic diapirism.

Other arguments against the possibility of magmatic diapiric upwellings in the crust involve field observations (Schwerdtner 1990), consideration of the relative ascent

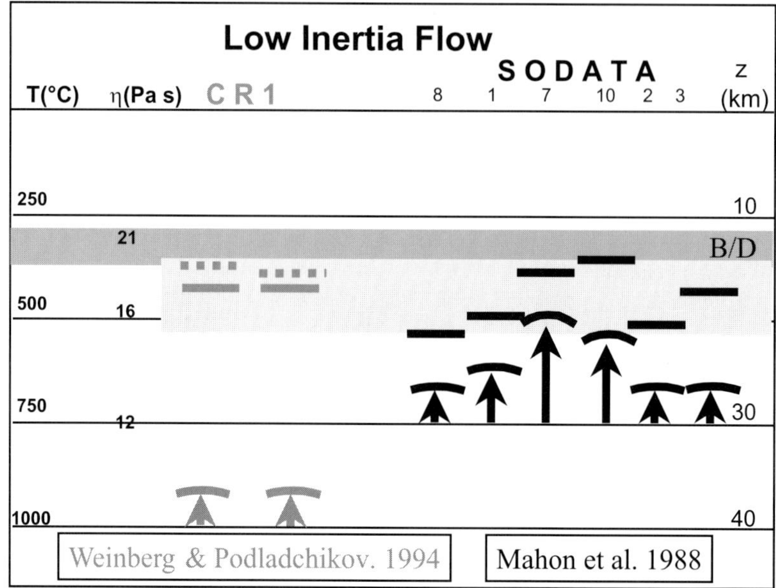

Fig. 1. Major results obtained by Weinberg & Podladchikov (1994) and by Mahon *et al.* (1988) using low-inertia flow modelling. Mahon *et al.* used so-called SODATA numbered experiments, whereas Weinberg & Podladchikov used CR1 models. The magma source is indicated by its origin, as well as by an arc that represents its radius. The maximum level reached by the rising body is indicated by a dash, or by small dashes when the body could rise further in a longer time. The temperature and the viscosity of the surrounding material used during these studies are also represented. The brittle–ductile transition (B/D) is assumed to occur at 300–350°C.

velocities compared with crystallization rates, the lack of the expected crustal structures (e.g. Clemens & Mawer 1992) or spacing between plutons using volumetric data for the emplaced magmas (Vigneresse 1995*b*). From all these arguments, we conclude that granitic diapirs could exist in the deep crust, but the lack of geological evidence for the passage of diapirs suggests that this may not occur. In any case, diapirism is unlikely to permit granitic magma ascent beyond the brittle–ductile transition.

Fracture propagation (dyking)

This mechanism incorporates elastic propagation of magma-filled cracks, solitary waves and dyke propagation in elasto-viscous rocks (Table 3). Although density contrast is the most important factor in the first case, viscosity contrast may be important for some kinds of dyking. In most cases, the density contrast between the magma and its wall rocks drives ascent. Thus, if no density contrast exists, this is commonly held to define a neutral buoyancy level (NBL), above which magma ascent is precluded.

In fracture propagation, ascent is controlled by the intrinsic properties of the magma that contribute to the maintenance of the upward propagation of a planar conduit. A model for this process was first formulated by Weertman & Chang (1977), who examined the forces at the propagating tip and at the bottom of a

Table 3. *Models for magma ascent by fracture propagation*

Fracture propagation Ascent rate controlled by the magma viscosity	
Liquid-filled crack	
Elastic crack	Weertman (1971)
+Pressure gradient	Pollard & Muller (1975)
	Takada (1990)
Waves in conduit	
Soliton	Whitehead (1986)
	Scott & Stevenson (1984)
Dyke propagation	
Elastic medium	Lister & Kerr (1991)
+Viscous	Rubin (1993)
Brittle–ductile (analogue)	Román–Berdiel *et al.* (1995)
Regional stresses	Parsons & Thompson (1991)
	Hogan & Gilbert (1995)

Models are subdivided into liquid-filled cracks, solitons and dyke propagation. In each case, major references are given for the important controlling factors.

liquid intrusion into an elastic medium (Weertman 1971, 1980). With time, an elongate, inverted teardrop shape evolves, as a result of the opening of the top of a crack, and its closure at the bottom. This model assumes that the driving force for ascent is a density contrast, inducing a vertical pressure gradient that overcomes the ambient stress field (gravity). Such pressure gradients are also inherent in models developed later by Pollard & Muller (1976) and experimentally reproduced by Takada (1989, 1990). This type of model implicitly assumes that a neutral buoyancy level exists, above which the density contrast between the magma and its wall rocks is insufficient to drive further ascent.

Modelling of dyke propagation can also encompass elastic rock behaviour and viscous pressure reduction in the magma (Lister & Kerr 1991; Rubin 1993). In these cases, the existence of a pre-existing fracture solves the problem of initial crack opening in the wall rock. Nevertheless, there are several ways in which magma source regions may become highly fractured at the time of partial melting (e.g. Clemens & Mawer 1992; Clemens & Droop 1998; Petford & Koenders 1998). The elastic response of the rock must be balanced, more or less, by internal magma pressure. In the case of a viscous wall rock, the viscosity ratio between the magma and its surroundings controls the viscous response of the wall rocks.

An important finding of these models is the development of solitary waves ('magmons') that can propagate without attenuation in the viscous medium (Scott & Stevenson 1984). Importantly, certain kinds of solitary waves can pass through one another, without interaction. Consequently, once a magma batch has begun its ascent, a solitary wave develops and propagates upward independently of the surrounding medium. Two separate waves (magma pulses) with different internal velocities could then cross and continue upward without major modification of their individual ascent velocities (Scott & Stevenson 1986; Scott *et al.* 1986).

The case of a fracture developing in an elasto-viscous material is similar to hydrothermal vein propagation. The mathematical treatment assumes that the elastic response of the matrix (country rock) depends on the internal magma pressure. Accordingly, the viscous response of the matrix depends on the viscosity contrast between the two materials, magma and country rock. Consequently, dykes of low-viscosity magmas (e.g. basalts) are preferentially observed in elastic rocks, whereas dykes of higher-viscosity magmas (e.g. granites) can develop in surrounding viscous rocks. This disparity in predicted behaviour is partly removed in models that assume viscoelastic behaviour of the wall rocks (Rubin 1993). With the intrusion of a low-viscosity magma (e.g. basalt) the medium deforms elastically, as the elastic displacement exceeds the viscous displacement at the centre of the dyke. Conversely, with intrusion of a higher-viscosity magma (e.g. rhyolite), the extent of viscous deformation greatly increases compared with the elastic response of the wall rock. This results in dyke aspect ratios (thickness/length) of around 10^{-2}, compared with 10^{-4} in purely elastic wall rocks.

A potential problem inherent in any model for viscosity-induced magma ascent is the exponential increase of magma viscosity as temperature falls. As shown by Clemens & Petford (1999), the effects of falling temperature will be largely offset by the rise in melt H_2O content that accompanies crystallization. Nevertheless, the growth of crystals will eventually increase the viscosity of the magma by a substantial amount, and finally preclude further ascent in any conduit of restricted width. For granites, computations indicate that conduit widths ranging from 2 to 10 m are sufficient to allow continued ascent, depending mainly on the thermal contrast with the wall rocks and the volumetric flow rate (Petford *et al.* 1994). The chemistry of magma may be far less influential than previously thought (Clemens & Petford 1999).

Competition between density and viscosity contrasts will restrict the development of the above processes. A theoretical neutral buoyancy level could exist for such a system, as the density contrast between magma and wall rocks is the main factor in the generation of internal magma pressure that maintains an open conduit. However, the question arises: is a neutral buoyancy level a reality in nature and, if so, does it present a barrier to continued magma ascent?

Arguments against a neutral buoyancy level (NBL)

The concept of an NBL offers several important contradictions to both theoretical considerations and observational data. The density of a granitic melt can be estimated using the partial molar volumes of oxide components, including H_2O (e.g. Lange & Carmichael 1987; Ochs & Lange 1997). For crystal-free granodioritic to leucogranitic magmas, at 100 MPa–1 GPa, 800–950°C and with 2–5 wt% H_2O in the melt, the calculated densities range from 2100 to 2500 kg m^{-3}. A survey of the densities of a wide range of crustal rock types shows that only vesicular volcanic rocks, porous sandstones, porous limestones and oil shales will have densities equal to or less than such magmas. The vast majority of compact crustal rocks (Fig. 2) have densities in the range 2550–3390 kg m^{-3}. Indeed, only shales reach as low as 2550 kg m^{-3}; the rest lie above 2630 kg m^{-3}. We add the caveat that this crude analysis of the problem is complicated if granitic magmas have crystal contents greater than 10–20%. Nevertheless, granitic magmas are commonly emplaced with rather low initial crystal contents (a few per cent). This suggests that a local NBL for granitic magmas could exist, in some rather specific cases, but that this cannot be a general feature

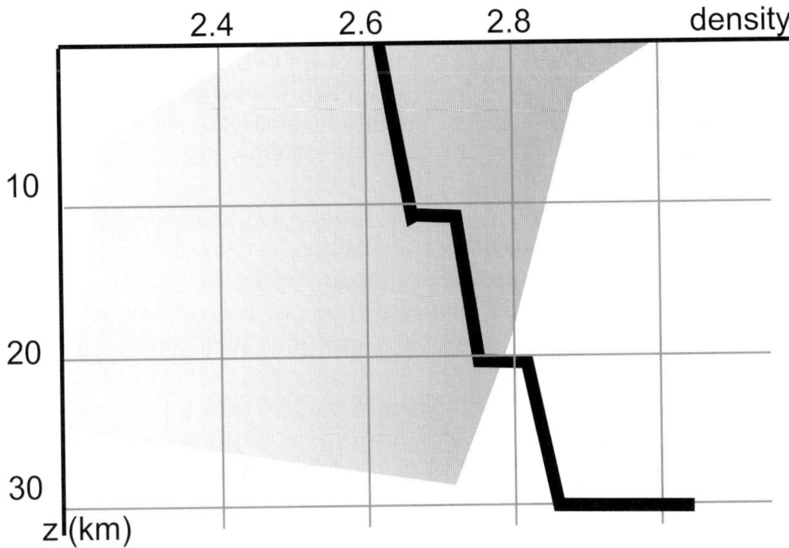

Fig. 2. Schematic bulk crustal density (in black) superimposed on the range of density of magmas. Granitic magmas are on the left, in light grey; gabbros are on the right in dark grey. The shaded sector encompasses material varying from molten at depth to fully crystallized at the top.

in the continental crust. If it did exist, it would commonly lie between 11 and 20 km depth (Fig. 2).

The ubiquitous occurrence of basaltic lavas (with melt densities of 2600–2750 kg m^{-3}) suggests that, at least for some magmas, the reality of a crust-wide NBL is immaterial to their ability to ascend all the way from the mantle to the Earth's surface. Indeed, Takada (1989, 1990) has shown, by experiment, that a propagating fracture can extend well beyond its level of neutral buoyancy, provided that the host medium (the crust in this case) is in extension. It is the buoyancy or magma pressure generated over the entire length of a magma column that drives ascent, not the local density difference at the top or at any point in the column.

Another argument against the existence of an NBL within the crust is the common observation of gravity anomalies over granitic intrusions. As liquid magmas are commonly less dense than their corresponding crystallized products (igneous rocks), magmas should ascend rapidly with respect to their surroundings and be emplaced only in rocks with densities lower than the magmas. On solidification of the magmas, this process should result in positive gravity anomalies over most granitic plutons. In fact, granitic intrusions commonly induce negative gravity anomalies, the amplitudes of which may reach −40 mgal. This observation also accords with the lower densities measured for granite samples (2630–2660 kg m^{-3}) compared with bulk crust (2670–2700 kg m^{-3}). Actually, lower densities and larger negative gravity anomalies correlate owing to the fact that granitic plutons have an average vertical thickness of 5 ± 2 km (Vigneresse 1995*b*). Clemens & Mawer (1992) concluded that emplacement levels for plutons are likely to be governed by the presence of structural discontinuities or easily deformable horizons that either divert or blunt the propagating tips of ascending magma fractures.

Deformation-induced ascent

We have argued that diapiric ascent should be restricted to the lower and ductile crust. We have also suggested that ascent through fractures is unlikely to be due to buoyancy forces alone. Nevertheless, granitic magmas are transferred from their deeper sources to the shallow upper crust. We therefore suggest that deformation actively participates in this process.

Deformation (other than purely magma-induced effects) must commonly occur during magma emplacement as the overall shapes of many intrusions, and fabrics within them, appear to be controlled by the ambient stress field (e.g. Vigneresse 1995*a*,*b*; Paterson *et al.* 1998). As observed from the inversion of gravity data on granitic plutons, two major types of intrusion exist, one with flat floors and restricted thickness, and another with wedge-like shapes, deeper in extent and with steep walls commonly controlled by fractures (Améglio *et al.* 1997). This shape dichotomy has been explained by a change of orientation of the intrusion plane (along which magma is delivered) as a result of a rotation of the local stress field (Parsons *et al.* 1992; Vigneresse *et al.* 1999).

During extension or strike-slip deformation, the extensional fracture planes are vertical and perpendicular to the least principal stress component (σ_3). When magma intrudes along this plane, the stress component perpendicular to this plane increases in proportion to the thickness of the intrusion (p_m). This is because the space taken by magma locally increases the stress field. Thus, the least principal stress grows ($\sigma_3 \rightarrow \sigma_3 + p_m$). The intermediate stress component also increases, but by a smaller amount because of the Poisson coefficient (ν) of the rocks ($\sigma_2 \rightarrow \sigma_2 + \nu p_m$). During this process the maximum principal stress component (σ_1) remains constant. An abrupt and major change may occur in the orientation of the stress field if the magnitude of the increasing minor stress component overtakes that of the former major component. This would drive the local stress field into compression, with the plane of opening horizontal (Vigneresse *et al.* 1999). Once the system had reached this stage, the horizontal magma sheet would form a structural trap for further intrusions of magma.

The mechanism above was modified from that of Parsons & Thompson (1991), which was valid only for extension. After adaptation to all tectonic environments, from extension to compression, this suggests that magma emplacement is strongly controlled by the stress field. This model also expands on the case of passive magma pressure driving emplacement in an extensional regime (Hogan & Gilbert 1995).

Before ascent and emplacement, deformation also appears to control segregation of melt in migmatites (Fig. 3). In migmatites, partial melts are commonly collected in shear zones (e.g. Collins & Sawyer 1996). The development of shear zones indicates a predominance of non-coaxial deformation, induced by the simultaneous presence of a liquid phase in a plastic matrix subjected to stress. Strain partitioning is a response to deformation in a two-phase material (Vigneresse & Tikoff 1999). In migmatites, it occurs on all scales, from millimetric and decimetric (the local shear zones that concentrate melt) to the massif scale. In granitic bodies, it is commonly observed when deformation is transpressional (e.g. Solar *et al.* 1998).

Given that deformation occurs during segregation and emplacement, it certainly also occurs during magma ascent (Fig. 3). Ascent of magma results from the flow of a liquid through a viscous matrix. Consequently, strain partitioning will occur,

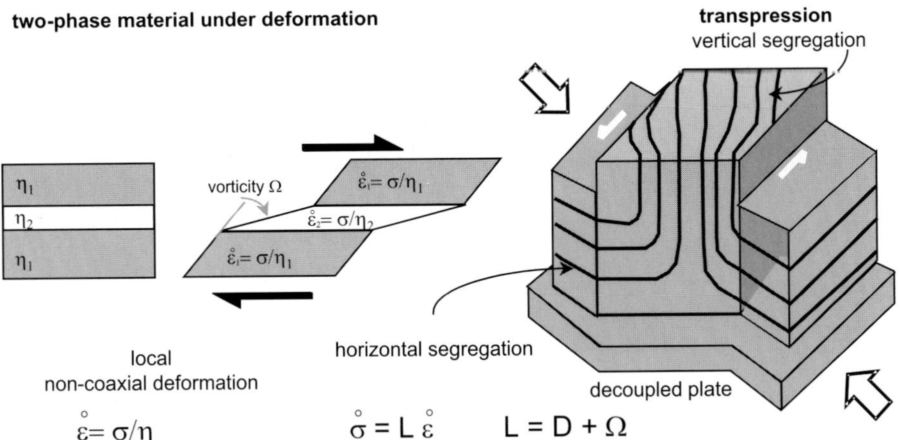

Fig. 3. Schematic presentation of strain partitioning. This occurs at a small scale (e.g. within a partly molten material) and at a large scale (e.g. in the case of a decoupled basal layer). In both cases, the strain rate is highest in the less viscous region. Consequently, melt is expelled horizontally from its matrix. At a larger scale, horizontal segregation of material is rotated into vertical movement (e.g. during transtensional or transpressional deformation).

leading to higher strain in the less viscous material, i.e. the magmatic body (de Saint Blanquat *et al.* 1998; Kisters *et al.* 1998). Because of the high viscosity contrast between magma and its matrix, vorticity will develop in the less viscous phase (magma), leading to non-coaxial deformation. This can easily overcome the buoyant gravity forces, as observed in migmatites. Thus, strain partitioning greatly assists magma ascent and provides the additional forces that allow magma transfer from the lower to the upper crust.

Deformation-driven ascent is not simply a theoretical mechanism. It is implicit in the most recent analogue modelling (Román-Berdiel *et al.* 1997; Benn *et al.* 1998). These experiments focused on transpressional deformation, which provides the flow vorticity to transform horizontal melt migration (during segregation at the magma source) to vertical migration during ascent in the shear and compressional regimes (Fig. 3). The predicted prevalence of such tectonic forcing suggests that a majority of granites could be classified, in some sense, as syntectonic.

Consequences for the chemical evolution of plutons

From the above analysis we have concluded that the most likely mechanism for granitic magma ascent, especially in the upper crust, is deformation-assisted fracture propagation. This mechanism implies that magma extraction (from the protolith) and magma delivery (to the emplacement site) will be pulsed, rather than a continuous flow. It is very difficult to comment on the rate of pulsed magma extraction, as it will be controlled by tectonics, as well as the intrinsic physics and chemistry of melting and segregation. On the other hand, given efficient melt supply to a fracture (dyke) system, it may be that the frequency of magma delivery to a growing pluton can be constrained by considering the volume of the future pluton in relation to the instantaneous volume of magma held in a feeder dyke. Weertman (1971) showed that any propagating, melt-filled fracture could remain open only up to some maximum length. Clemens & Mawer (1992) used this model to predict that granitic dykes could remain open only for lengths of a few kilometres, implying that ascent over crustal distances (tens of kilometres) would result in pulsed magma delivery. Those workers calculated a minimum pulse frequency of 1.3×10^{-6} pulses per year, for their modelled batholith. Perhaps the controls here are also rather too complicated, and we should not take such a value too seriously. However, the calculations suggest that batholiths could have been constructed from a large number of individual magma pulses ($>10^4$ in this case). The other important point is that these pulses (magmons) may well be able to cross without interaction. These two conclusions have some important implications for the chemical and isotopic characteristics of granites.

Pulsed magma extraction

Granitic melt is probably evacuated from its partially molten protolith through a self-focusing network of fractures. Ito *et al.* (1997) examined one mechanism for such focusing (see also Clemens *et al.* (1997) and Weinberg (1999)). Local physical and chemical segregation of the melt phase may also be controlled by the initiation of fracture networks (e.g. Clemens & Droop 1998). If this is true, it means that the rate of melt withdrawal may influence the rate of melt production. This is an example of Le Chatelier's principle; as melt is a product of the reaction, the higher its concentration, the slower the reaction rate will be. Thus, the high efficiency of melt tapping by fractures (Rutter & Neumann 1995) suggests that this mechanism would promote rapid melting and high degrees of reaction while impeding the attainment of equilibrium between restite and melt. Especially for trace elements (e.g. Y, Zr, Ce, LREE) that form essential structural constituents of common accessory phases (e.g. zircon, monazite, xenotime and allanite), this would result in gross disequilibrium between magma fractions withdrawn from the same protolith. Furthermore, as zoned, refractory phases commonly carry the isotopic signals used in geochronology, this model implies that studies of isotopic inheritance may be less reliable than commonly supposed for the identification of crustal protoliths. Sawyer (1994) and Bea (1996) discussed rapid chemical isolation of melt from restite (through dissolution kinetics or by physical segregation) as a cause of chemical and isotopic disequilibrium. The full extent of this potential problem has yet to be investigated.

Pulsed magma delivery

The delivery of contrasting magma batches to a growing chamber suggests the possibility that portions of such a composite magma may have a common cooling history but show contrasting chemical and/or isotopic evolutionary patterns, particularly for some trace elements. As pointed out by Clemens & Mawer (1992), dyke ascent of granitic magma would be sufficiently rapid to preclude any great degree of interaction between magma and wall rocks. Thus, source-related heterogeneities should be largely preserved, at least until a large magma chamber is formed.

Following the formation of a magma chamber, the degree to which chemical and isotopic heterogeneities will be preserved must depend partly on the longevity of the chamber (future pluton). The time that a magma chamber spends in a sufficiently fluid state to promote homogenization is potentially very variable. However, with the exception of high-level rhyolite chambers, periodically replenished from below by hot basaltic magma, cooling calculations suggest that most granitic magmas will spend between 10^3 and 10^5 years in a partially liquid state. The few detailed spatial studies of chemical and especially isotopic variations within single granitic plutons reveal preservation of relatively small-scale heterogeneities, some of which are interpreted as being source related rather than caused by fractionation (e.g. Stussi & Cuney 1993). It seems that there is generally insufficient time for processes in granitic magmas to erase a range of isotopic heterogeneities, some of which are likely to be source related.

Pulsed delivery to high-level magma chambers need not always involve transport of source-derived heterogeneities. Granitic magmas emplaced as pyroclastic deposits commonly contain phenocryst phases formed at pressures that suggest significant magma evolution at deeper levels than the final volcanic magma chamber (pluton analogue). An example is the Tolmie Highlands Igneous Complex, in which rhyolites contain an early phenocryst assemblage, equilibrated at around 600 MPa, and a later one formed at about 150 MPa (Clemens 1982). Also, Clemens & Wall (1981) showed that many such magmas must have cooled and crystallized more than would be expected for adiabatic ascent paths. This evokes the idea of stalled ascent and the existence of deep magma chambers that eventually feed the shallower ones from which eruptions take place. Thus, a final pluton might well be fed with heterogeneous magma batches resulting from crystal fractionation, magma mixing, etc. Pitcher (1979, p. 637) stated that some plutons show abrupt steps in compositional gradients, 'even represented by internal contacts, with sufficient chemical difference across them to suggest that the plutons were constructed by multiple injection of magmas which differentiated elsewhere'. The persistence of the identity of separate magmas that had a common chamber is evident in the occurrence of some kinds of zoned plutons and the abundance of magmatic enclaves in some granites.

Although many crystals forming within a magma might remain within the bounds of the magma batch that gave them birth, in some cases refractory crystals may cross boundaries between magma batches. Such circulations, between magma batches that may even have had a common protolith, could cause some of the complex zoning (major compositional steps as opposed to fine oscillatory zoning) in certain magmatic minerals (e.g. plagioclase). This type of variation may well be the signature of pulsed magma withdrawal and ascent, and need not always result from the mixing of magmas derived from disparate sources (e.g. crust and mantle), as is commonly assumed.

Conclusions

We have attempted to form an overview of the major possible mechanisms for the ascent of granitic magma, from its source in the lower to intermediate crust toward the upper crust. Diapirism, or low-inertia flow, is controlled by the viscous response of the wall rocks. It implies that magma ascent is restricted to the more mobile part of the crust and cannot operate in the upper brittle crust. Buoyant magma ascent results from the internal magma forces induced by its low density. On the face of it, this implies that magma should cease ascent when entering a zone with which it has no density contrast. Such a neutral buoyancy level is never actually observed in the crust.

We suggest that the regional deformation field may largely control magma ascent. Strain partitioning occurs between the ductile crust and the still less viscous liquid magma. Consequently, strain preferentially concentrates in magma and promotes magma ascent in zones of restricted thickness but high strain (Kisters *et al.* 1998). Magma ascent is not peculiar to any specific deformational regime, but occurs in all tectonic environments (extensional, strike-slip and compressional). Development of vorticity, as a consequence of strain partitioning, implies that non-coaxial deformation, rather than gravity driven forces, acts to segregate melt from solid residue (see McKenzie 1984). Strain partitioning also favours local movement of melt as soon as is formed. Ponding of magma near the protolith, before delivery to the upper crust, is unlikely. The fact that some deep magma evolution clearly occurs must relate to temporary storage of magmas part-way through the crust. Such storage, and final emplacement, cannot relate to neutral buoyancy levels, and probably reflects ponding by pre-existing geological structures. Finally, the existence of small channels through which magma is transported toward the upper crust implies that many of these may not reach higher crustal levels. This suggests that the observed volume of granitic magma underestimates the real volume of partial melt material produced during magma genesis.

We thank B. Vendeville for stimulating discussions about mechanisms that lead to salt and igneous intrusions. S. Cruden and K. Benn are thanked for critical but impartial reviews of a paper that, in certain points, dissents from accepted hypotheses. Funding for part of this work was from the CREGU programme at Nancy.

References

AHRENS, T. J. (ed.) 1995. *Rock Physics and Phase Relations. A Handbook of Physical Constants.* American Geophysical Union, Washington, DC.

AMÉGLIO, L., VIGNERESSE, J. L. & BOUCHEZ, J. L. 1997. Granite pluton geometry and emplacement mode inferred from combined fabric and gravity data. *In*: BOUCHEZ, J. L., HUTTON, D. H. W. & STEPHENS, W. E. (eds) *Granite: from Segregation of Melt to Emplacement Fabrics.* Kluwer, Dordrecht, 199–214.

ANMA, R. & SOKOUTIS, D. 1997. Oblique diapirism in the Ryukyu arc, Japan. *In*: BOUCHEZ, J. L., HUTTON, D. H. W. & STEPHENS, W. E. (eds) *Granite: from Segregation of Melt to Emplacement Fabrics*, Kluwer, Dordrecht, 295–318.

BARNICHON, J. D., HANEVITH, H. D., HOFFER, B., CHARLIER, R., JONGMANS, D. & DUCHESNE, J. C. 1999. The deformation of the Egersund–Ogna anorthosite massif, south Norway: finite element modelling of diapirism. *Tectonophysics*, **303**, 109–130.

BEA, F. 1996. Controls on the trace element composition of crustal melts. *Transactions of the Royal Society of Edinburgh: Earth Sciences*, **87**, 33–41.

BENN, K., ODONNE, F. & DE SAINT BLANQUAT, M. 1998. Pluton emplacement during transpression in brittle crust: new views from analogue experiments. *Geology*, **26**, 1079–1082.

BOTTINGA, Y., RICHET, P. & WEILL, D. F. 1983. Calculation of the density and thermal expansion coefficient of silicate liquids. *Bulletin de Minéralogie*, **106**, 129–138.

BROWN, M. 1994. The generation, segregation, ascent and emplacement of granite magma: the migmatite-to-crustally-derived granite connection in thickened orogens. *Earth-Science Reviews*, **36**, 83-130.

CLEMENS, J. D. 1982. The Tolmie Igneous Complex, Australia: high-*T* S-type rhyolites with polybaric crystallization histories. *Geological Society of America, Abstracts with Programs*, **14**, 464–465.

—— & DROOP, G. T. R. 1998. Fluids, *P-T* paths and the fates of anatectic melts in the Earth's crust. *Lithos*, **46**, 21–36.

—— & MAWER, C. K. 1992. Granitic magma transport by fracture propagation. *Tectonophysics*, **204**, 339–360.

& PETFORD, N. 1999. Granitic melt viscosity and silicic magma dynamics in contrasting tectonic settings. *Journal of the Geological Society, London*, **156**, 1057–1060.

—— & WALL, V. J. 1981. Origin and crystallization of some peraluminous (S-type) granitic magmas. *Canadian Mineralogist*, **19**, 111–131.

——, PETFORD, N. & MAWER, C. K. 1997. Ascent mechanisms of granitic magmas: causes and consequences. *In*: Holness, M. (ed.) *Deformation-enhanced Fluid Transport in the Earth's Crust and Mantle*. Chapman & Hall, London, 144–171.

COLLINS, W. J. & SAWYER, E. W. 1996. Pervasive granitoid magma transfer through the lower–middle crust during non-coaxial compressional deformation. *Journal of Metamorphic Geology*, **14**, 565–579.

CRUDEN, A. R. 1988. Deformation around a rising diapir modeled by creeping flow past a sphere. *Tectonics*, **7**, 1091–1101.

—— 1990. Flow and fabric development during the diapiric rise of magma. *Journal of Geology*, **98**, 681–698.

DE SAINT BLANQUAT, M., TIKOFF, B., TEYSSIER, C. & VIGNERESSE, J. L. 1998. Transpressional kinematics and magmatic arcs. *In*: HOLDSWORTH, R. E., STRACHAN, R. A. & DEWEY, J. F. (eds). *Continental Transpressional and Transtensional Tectonics*. Geological Society of London, Special Publications, **135**, 327–340.

DIXON, J. M. 1975. Finite strain and progressive deformation in models of diapiric structures. *Tectonophysics*, **28**, 89–124.

FINGER, F. & CLEMENS, J. D. 1995. Migmatization and 'secondary' granitic magmas: effects of emplacement of 'primary' granitoids in Southern Bohemia, Austria. *Contributions to Mineralogy and Petrology*, **120**, 315–326.

HANMER, S. K. & VIGNERESSE, J. L. 1983. Mise en place de diapirs syntectoniques dans la chaîne hercynienne. Exemple des massifs granitiques de Locronan et Pontivy (Bretagne). *Bulletin de la Société Géologique de France*, **22**(7), 193–202.

HESS, K.-U. & DINGWELL, D. B. 1996. Viscosities of hydrous leucogranitic melts: a non-Arrhenian model. *American Mineralogist*, **81**, 1297–1300.

HOGAN, J. P. & GILBERT, M. C. 1995. The A-type Mount Scott granite sheet: importance of crustal magma traps. *Journal of Geophysical Research*, **B100**, 15779–15792.

ITO, G., MARTEL, S. J. & BERCOVICI, D. 1997. Magma transport in oceanic lithosphere through interacting dikes. *Eos Transactions, American Geophysical Union*, **78**, F694.

KIRBY, S. & KRONENBERG, A. K. 1987. Rheology of the lithosphere: selected topics. *Reviews of Geophysics*, **25**, 1219–1244.

KISTERS, A. F., GIBSON, R. L., CHARLESWORTH, E. G. & ANHAEUSSER, C. R. 1998. The role of strain localization in the segregation and ascent of anatectic melts, Namaqualand, South Africa. *Journal of Structural Geology*, **20**, 229–242.

KUKOWSKI, N. & NEUGEBAUER, H. J. 1990. On the ascent and emplacement of granitoid magma bodies. Dynamic–thermal numerical models. *Geologische Rundschau*, **79**, 227–239.

—— & —— 1994. Mathematical modelling of granite emplacement and forms on basis of the gravimetric, petrographic and structural data. *In*: HASLAM, H. W. & PLANT, J. A. (eds) *Granites, Metallogeny, Lineaments and Rock–Fluid Interactions*. British Geological Survey Research Report **SP/94/1**, 64–72.

LANGE, R. A. & CARMICHAEL, I. S. E. 1987. Densities of Na_2O–K_2O–CaO–MgO–FeO–Fe_2O_3–Al_2O_3–TiO_2–SiO_2 liquids: new measurements and derived partial molar properties. *Geochimica et Cosmochimica Acta*, **51**, 2931–2946.

LISTER, J. R. & KERR, R. C. 1991. Fluid-mechanical models of crack propagation and their application to magma transport in dykes. *Journal of Geophysical Research*, **B96**, 10049–10077.

MAHON, K. I., HARRISON, T. M. & DREW, D. A. 1988. Ascent of a granitoid diapir in a temperature varying medium. *Journal of Geophysical Research*, **B93**, 1174–1188.

MATTHEWS, D. M. 1987. Can we see granites on reflection profiles? *Annales Geophysicae*, **5B**, 353–356.

MCCAFFREY, K. J. W. & PETFORD, N. 1997. Are granitic intrusions scale invariant? *Journal of the Geological Society, London*, **154**, 1–4.

McKENZIE, D. 1984. The generation and compaction of partially molten rocks. *Journal of Petrology*, **25**, 713–765.

MEISSNER, R. 1986. *The Continental Crust: a Geophysical Approach*. Academic Press, Orlando, FL.

NEUGEBAUER, H. J. & REUTHER, C. 1987. Intrusion of igneous rocks—physical aspects. *Geologische Rundschau*, **76**, 89–99.

OCHS, F. A. & LANGE, R. A. 1997. The partial molar volume, thermal expansivity, and compressibility of H_2O in $NaAlSi_3O_8$ liquid: new measurements and an internally consistent model. *Contributions to Mineralogy and Petrology*, **129**, 155–165.

PARSONS, T. & THOMPSON, G. A. 1991. The role of magma overpressure in suppressing earthquakes and topography: worldwide examples. *Science*, **253**, 1399–1402.

——, SLEEP, N. H. & THOMPSON, G. A. 1992. Host rock rheology controls on the emplacement of tabular intrusions: implications for underplating of extending crust. *Tectonics*, **11**, 1348–1356.

PATERSON, S., FOWLER, T., SCHMIDT, K., YOSHINOBU, A., YUAN, E. & MILLER, R. 1998. Interpreting magmatic fabric patterns in plutons. *Lithos*, **44**, 53–82.

PETFORD, N. & KOENDERS, M. A. 1998. Self-organisation and fracture connectivity in rapidly heated continental crust. *Journal of Structural Geology*, **20**, 1425–1434.

——, LISTER, J. R. & KERR, R. C. 1994. The ascent of felsic magmas in dykes. *Lithos*, **32**, 161–168.

PITCHER, W. S. 1979. The nature, ascent and emplacement of granitic magmas. *Journal of the Geological Society, London*, **136**, 627–662.

POLLARD, D. D. & MULLER, O. H. 1976. The effect of gradients in regional stress and magma pressure on the form of sheet intrusions in cross section. *Journal of Geophysical Research*, **B81**, 975–984.

RAMBERG, H. 1981. *Gravity, Deformation and the Earth's Crust in Theory, Experiments and Geological Applications*. Academic Press, London.

ROMÁN-BERDIEL, T., GAPAIS, D. & BRUN, J. P. 1995. Analogue models of laccolith formation. *Journal of Structural Geology*, **17**, 1337–1346.

——, —— & —— 1997. Granite intrusion along strike-slip zones in experiment and nature. *American Journal of Science*, **297**, 651–678.

RUBIN, A. M. 1993. Dikes vs. diapirs in viscoelastic rock. *Earth and Planetary Science Letters*, **119**, 641–659.

RUTTER, E. H. & NEUMANN, D. H. K. 1995. Experimental deformation of partially molten Westerly Granite under fluid-absent conditions, with implications for the extraction of granitic magmas. *Journal of Geophysical Research*, **B100**, 15697–15715.

SAWYER, E. W. 1994. Melt segregation in the continental crust. *Geology*, **22**, 1019–1022.

SCHMELLING, H., CRUDEN, A. R. & MARQUART, G. 1988. Finite deformation in and around a fluid sphere moving through a viscous medium: implications for diapiric ascent. *Tectonophysics*, **149**, 17–34.

SCHWERDTNER, W. M. 1990. Structural tests of diapir hypotheses in Archean crust of Ontario. *Canadian Journal of Earth Science*, **27**, 387–402.

SCOTT, D. R. & STEVENSON, D. J. 1984. Magma solitons. *Geophysical Research Letters*, **11**, 1161–1164.

—— & —— 1986. Magma ascent by porous flow. *Journal of Geophysical Research*, **B91**, 9283–9296.

——, —— & WHITEHEAD, J. 1986. Observations of solitary waves in a viscously deformable pipe. *Nature*, **319**, 759–761.

SHAW, H. R. 1972. Viscosities of magmatic silicate liquids: an empirical method of prediction. *American Journal of Science*, **272**, 870–893.

SOLAR, G. S., PRESSLEY, R. A., BROWN, M. & TUCKER, R. D. 1998. Granite ascent in convergent orogenic belts: testing a model. *Geology*, **26**, 711–714.

STUSSI, J. M. & CUNEY, M. 1993. Chemical evolution models in peraluminous granitoids—examples from the Millevaches plutonic complex (French Massif-Central). *Bulletin de la Société Géologique de France*, **164**, 585–596.

TAKADA, A. 1989. Magma transport and reservoir formation by a system of propagating cracks. *Bulletin of Volcanology*, **52**, 118–126.

—— 1990. Experimental study on propagation of liquid-filled crack in gelatin: shape and velocity in hydrostatic stress condition. *Journal of Geophysical Research*, **B95**, 8471–8481.
VIGNERESSE, J. L. 1995*a*. Control of granite emplacement by regional deformation. *Tectonophysics*, **249**, 173–186.
—— 1995*b*. Crustal regime of deformation and ascent of granitic magma. *Tectonophysics*, **249**, 187–202.
—— & TIKOFF, B. 1999. Strain partitioning during partial melting and crystallizing felsic magmas. *Tectonophysics*, **312**, 117–132.
——, —— & AMÉGLIO, L. 1999. Modification of the regional stress field by magma intrusion and formation of tabular granitic plutons. *Tectonophysics*, **302**, 203–224.
WEERTMAN, J. 1971. Theory of water-filled crevasses in glaciers applied to vertical magma transport beneath ocean ridges. *Journal of Geophysical Research*, **B76**, 1171–1183.
—— 1980. The stopping of a rising, liquid-filled crack in the Earth's crust by a freely slipping horizontal joint. *Journal of Geophysical Research*, **B85**, 967–976.
—— & CHANG, S. P. 1977. Fluid flow through a large vertical crack in the Earth's crust. *Journal of Geophysical Research*, **B82**, 929–932.
WEIJERMARS, R. 1986. Flow behaviour and physical chemistry of bouncing putties and related polymers in view of tectonic laboratory applications. *Tectonophysics*, **124**, 325–358.
WEINBERG, R. F. 1993. Drops, surface tension and diapirs models. *Journal of Structural Geology*, **15**, 227–232.
—— 1996. The ascent mechanism of felsic magmas: news and views. *Transactions of the Royal Society of Edinburgh: Earth Sciences*, **87**, 95–103.
—— 1999. Mesoscale pervasive felsic magma migration: alternative to dykes. *Lithos*, **46**, 393–410.
—— & PODLADCHIKOV, Y. Y. 1994. Diapiric ascent of magmas through power law crust and mantle. *Journal of Geophysical Research*, **B99**, 9543–9559.
—— & —— 1995. The rise of solid-state diapirs. *Journal of Structural Geology*, **17**, 1183–1195.
WHITEHEAD, J. A. 1986. Buoyancy-driven instabilities of low-viscosity zones as models of magma-rich zones. *Journal of Geophysical Research*, **B91**, 9303–9314.
—— & HELFRICH, K. R. 1991. Instability of flow with temperature-dependent viscosity: a model of magma dynamics. *Journal of Geophysical Research*, **B96**, 4145–4155.

Experiments on granite intrusion in transtension

T. ROMÁN-BERDIEL[1], A. ARANGUREN[1],
J. CUEVAS[1], J. M. TUBÍA[1], D. GAPAIS[2] & J.-P. BRUN[2]

[1] *Departamento de Geodinámica, Universidad del Pais Vasco, Apartado 644, 48080 Bilbao, Spain (e-mail: acasas@posta.unizar.es)*

[2] *Géosciences Rennes (UPR 4661, CNRS), Université de Rennes 1, 35042 Rennes Cédex, France*

Abstract: Granite intrusion in transtensional regime is modelled by injecting a Newtonian fluid into a sand pack containing a ductile layer. The transtensional regime is obtained using two plastic sheets sliding along two rigid horizontal plates, and diverging from two narrow spaces (two fixed velocity discontinuities). The injection tube is located in a central space between these plates. Both symmetric experiments (when the two sheets were displaced with equal and opposite velocity vectors) and asymmetric experiments (in which only one sheet was displaced) were performed. Transtension was applied with a systematic variation (every 15°) of the divergence angle (α), between 15° and 90°. Experiments showed that: intrusions localize strain from the first stages of deformation; intrusions result in partially conformable laccoliths with bowler-hat geometry in cross-section; intrusions show an important offset towards the mobile basal plate for asymmetric transtensional regime, and are more symmetric and centred on the injection point for symmetric transtensional regime; the geometry of intrusions is controlled by the faults developed in the overburden. The significance of this control depends upon the angle of divergence α. Examples of the Hombreiro and Los Pedroches granites of the Variscan belt of Spain have been addressed to test the applicability of these experimental results.

Transtension is a strike-slip deformation that deviates from simple shear because of a component of extension orthogonal to the deformation zone (Harland 1971; Dewey *et al.* 1998). It is caused by the obliquity between the shear direction and the main faults, or by the existence of segments with different directions along the main wrench fault. The kinematics and dynamics associated with transtension have been extensively addressed in the literature (McKenzie & Jackson 1983, 1986, 1989; Jackson & McKenzie 1989; Fossen & Tikoff 1998; Lin *et al.* 1998). They are documented by several sets of small-scale modelling (Withjack & Jamison 1986; Cobbold *et al.* 1989; Tron & Brun 1991; Schreurs & Colletta 1998; Basile & Brun 1999), and by analyses of natural examples (Pavlides *et al.* 1990; Chorowicz & Sorlien 1992; Allen *et al.* 1998; Dokka *et al.* 1998; Krabbendam & Dewey 1998; Watkeys & Sokoutis 1998). Brittle–ductile analogue experiments by Tron & Brun (1991) have documented changes in fault patterns according to boundary conditions from graben patterns in pure divergence conditions to Riedel-type patterns in pure strike-slip conditions.

Structural controls on granite emplacement have been extensively discussed and documented in the literature (see reviews by Castro (1987), Hutton (1988) and Pitcher

From: VENDEVILLE, B., MART, Y. & VIGNERESSE, J.-L. (eds) *Salt, Shale and Igneous Diapirs in and around Europe*. Geological Society, London, Special Publications, **174,** 21–42. 1-86239-066-5/00/$15.00

(1992)). Close relationships between plutons and crustal-scale shear zones are particularly well documented. For strike-slip faults, many emplacement models have invoked local sites of extension, such as releasing bends (McCaffrey 1992; Lacroix *et al.* 1998) or pull-apart structures (Schmidt *et al.* 1990). Other models invoke regional transtensional regime because stretching lineations and the trend of the shear zones are oblique (Aranguren *et al.* 1997*a*). On the other hand, recent brittle–ductile analogue models have shown that pluton emplacement does not require the creation of holes, even under pure strike-slip boundary conditions (Román-Berdiel *et al.* 1997). These experiments have further shown that intrusion geometries and their relationships with faults developed in the brittle layers were diagnostic of both crustal rheological profile and boundary conditions. Field studies by Corriveau *et al.* (1998) and Benn *et al.* (1999) concluded that the nature of the host rock and the crustal structure–rheology and not the tectonic regime control the mechanics of emplacement. Moreover, pluton emplacement can also control the kinematics of deformation and the rheology of the host rocks. Crawford *et al.* (1987), Davidson *et al.* (1992), Karlstrom *et al.* (1993) and Chardon *et al.* (1999) explained the weakening of the crust and the high strain rates along ductile thrust zones by the presence of melted materials.

In previous experiments, we have examined intrusion within the upper brittle crust under static conditions or extensional regime (Román Berdiel *et al.* 1995), as well as during regional strike-slip (Román-Berdiel *et al.* 1997). These works are amongst the rare ones that incorporate an uppermost brittle layer where faulting of country rocks could accompany intrusion (Merle & Vendeville 1992; Román-Berdiel *et al.* 1995, 1997; Benn *et al.* 1998), which better fits the conditions of the continental crust. In the present paper, extension is combined with strike-slip to model intrusion in a brittle crust undergoing overall transtension. In contrast, during progressive injection compressional conditions developed at local scale over the roof of the intrusions.

Experimental procedure

Experiments on intrusions in a brittle crust have been performed using silicone putty and sand by Merle & Vendeville (1992, 1995), Román-Berdiel *et al.* (1995, 1997), and Benn *et al.* (1998). Román-Berdiel *et al.* (1995) have shown that properly scaled sand–silicone putty models can be used to study intrusion processes in the upper crust. In particular, they have shown that the occurrence of a ductile weak layer interbedded within the brittle crust was the first-order factor controlling the formation of a conformable laccolith versus a piercing diapir. Several previous experiments showed that weak layers, such as shales, pelites or salt, within the brittle crust were conveniently modelled using silicone layers (Ballard *et al.* 1987; Vendeville 1987; Richard *et al.* 1989, 1991; Basile 1990).

Materials, models and scaling used in this paper are similar as those previously used by Román-Berdiel *et al.* (1995, 1997).

Analogue materials and scaling

The sand used for the experiments is pure quartz with rounded grains with a maximum size of 500 μm. The sand shows Coulomb behaviour with negligible cohesion and frictional angle of 30–32°. Its density is *c.* 1500 kg m^{-3}. For ductile materials, we used silicone putty (GS1R gum from Rhône-Poulenc), which is almost perfectly Newtonian (Weijermars & Schmeling 1986; Vendeville *et al.* 1987). Two types of silicone have been used: a standard, high-viscosity silicone ($\mu = 1.5 \times 10^4$ Pa s, $\rho = 1231$ kg m^{-3}) was used to introduce soft ductile layers within the sand pack, and a low-viscosity silicone ($\mu = 1.3 \times 10^4$ Pa s, $\rho = 1151$ kg m^{-3}) was used for the intrusion. In the models, the intrusion is pressurized by moving a piston. Because in these experiments rigid plates supported the overburden, the density difference between injected silicone and sand is irrelevant, as it does not affect the pressure through differential loading. Viscosity of silicone putty depends strongly on temperature, decreasing with temperature rise (Nalpas & Brun 1993). All our models are run at a controlled room temperature of 28°C.

Our models were scaled for length, viscosity and time by means of methods discussed by Hubbert (1937) and Ramberg (1981). Scale ratios between models and natural examples are 10^{-5} for length (1 cm in the model represents 1 km in nature), 7.7×10^{-13} for viscosity (implying a viscosity of 10^{18} Pa s for crystallizing magma) and 3.6×10^{-8} for time (1 h of experiment represents 53 300 years in nature). The remaining magnitudes are imposed by the scaling. Linear injection rates of the order of 41 cm h^{-1} scale to 2.2×10^{-8} m s^{-1} (77 cm a^{-1}) in nature, and displacement velocity of 5 cm h^{-1} scales to 2.8×10^{-9} m s^{-1} (9 cm a^{-1}). The magma ascent rate is within the values for granitic magma given by various researchers (Clemens & Mawer 1992; Paterson & Tobisch 1992). The high displacement velocity obtained is characteristic of regions undergoing high strain rates favoured by weakening of the crust in the presence of melt (Karlstrom *et al.* 1993). In models, high displacement velocities are required to obtain an overall syn-intrusion deformation sufficiently large and a localized deformation pattern, for the brittle–ductile ratio of the rheological profile used in models. In scaling, limitations are imposed because some factors are either not well known or are difficult to model. In our experiments factors such as changes of magma viscosity during rising and cooling are not considered. In nature, viscosity in partially molten granitic systems can be rather high during the segregation process (10^{17} Pa s, Cruden 1990), decreasing during the rise through fractures (to 10^6 Pa s) because of high-pressure conditions (Marsh 1984), and increasing again when magma reaches the intrusion site and cools during ballooning (to 10^{19} Pa s, Wickham 1987; Kukowski & Neugebauer 1990). In experimental models limitations are imposed by analogue materials: injected silicone has a viscosity of 1.3×10^4 Pa s, implying a very high viscosity for crystallizing magma (10^{18} Pa s). No better analogue material for cristallizing magma is find at this point. Our experiments model magma when it has reached the intrusion site. In our experiments the motion was applied only during injection, to avoid post-emplacement deformation.

Despite their limitations, these models can be very useful in recognizing and examining the effects of critical factors, such as the crustal rheological profile and the tectonic regime, which control the geometry of intrusions.

Apparatus

The experiments were performed in the Analogue Modelling Laboratory of Géosciences, Rennes. The deforming box consists of two plastic sheets attached to two vertical end walls, sliding along two rigid horizontal plates, and diverging from

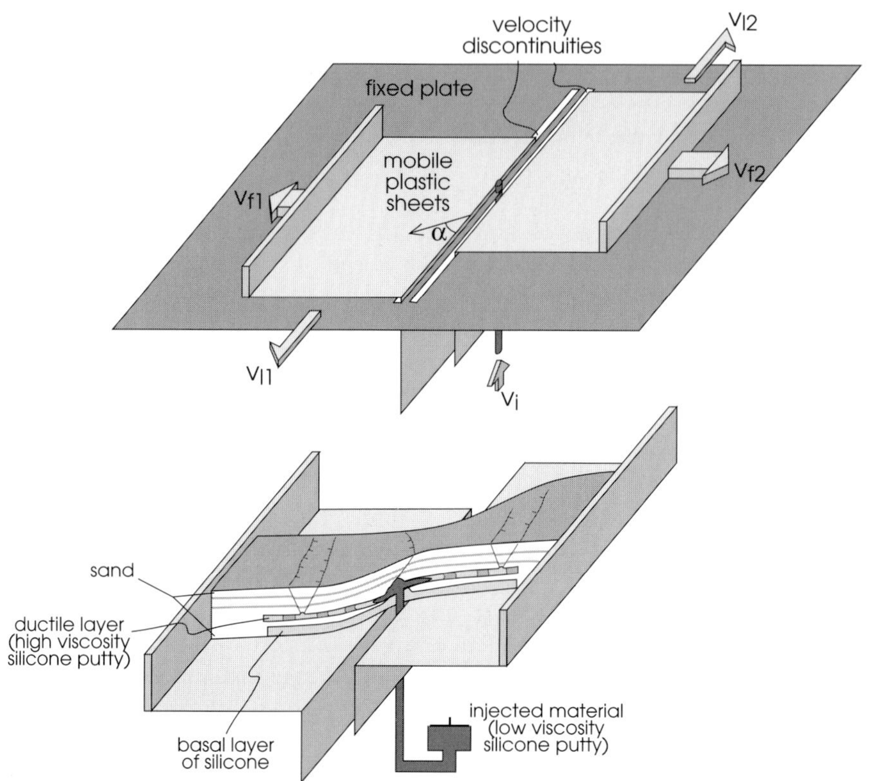

Fig. 1. Sketch of experimental setting (modified after Tron & Brun 1991). v_f is the frontal velocity, v_l is the lateral velocity, v_i is the injection velocity, α is the angle of divergence.

two narrow spaces (two velocity discontinuities, VD). An injection tube is located in a central space between these plates (Fig. 1). The vertical end walls are pulled perpendicular to the VD by screw jacks with a frontal velocity (v_f) equivalent to a pure shear component. The rigid horizontal plates are pushed by two other screw jacks at a lateral velocity (v_l) or simple shear component, giving a sinistral strike-slip displacement along the VD. Screw jacks are powered by stepping motors controlled by a computer. The relative magnitudes of (v_f) and (v_l) determine the obliquity α of the displacement direction with respect to the VD (Fig. 1 and Table 1). During motion, silicone putty was pushed in the tube by a piston driven at constant injection velocity (v_i) by another stepper motor (Fig. 1).

Symmetric transtensional injection (STTI) models are obtained when the two sheets are displaced at equal and opposite velocities ($v_{f1} = v_{f2}$ and $v_{l1} = v_{l2}$), with respect to the injection point. If only one basal sheet has been displaced ($v_{f2} = v_{l2} = 0$), asymmetric transtensional injection (ATTI) models develop with respect to the feeding pipe and silicone putty is injected beside the basal VD. Both symmetric VD experiments and asymmetric VD experiments have been performed. The final dimensions of both sheets are the same in the symmetric experiments, but in the asymmetric models only the moving sheet increases its size as deformation progresses (Figs 2 and 3).

Table 1. *Model characteristics and conditions for the experiments*

Series	Model	Brittle–ductile ratio (cm)	α (°)	Frontal deformation velocity ($cm\,h^{-1}$)	Lateral deformation velocity ($cm\,h^{-1}$)	Total deformation velocity ($cm\,h^{-1}$)	Injection rate ($cm^3\,h^{-1}$)	Duration (min)	Frontal/lateral displacement (cm)	Total displacement (cm)	Volume of injected silicone (cm^3)	Room temperature (°C)
Asymmetric	ATTI90	4.5/1.6	90	5.00	0.00	5	32	60	5.00/0.00	5	32	23.50
Asymmetric	ATTI75	4.5/1.6	75	4.82	1.28	5	32	60	4.82/1.28	5	32	29.85
Asymmetric	ATTI60	4.5/1.6	60	4.32	2.50	5	32	60	4.32/2.50	5	32	23.50
Asymmetric	ATTI45	4.5/1.6	45	3.54	3.54	5	32	60	3.54/3.54	5	32	27.35
Asymmetric	ATTI30	4.5/1.6	30	2.50	4.32	5	32	60	2.50/4.32	5	32	29.15
Asymmetric	ATTI15	4.5/1.6	15	1.28	4.82	5	32	60	1.28/4.82	5	32	27.85
Symmetric	STTI90	4.5/1.6	90	2.50 + 2.50	0.00 + 0.00	5	32	60	5.00/0.00	5	32	29.65
Symmetric	STTI75	4.5/1.6	75	2.41 + 2.41	0.64 + 0.64	5	32	60	4.82/1.28	5	32	29.20
Symmetric	STTI60	4.5/1.6	60	2.16 + 2.16	1.25 + 1.25	5	32	60	4.32/2.50	5	32	28.65
Symmetric	STTI45	4.5/1.6	45	1.77 + 1.77	1.77 + 1.77	5	32	60	3.54/3.54	5	32	27.90
Symmetric	STTI30	4.5/1.6	30	1.25 + 1.25	2.16 + 2.16	5	32	60	2.50/4.32	5	32	27.10
Symmetric	STTI15	4.5/1.6	15	0.64 + 0.64	2.41 + 2.41	5	32	60	1.28/4.82	5	32	27.30

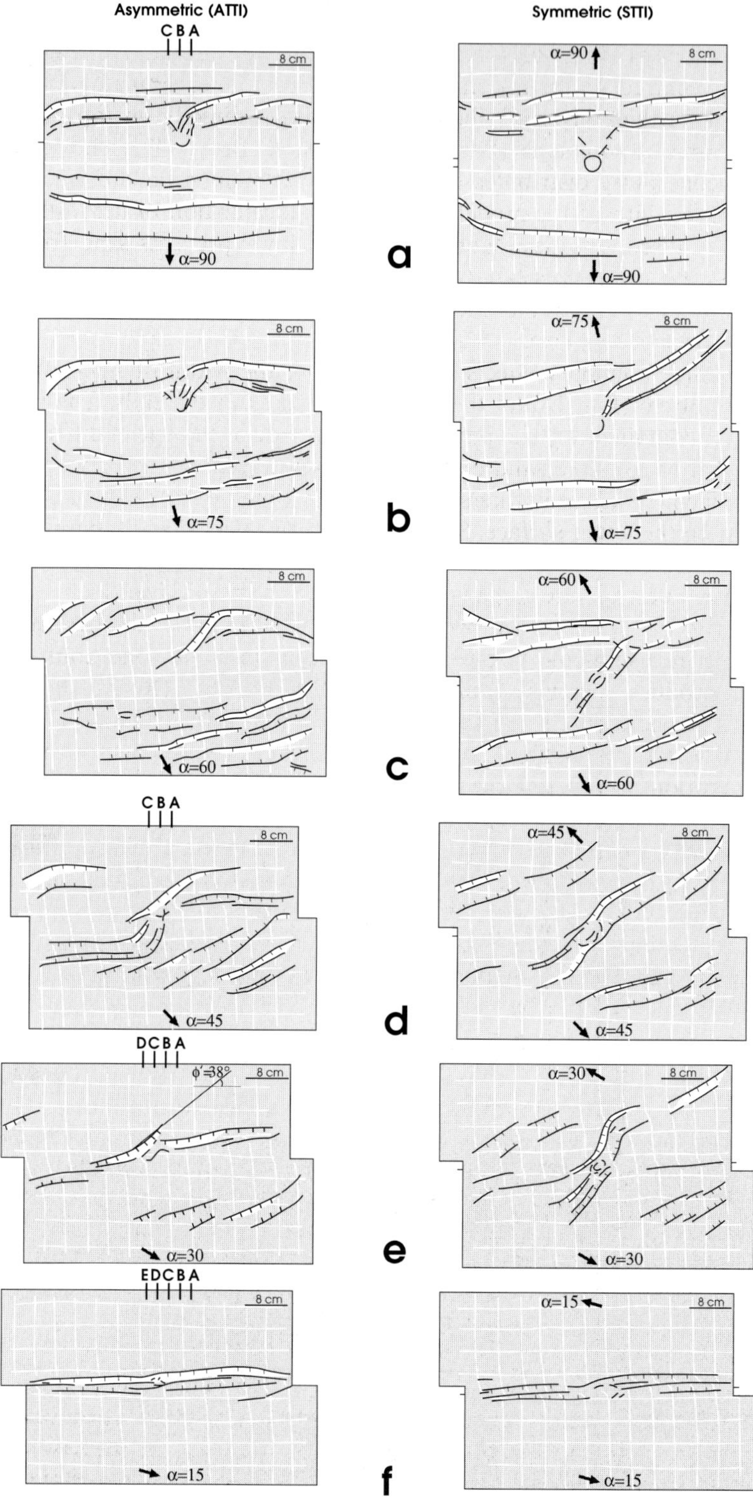
Asymmetric (ATTI)
Symmetric (STTI)
C B A
8 cm
α=90
a
α=75
b
α=60
c
d
α=45
D C B A
φ'=38°
α=30
e
E D C B A
α=15
f

Types of models and experiments

Total dimensions of the models are 56 cm length, 40 cm width and 6.1 cm height. The models were centred on the injection point. The base of each analogue model consisted of a 1 cm thick layer of silicone putty, placed directly on top of the plastic sheets. Previous experiments (Basile 1990; Tron & Brun 1991) showed that this layer of silicone putty distributes the imposed strike-slip shear deformation evenly over the entire width of the model and prevents localization of deformation above basal velocity discontinuities. Moreover, such a layer prevents sand from pouring down into the narrow spaces. Sand of various colours with the same mechanical properties was poured on top of the silicone to produce a 4.5 cm thick sand pack. A interbedded ductile (silicone) layer 0.6 cm thick was located in the sand pack at a depth of 3 cm. To reduce boundary effects to a minimum, the width of silicone layers was restricted to 30 cm (Fig. 1) and they were unconfined at the ends. The interbedded silicone layer contained vertical passive markers made of silicone of different colours. A square grid of coloured sand markers was then traced on the upper free surface. Sand and silicone passive markers allowed us to observe the deformation in cross-section and on the upper surface. Silicone putty was injected at the base of the interbedded ductile layer.

For a given injection rate and a given injected volume, the layer thicknesses chosen in the experiments of the present work correspond to intermediate situations, where conformable intrusions of laccolithic shapes occur in static conditions, but where intrusions produce incipient faulting in the overburden (Román-Berdiel *et al.* 1995). This choice ensured that interactions between intrusions and faulting in the overburden would be possible, and that the effects of faulting as a result of transtensional motion would not be totally overprinted by deformations induced by piercement of the overburden.

In all models, the thicknesses of sand and silicone layers are in the ratio 4.5:1.6. Silicone was injected at a depth of 3.6 cm, which scales with 3.6 km in nature. All experiments are performed with a total displacement velocity $v_t = 5\,\mathrm{cm\,h^{-1}}$ (Table 1). In all experiments, the injection velocity is $v_i = 1.64\,\mathrm{cm\ h^{-1}}$ at the piston ($41.5\,\mathrm{cm\,h^{-1}}$ at the bottom of the model) and the injected volume was $32\,\mathrm{cm^3}$, so the injection rate is $32\,\mathrm{cm^3\,h^{-1}}$. Injection and displacement start and end at the same time. Two series of experiments have been performed, for symmetric and asymmetric VD, with $\alpha = 15°, 30°, 45°, 60°, 75°$ and $90°$ (Table 1). The total displacement (incorporating extension and strike-slip components) was 5 cm in all experiments. Top surface photographs were taken at regular time intervals during experiments, which allowed us to observe progressive deformation in surface view. At the end of the experiments, models were cut serially into sections perpendicular to the VD.

Fig. 2. Line drawing from top view photographs for experiments of both asymmetric and symmetric series. **(a)** $\alpha = 90°$; **(b)** $\alpha = 75°$; **(c)** $\alpha = 60°$; **(d)** $\alpha = 45°$; **(e)** $\alpha = 30°$; **(f)** $\alpha = 15°$. ϕ' is the angle of the strike of the T fault with respect to the VD. A–E refer to serial cross-sections made perpendicular to the VD at the final stage, which were used to make the horizontal sections in Figs 5 and 6.

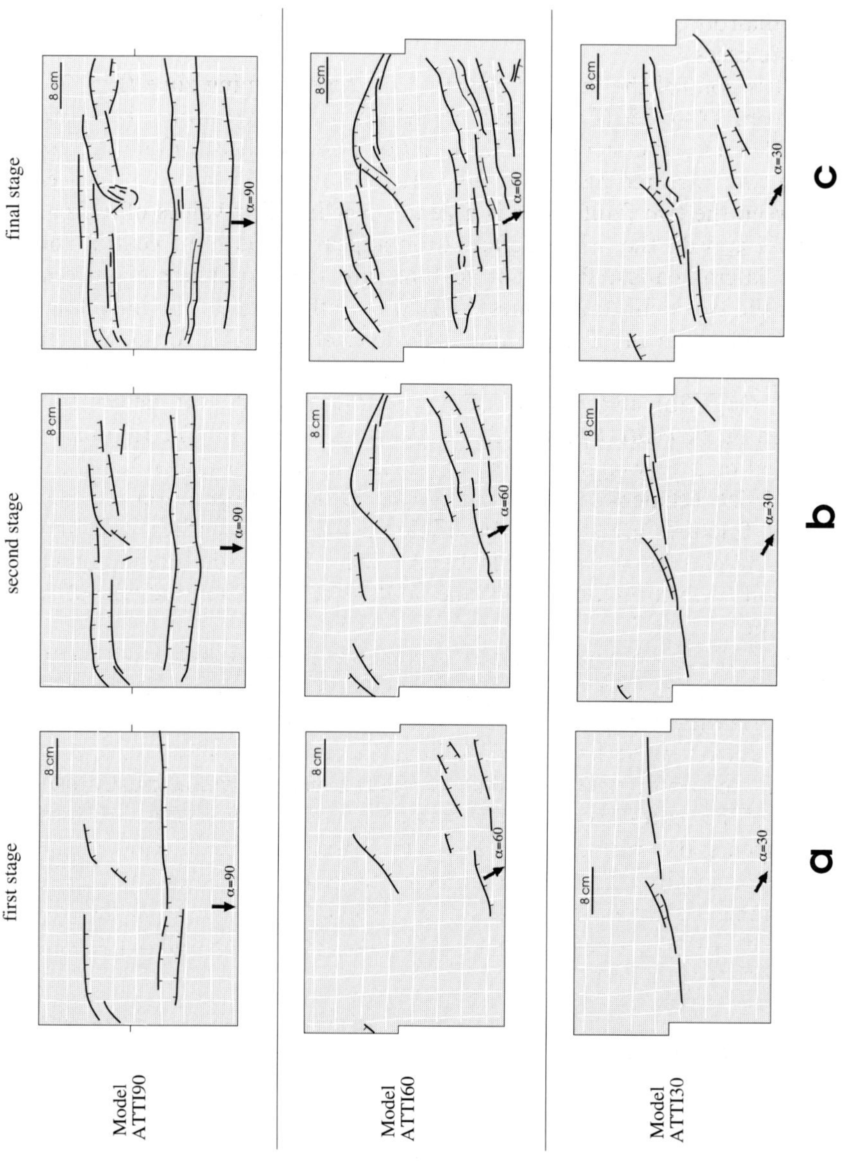

Fig. 3. Top views showing fault development in asymmetric experiments for $\alpha = 30°$, $\alpha = 60°$ and $\alpha = 90°$. Bulk displacement is 1.64 cm (first stage), 2.5 cm (second stage) or 5 cm (final stage).

Experimental results

Fault pattern

At the final stage of the experiments, in surface view, the deformation zone is as wide as the effective width of the model (width of the basal silicone layer) for α values ranging from 90° to 30° (Fig. 2a–2e). The deformation zone is narrower for α values lower than 30° (Fig. 2f). These observations were made in both the asymmetric and symmetric cases.

The α values control the internal structure of the deforming zone (Fig. 2). For values of α from 90° to 75° (Fig. 2a and b), the surface deformation is concentrated in two parallel bands, on either side of the injection point, with an unfractured central domain. The two fractured bands are characterized by graben systems. The central domain between the two fault zones is a gentle syncline. For values of α from 60° to 45° (Fig. 2c and d), the brittle deformation appears in both the central domain and the two lateral bands. Particular features are the occurrence of en echelon faults within the two main fault zones, located on either side of the intrusion, and an oblique band made of 'T type' faults (Wilcox *et al.* 1973) forming a graben above the injection point. The onset of this oblique zone can be seen in models with $\alpha = 75°$ (Fig. 2b) and $\alpha = 30°$ (Fig. 2e). For $\alpha < 30°$, the deformation is mainly located in the central part of the model (Fig. 2e and f). The localization of brittle deformation is very high. Its geometry is close to strike-slip, with R and Y fractures (Riedel 1929; Tchalenko 1970; Bartlett *et al.* 1981) developed over the VD. For $\alpha = 30°$ the T fractures are still clearly visible on the model surface, and some en echelon faults develop in the two parallel bands. The strike-slip deformation becomes more important, with formation of oblique faults over the VD in the case of asymmetric experiments.

Differences between the symmetric and asymmetric experiments are systematically observed except when $\alpha = 15°$. In general, in symmetric experiments the deformation is also symmetric with respect to the VD, and in asymmetric series the fractured band located above the diverging basal sheet is the more developed (Fig. 2). For α values from 90° to 75°, in symmetric experiments, the fault zones are symmetrically disposed on both sides of the central VD, and intrusion occurs in the central domain. In asymmetric experiments, the fractured zone above the basal fixed plate is close to the VD and the intrusion localizes small-scale faults in this zone (Fig. 2a and b).

In some symmetric experiments, the strike of some faults changes at their tips to become parallel to the unconfined borders of the model, probably as a result of boundary effects (Fig. 2b, d and e).

Progressive deformation

Some general features reflect the progressive development of fault patterns during increasing deformation in both symmetric and asymmetric series. Here, we present illustrations for the asymmetric case only (Fig. 3). In general, faults initiate with the creation of half-grabens (Fig. 3a). They rapidly evolve to full grabens with increasing deformation (Fig. 3b). Once formed, the grabens do not widen and new subparallel grabens form (Fig. 3c). This general evolution is consistent with experimental results of Tron & Brun (1991) in oblique rifting without injection. In our models the intrusion does not control the general deformation pattern, but deflection

ATTI90
a

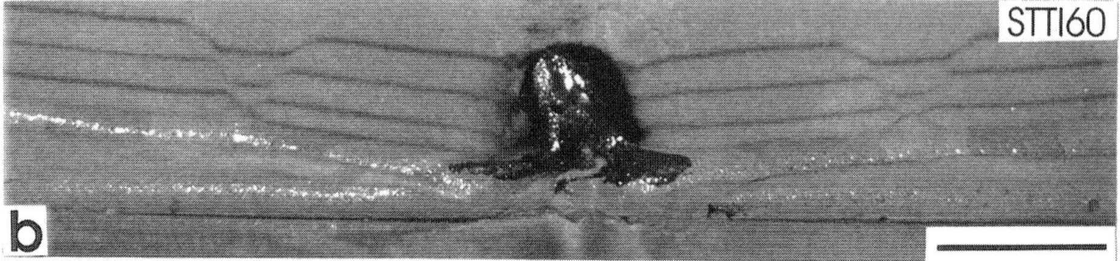
STTI60
b

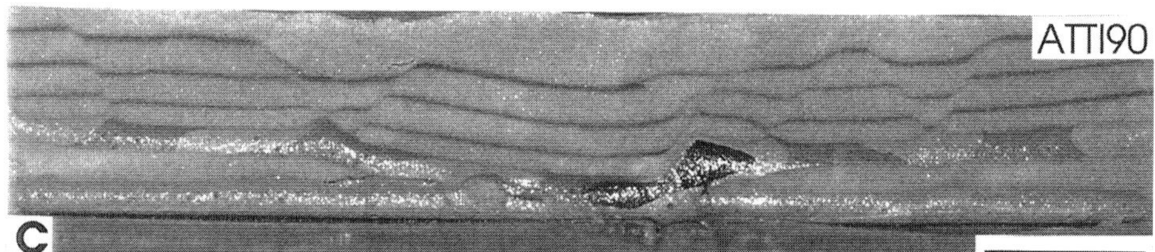
ATTI90
c

STTI75
d

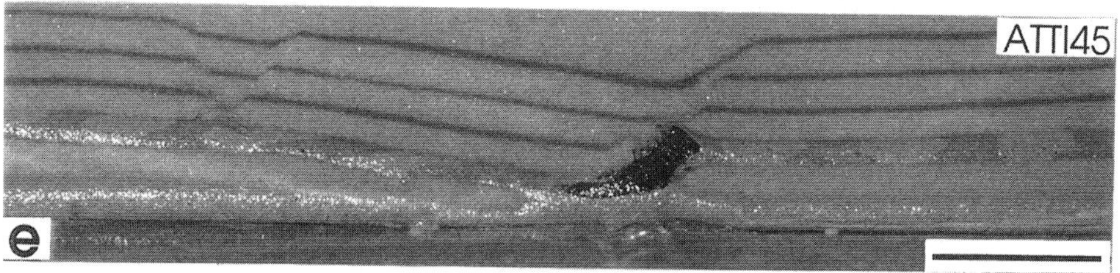
ATTI45
e

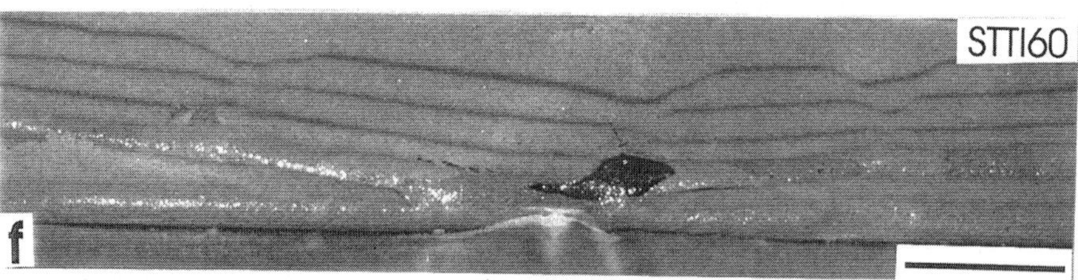
STTI60
f

of faults developed in the central part of the models occurs. For α values of 90° the two parallel bands to the side of the VD initiate at the same time. The intrusion deflects the strike of normal faults, from the first stages (Fig. 3, Model ATTI90). For $\alpha = 60°$ the first structure to appear is a listric normal fault (cross-section, Fig. 4e) above the intrusion with T-fracture orientation. Later, en echelon normal faults develop on both sides of the VD (Fig. 3, Model ATTI60). For $\alpha = 30°$ the first structure to appear is a T fracture above the intrusion, as observed for $\alpha = 60°$. Then, an oblique strike-slip fracture developed above the VD (Fig. 3, Model ATTI30). In a third stage, en echelon normal faults appear. For $\alpha = 15°$ the deformation is located in the central part of the model (Fig. 2f). Strike-slip faults parallel to the VD appeared above it from the first stages.

Geometry of intrusions

For both symmetric (STTI) and asymmetric (ATTI) experiments, vertical sections perpendicular to the VD and across the central part of the models show an intrusive body that pierces the overburden, with partial spreading of the injected silicone within the ductile layer. In general, the geometry is that of a bowler hat (Fig. 4a and b). Piercing occurs only in the central section of the models, and is favoured by thinning and fracturing of the overburden, produced by the transtensional tectonic regime. With a similar rheological profile under static conditions (no active tectonic regime) failure of the overburden is produced by the intrusion stress alone, but there is no piercement, and the injected silicone produces concordant lens-shaped intrusions (Román-Berdiel *et al.* 1995). In sections across the central part of intrusions, differences between symmetric and asymmetric transtensional regimes are expressed by an important offset of the intrusion toward the mobile basal plate in the case of an asymmetric regime (Fig. 4a), and a more symmetric intrusion centred on the injection point for a symmetric transtensional regime (Fig. 4b). In sections perpendicular to the VD and across the lateral part of intrusions, injected material remains confined in the soft level, and the interaction between the intrusion process and faults can be better observed (Fig. 4c–f).

Experiments were grouped in three types depending on their surface deformation pattern whatever the symmetric or asymmetric character of the deformation. The three groups have different shape of intrusion, as seen in vertical cross-sections (Fig. 4c–f) and in horizontal sections (Fig. 5). The first group corresponds to α values of 90° and 75°, for which structures mainly reflect extension. The main extensional structures developed in the overburden are two graben systems localized in two domains, on each side of the intrusion (Fig. 2a and b). In cross-section the overburden is affected by steep-dipping normal faults, which evolve to sub-horizontal ductile shear zones at the deepermost ductile layers. Also, intrusions are affected by normal faults and the injected material spreads into the soft layer in both walls of the fault, with a greater volume intruded in the footwall (Fig. 4c, lateral section of

Fig. 4. Photographs showing some sections of experiments for asymmetric (**a**, **c** and **e**) and symmetric VD (**b**, **d** and **f**), cut across intrusions and perpendicular to the VD. (**a**) and (**c**) correspond to sections B and A (see Fig. 2a and 5a), respectively, of Model ATTI90; (**b**) and (**f**) central and lateral sections, respectively, of Model STTI60; (**d**) lateral section of Model STTI75; (**e**) section A (see Fig. 2d and 5b) of Model ATTI45. The scale bars represent 5 cm.

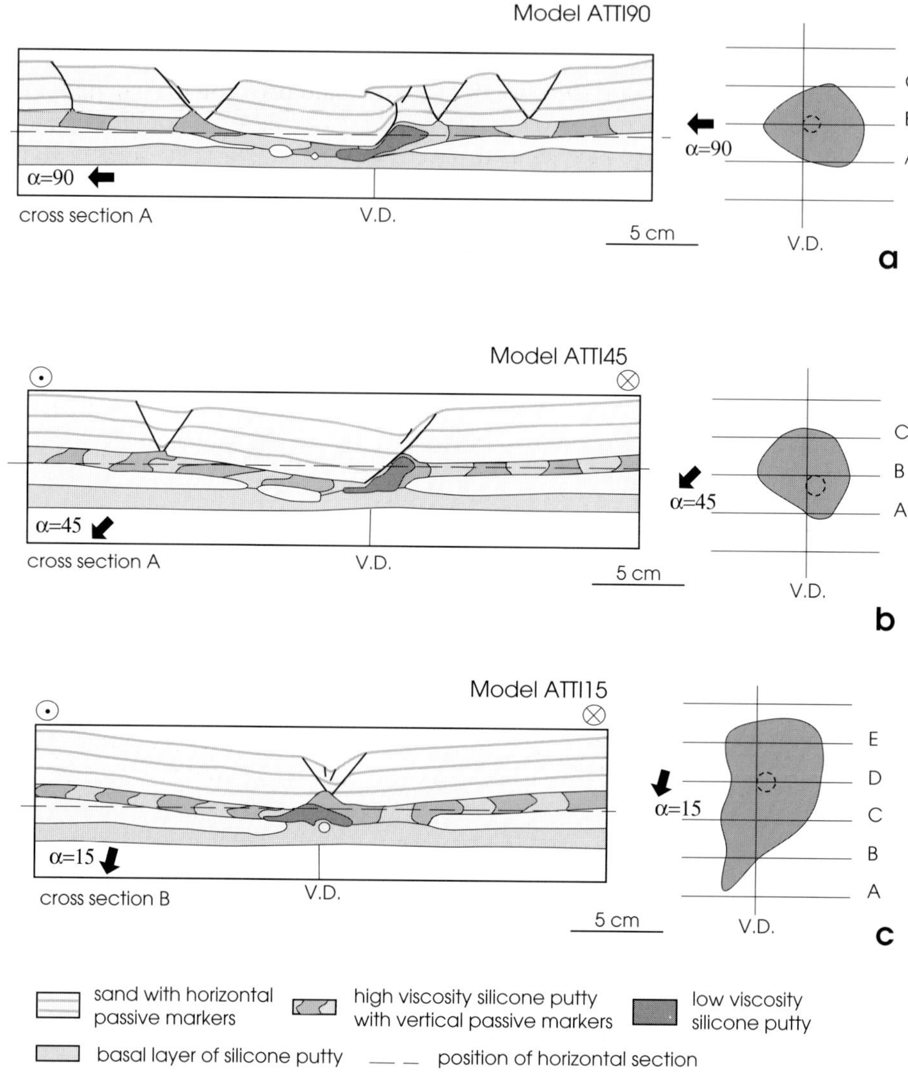

Fig. 5. Representative vertical and horizontal sections for transtensional experiments performed with asymmetric velocity discontinuity. (**a**) $\alpha = 90°$; (**b**) $\alpha = 45°$; (**c**) $\alpha = 15°$. Dotted circle on maps is the trace of underlying feeding pipe. The horizontal line is the trace of the basal velocity discontinuity. A–E refer to serial cross-sections made perpendicular to the VD at the final stage and used to make the drawing.

Model ATTI90; Fig. 4d, lateral section of Model STTI75). In horizontal view, intrusions show elliptical geometry gently elongated in the stretching direction (Fig. 5a, Model ATTI90).

For α values of 60° and 45° there is an interaction between the injected material and the listric faults developed above the intrusion. From surface photographs taken at regular time intervals during experiments (Fig. 3, Model ATTI60) we infer that, in the first deformation stages, the injected material controls the position of the listric fault that forms on top of it. As deformation proceeds, the injected fluid spreads in the soft level but only in the footwall of the listric faults (Fig. 4e and f). Intrusion does not spread in the hanging wall despite the existence of the soft level. The fault acts as a barrier, preventing the injection of silicone in the ductile layer. This can be explained by considering that the fault is the limit between the sand pack (on the hanging wall) and the silicone layer (in the footwall). The sand pack behaves as the limit for the expansion of the intruding silicone, as it occurs with the upper sand layer in static experiments (Román-Berdiel *et al.* 1995). From this we infer that, once formed, the listric fault, together with the rheological profile, controls the site of intrusion. In horizontal view, intrusions show subcircular geometry (Fig. 5b).

For $\alpha = 30°$, the T fractures are still clearly visible in surface view (Fig. 2e) and control the shape of intrusion at depth (Fig. 6a), although the strike-slip deformation is dominant. Comparing the horizontal section of Model ATTI30 (Fig. 6a) with its surface view (Fig. 2e), it can be observed that the left boundary of the intrusion is controlled by the T fault (compare the angle between the T fault and the VD, ϕ', in Fig. 2e with the angle between the local boundary of intrusion and the VD, ϕ, in Fig. 6a). For

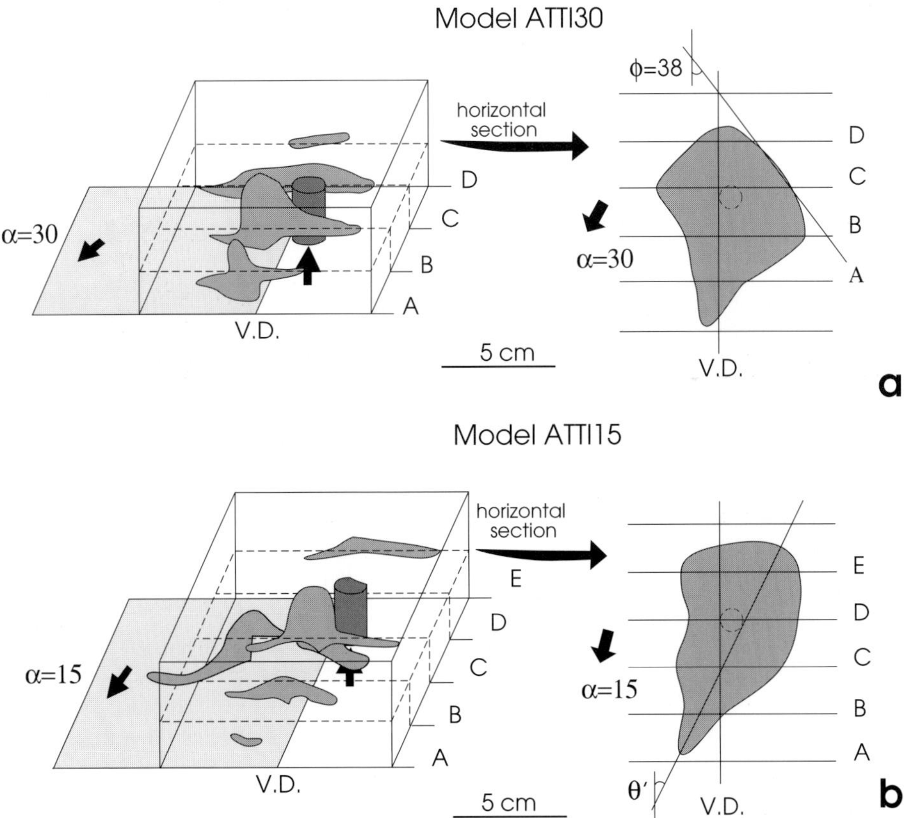

Fig. 6. Block diagrams and horizontal sections for transtensional experiments performed with asymmetric velocity discontinuity. (**a**) $\alpha = 30°$; (**b**) $\alpha = 15°$. Dotted circle on maps is the trace of underlying feeding pipe. The horizontal line is the trace of the basal velocity discontinuity. A–E refer to serial cross-sections made perpendicular to the VD at the final stage and used to make the drawing. ϕ, local angle between intrusion boundary and VD direction; θ', angle between long axis of intrusion and VD direction.

$\alpha = 15^\circ$ the intrusion shows an asymmetric drop shape marked by a sheared tail trailing behind the intrusion (Fig. 5c and b). This result is similar to those obtained for pure strike-slip ($\alpha = 0^\circ$, Román-Berdiel *et al.* 1997). In pure strike-slip experiments, the theoretical angle θ between the principal stretching direction and the shear zone boundary ($\tan 2\theta = 2\gamma$) is close to the angle θ' measured in models between the long axes of intrusions and the shearing direction (Fig. 6b). From this consistency between theoretical and experimental values, Román-Berdiel *et al.* (1997) inferred that the long axis of model intrusions tends to track the long axis of the bulk strain ellipsoid.

In summary, intrusion geometries can be grouped into three types according to the angle of divergence α (Fig. 5). For α values of 90° and 75° (Fig. 5a) intrusions are sub-circular to gently elongate in the principal stretching direction in horizontal view. In cross-section intrusions are affected by listric normal faults and the injected material spreads into the ductile layer in both walls of the fault. For α values of 60° and 45° (Fig. 5b) intrusions are located in the footwall of the listric fault. In horizontal view intrusions show sub-circular geometry. For $\alpha = 15^\circ$ (Fig. 5c) intrusions are elongate and show asymmetric drop shapes with a sheared tail trailing behind the intrusion.

Discussion and comparison with some natural examples

Our experiments simulate granite intrusions in a transtensional regime at upper-crustal levels. The main features of fault patterns are consistent with experimental results on oblique rifting reported by Tron & Brun (1991). In these experiments, oblique rifting is characterized by en echelon fault patterns and the mean fault trends are not perpendicular to the displacement direction (this angle decreases as α decreases). For high α values (90° and 60°) deformation localizes in two parallel subzones. For medium α values (45°, 30°) the major faults crosscut the whole deforming zone, and a partition arises creating distinct families of oblique-slip faults and strike-slip faults. Our models provide two new structural features: deformation is much more concentrated in two parallel corridors with normal faults than in the models by Tron & Brun (1991) because we used a higher brittle–ductile ratio (Davy *et al.* 1995); the central domain between these two corridors is deformed by an oblique graben at high angle to the displacement direction (Fig. 2). This angle decreases as α decreases in the same way as the en echelon fault pattern of Trons' experiments. These oblique grabens were probably induced by the incorporation of the intrusive materials in the experiments because they are systematically developed above the injection point.

In our models with $\alpha = 45^\circ$ and 60°, intrusions spread into the ductile layer along the footwalls of the listric normal faults developed in the brittle cover and emplacement occurs in fault-bounded laccoliths (see Fig. 4e and f). Similar relationships have been observed on 3D seismic images of normal fault related salt structures, and compared with granitic intrusions that appear to have been emplaced during extension on the footwall side of active normal faults (Quirk *et al.* 1998).

In the Variscan belt of Spain the Hombreiro granite (Aranguren & Tubía 1992; Aranguren 1997) and the Los Pedroches batholith (Aranguren *et al.* 1997*a*) were emplaced in the footwall side of the Vivero and the Conquista faults, respectively, two extensional shear zones (Martínez Catalán 1985; Aranguren *et al.* 1997*b*).

The Vivero fault is a normal shear zone with a subordinate dextral motion (Martínez Catalán 1985). Its curved trace is outlined by a row of synkinematic granites of Carboniferous age, which were highly deformed by the fault and are limited to the footwall block (Fig. 7). At first interpreted as a thrust fault (Ries & Shackleton 1971; Marcos 1973; Matte & Ribeiro 1975; Martínez Catalan *et al.* 1977; Bard 1978; Pérez Estaún 1978), the extensional nature of the Vivero fault is now widely accepted but its origin is still the subject of some controversy (Pérez-Estaún *et al.* 1991; Aranguren & Tubía 1992; Martínez Catalán *et al.*, 1992; Martínez *et al.* 1996). It could be a reactivated fault; either a Hercynian thrust related to the

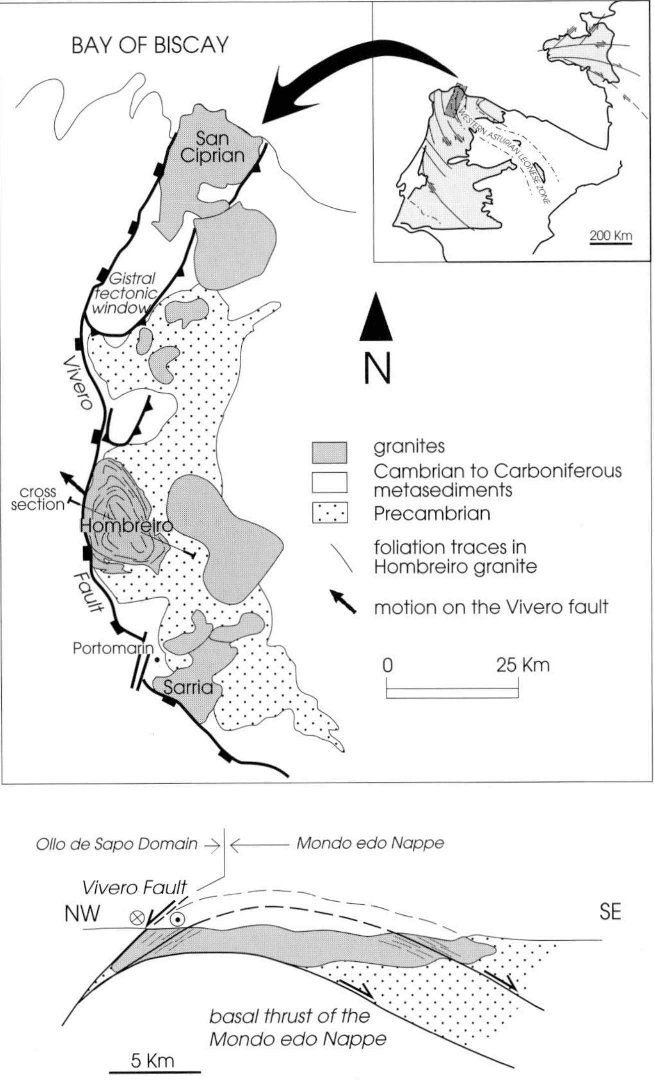

Fig. 7. Geological and structural sketch map and cross-section of the area around the Hombreiro granite (modified after Aranguren & Tubía 1992).

previous crustal thickening or an even older (Ordovician to Lower Silurian) synsedimentary normal fault (Martínez Catalán *et al.* 1992). It has also been interpreted as a newly formed extensional fault developed at the rear of a Hercynian antiformal stack of thrust sheets (Pérez-Estaún *et al.* 1991; Aranguren & Tubía 1992; Martínez *et al.* 1996). Whatever its origin may be, the Vivero fault fits the main requirement of our analogue models, as both faulting and intrusion evolved simultaneously. It is closely associated with the intrusion of the Hombreiro granite, whose structure and 3D shape were well constrained from field, anisotropy of magnetic susceptibility and gravity data (Aranguren & Tubía 1992; Vigneresse 1995; Aranguren 1997). This massif is sheet like extending down to a depth of only 2 km. It displays an ellipsoidal contour, elongate toward the stretching lineation in map view (Fig. 7). This pluton intruded Precambrian slates, shales and gneiss, and the emplacement level can be related to the basal thrust of the Mondoñedo Nappe. The petrofabric data clearly show that the granites deformed as a result of the oblique motion of the Vivero fault during temperature change from sub-solidus to low temperatures. These data imply that the cooling of the Hombreiro granite and the motion of the Vivero fault were coeval (Aranguren & Tubía 1992).

Given the obliquity between the strike of the Vivero fault and the trend of the stretching lineation (Fig. 7), this natural example can be compared with our models with α values of 45° and 60°. In these analogue models the main structure developed in the overburden is a listric fault above the intrusion (see Fig. 2c and d). The injected silicone pierces the footwall of this structure and spreads into the soft level, generating a flat intrusive body (see Fig. 4e and f). In map view, plutons associated with the Vivero fault have rather circular shapes (Fig. 7). This feature is also consistent with a dominant normal component along the fault, compared with the strike-slip component (large value of α) (Fig. 5a and b). Another important result of our models is that intrusions localize strain from the first stages of deformation, and therefore nucleate the listric fault (see Fig. 3). This result corroborates the hypothesis of Aranguren & Tubía (1992) that granites served as the nucleation site for the Vivero fault.

The Los Pedroches batholith, located in the southern part of the Central Iberian zone (Variscan belt) in southern Spain (Fig. 8), is an elongate batholith (>200 km long) composed of a large granodiorite massif and several late granite bodies emplaced in very low grade metamorphic rocks: slates, phyllites and chlorite-bearing schists of Precambrian to Viséan ages (Quesada 1987). This batholith was emplaced along the boundary between the mechanically strong pre-Carboniferous country rocks and the overlying Carboniferous plastic shales. In the same way as the Hombreiro pluton, this pluton crops out in the footwall side of an extensional shear zone, the Conquista fault. The emplacement of the Los Pedroches batholith was clearly controlled by a blind dextral transtensional shear zone at crustal scale (Aranguren *et al.* 1997*a*). A subordinate transtensional shear zone, the Conquista fault (Fig. 8), developed along more than 50 km in the northern contact of the batholith during its intrusion at shallow crustal levels (Aranguren *et al.* 1997*b*). The Conquista fault deforms the Los Pedroches granodiorite but is sealed by the subsequent intrusion of the Cerro Mogabar granite (Fig. 8). This observation implies that its activity was coeval with the emplacement of the Los Pedroches batholith. Deformation related to the Conquista fault leads to strongly foliated granodiorites with S–C microstructures. Stretching lineations are oblique to the trend of the Conquista fault and plunge 35° to the ENE (Aranguren *et al.* 1997*b*). The dominant

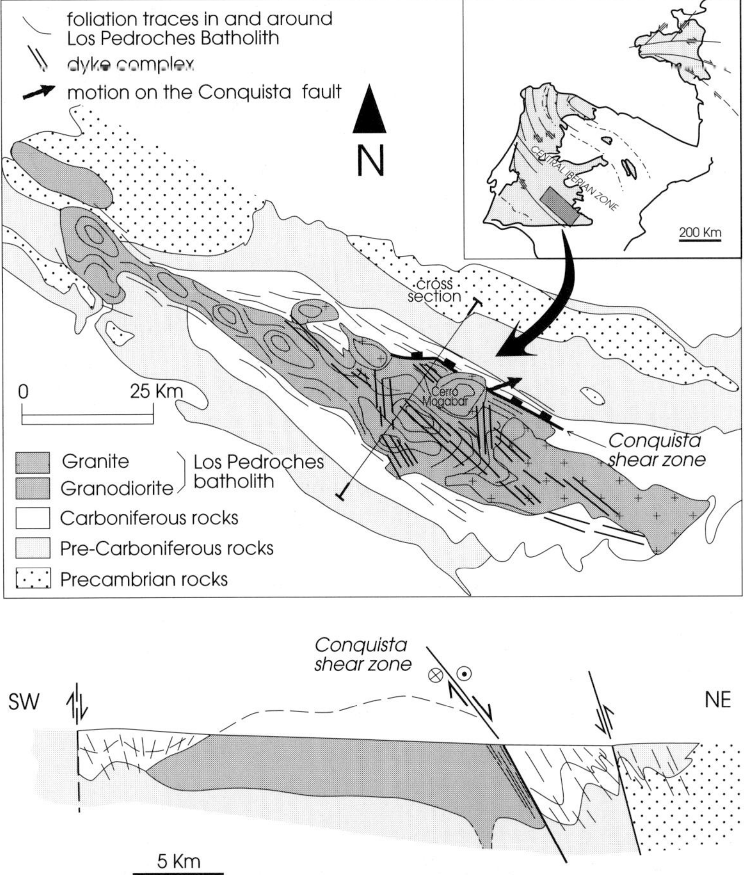

Fig. 8. Geological and structural sketch map and cross-section of the area around the Los Pedroches batholith (modified after Aranguren *et al.* 1997*a*).

orientation of the magmatic foliation of the Los Pedroches batholith is sub-horizontal and the batholith has been interpreted as a sub-horizontal sheeted granite (Aranguren *et al.* 1997*a*).

Given the obliquity between the strike of the Conquista fault and the trend of the stretching lineation, this batholith can be compared with our models with $\alpha = 45°$ and $\alpha = 60°$. However, the Los Pedroches batholith shows a very elongate shape parallel to the Conquista fault, a feature observed for experimental intrusions associated with dominant strike-slip components (see Fig. 5c and Román-Berdiel *et al.* (1997)). The elongate shape of the Los Pedroches batholith contrasts with the sub-circular geometry of intrusions developed during our transtensional experiments (Fig. 5a and b). This apparent contradiction may reflect that our models have only one fixed feeding pipe, whereas the Los Pedroches batholith is a composite laccolith consisting of at least five coalescent plutons, with different feeding pipes aligned in parallel with a blind, 120°E-trending dextral shear zone (Aranguren *et al.* 1997*a*). On the other hand, a direct comparison of the shape of these two natural examples is not straightforward, as the outcrop shape of a granite is also dependent on the erosional level.

The granodiorites of the Los Pedroches batholith are intruded by a well-developed dyke swarm of granites and granodiorites (Fig. 8). The orientation of the oldest N–S-trending dyke swarm is also consistent with the subordinate fault pattern developed at high angle to the VD (compare Fig. 8 with Fig. 2c, symmetric case).

In summary, the Vivero and the Conquista faults, two major dextral extensional shear zones of the Iberian Massif (Variscan belt of Spain) can be compared with the listric faults developed above the intrusions in our models with $\alpha = 45^\circ$ and 60°. The emplacement site of the Hombreiro pluton (basal thrust of the Mondoñedo Nappe and Precambrian slates, shales and gneiss) and the Los Pedroches batholith (boundary between the pre-Carboniferous and the Carboniferous country rocks) can be compared with the ductile layer of our models interpreted as a major discontinuity within the brittle crust. Our models suggest that the intrusion of most leucogranites related to the Vivero fault (San Ciprian, Hombreiro and Sarria) and the Los Pedroches batholith played a major role in controlling the location of both the Vivero and the Conquista faults. On the other hand, these faults have acted as barriers for magma ascent, limiting their spreading in the hanging wall and controlling their shape and site of emplacement.

Conclusions

Experiments with a soft level interbedded into the brittle cover favour localization and horizontal spreading of the igneous body in the incompetent level. This condition has been shown to be a first-order factor controlling the formation of either a conformable laccolith or a piercing diapir (Román-Berdiel *et al.* 1995). The more important results of the present study are as follows:

(1) Intrusions localize strain from the first stages of deformation.

(2) Intrusions result in partially conformable laccoliths with a bowler-hat geometry, whereas entirely conformable laccoliths were obtained in static conditions for the same rheological profile (Román-Berdiel *et al.* 1995). This difference is due to the thinning and fracturing of the overburden produced by the extension, which decreases the overall brittle–ductile ratio of the crustal section.

(3) Intrusions show an important offset towards the mobile basal plate for an asymmetric transtensional regime. They are more symmetric and centred on the injection point for a symmetric transtensional regime.

(4) The geometry of intrusions is controlled by the faults developed in the overburden. The significance of this control depends upon the angle of divergence α. For low α values (15°) the dominant tectonic regime is strike-slip. There, faults localize in the central part of the model, above the VD, and allow local rising of the injected fluid across the overburden; but intrusions elongate in the principal extension direction determined by the bulk strain field and have an asymmetric drop shape in horizontal section. For α values of 45° and 60°, the main structure developed in the overburden is a listric fault above the intrusion. This fault controls strongly the geometry of intrusions. The injected silicone intrudes the footwall of the listric fault. The injected fluid does not spread in the hanging wall despite the existence of the soft level because the listric fault acts as a barrier. For α values of 75° and 90° the dominant tectonic regime is extensional, and the main extensional structures developed in the overburden concentrate in domains localized on both sides of the

intrusion. The igneous body is circular in horizontal view and is deformed by normal faults, but the injected material spreads in both walls of the faults.

Special thanks are due to J.-J. Kermarrec for invaluable technical assistance. A. Puzo is thanked for her help in the laboratory. This paper has benefited greatly from discussion with A. Casas and O. Dauteuil, to whom we are grateful. J. L. Vigneresse, F. Odonne and D. Grujic are thanked for their constructive reviews. This work forms part of research project PB96-1452-C03-03 (DGICYT).

References

ALLEN, M. B., MCDONALD, D. I. M., ZHAO XUN, VINCENT, S. J. & BROUET-MENZIES, C. 1998. Transtensional deformation in the evolution of the Bohai Basin, northern China. *In*: HOLDSWORTH, R. E., STRACHAN, R. A. & DEWEY, J. F. (eds) *Continental Transpressional and Transtensional Tectonics*. Geological Society, London, Special Publications, **135**, 215–229.

ARANGUREN, A. 1997. Magnetic fabric and 3D geometry of the Hombreiro–Sta Eulalia pluton: implications for the Variscan structures of eastern Galicia, NW Spain. *Tectonophysics*, **273**, 329–344.

—— & TUBÍA, J. M. 1992. Structural evidence for the relationship between thrusts, extensional faults and granite intrusions in the Variscan belt of Galicia (Spain). *Journal of Structural Geology*, **14**, 1229–1237.

——, CUEVAS, J., TUBÍA, J. M., CARRACEDO, M. & LARREA, F. J. 1997*b*. La faille de Conquista: caractérisation structurale et cinématique d'un cisaillement ductile dextre B composante normale dans le domaine sud de l'Arc Ibéro-armoricain. *Comptes Rendus de l'Académie des Sciences*, **325**, 601–606.

——, LARREA, F. J., CARRACEDO, M., CUEVAS, J. & TUBÍA, J. M. 1997*a*. The Los Pedroches Batholith (Southern Spain): polyphase interplay between shear zones in transtension and setting of granites. *In*: BOUCHEZ, J. L., HUTTON, D. H. W. & STEPHENS, W. E. (eds) *Granite: from Segregation of Melt to Emplacement Fabrics*. Kluwer, Dordrecht, 215–229.

BALLARD, J. F., BRUN, J. P., VAN-DEN-DRIESSCHE, J. & ALLEMAND, P. 1987. Propagation des chevauchements au-dessus des zones de décollement: modèles experimentaux. *Comptes Rendus de l'Académie des Sciences*, **305**, 1249–1253.

BARD, J. P. 1978. Contribution au problème de la signification des ceintures métamorphiques dans les orogènes antémésozoïques: éléments d'approche dans la virgation Galicio-Armoricaine. *In*: *Geología de la parte Norte del Macizo Ibérico*. Cuadernos del Seminario de estudios Cerámicos de Sargadelos, Sada–La Coruña, 73–92.

BARTLETT, W. L., FRIEDMAN, M. & LOGAN, J. M. 1981. Experimental folding and faulting of rocks under confining pressure—IX. Wrench faults in limestone layers. *Tectonophysics*, **79**, 255–277.

BASILE, C. 1990. *Analyse structurale et modélisation analogique d'une marge transformante. Exemple de la marge de Côte d'Ivoire–Ghana*. Mémoires et Documents du CAESS, **39**, 220.

—— & BRUN, J. P. 1999. Transtensional faulting patterns ranging from pull-apart basins to transform continental margins: an experimental investigation. *Journal of Structural Geology*, **21**, 23–37.

BENN, K., ODONNE, F. & DE SAINT BLANQUAT, M. 1998. Pluton emplacement during transpression in brittle crust: new views from analogue experiments. *Geology*, **26**, 1079–1082.

——, ROEST, W. R., ROCHETTE, P., EVANS, N. G. & PIGNOTTA, G. S. 1999. Geophysical and structural signatures of syntectonic batholith construction: the South Mountain Batholith, Meguma Terrane, Nova Scotia. *Geophysical Journal International*, **136**, 144–158.

CASTRO, A. 1987. On granitoid emplacement and related structures. A review. *Geologische Rundschau*, **76**, 101–124.

CHARDON, D., ANDRONICOS, C. L. & HOLLISTER, L. S. 1999. Large-scale transpressive shear zone patterns and displacement within magmatic arcs: the Coast Plutonic Complex, British Columbia. *Tectonics*, **18**, 278–292.

Chorowicz, J. & Sorlien, C. 1992. Oblique extensional tectonics in the Malawi Rift, Africa. *Geological Society of America Bulletin*, **104**, 1015–1023.

Clemens, J. D. & Mawer, C. K. 1992. Granitic magma transport by fracture propagation. *Tectonophysics*, **204**, 339–360.

Cobbold, P. R., Brun, J. P., Davy, P., Fiquet, G., Basile, C. & Gapais, D. 1989, Some experiments on block rotation in the brittle crust. *In*: Kissel, C. & Laj, C. (eds) *Paleomagnetic Rotations and Continental Deformation*. Kluwer, Dordrecht, 145–155.

Corriveau, L., Rivard, B. & Breemen, O. V. 1998. Rheological controls on Grenvillian intrusive suites: implications for tectonic analysis. *Journal of Structural Geology*, **20**, 1191–1204.

Crawford, M. L., Hollister, L. S. & Woodsworth, G. J. 1987. Crustal deformation and regional metamorphism across a terrane boundary, Coast Plutonic Complex, British Columbia. *Tectonics*, **6**, 343–361.

Cruden, A. 1990. Flow and fabric development during the diapiric rise of magma. *Journal of Geology*, **98**, 681–698.

Davidson, C., Hollister, L. S. & Schmid, S. M. 1992. Role of melt in the formation of a deep-crustal compressive shear zone: the Maclaren glacier metamorphic belt, South Central Alaska. *Tectonics*, **11**, 348–359.

Davy, P., Hansen, A., Bonnet, E. & Zhang, S.-Z. 1995. Localization and fault growth in layered brittle–ductile systems: implications for deformations of continental lithosphere. *Journal of Geophysical Research*, **100**, 6281–6294.

Dewey, J. F., Holdsworth, R. E. & Strachan, R. A. 1998, Transpression and transtension zones. *In*: Holdsworth, R. E., Strachan, R. A. & Dewey, J. F. (eds) *Continental Transpressional and Transtensional Tectonics*. Geological Society, London, Special Publications, **135**, 1–14.

Dokka, R. K., Ross, T. M. & Lu, G. 1998. The Trans Mojave–Sierran shear zone and its role in Early Miocene collapse of southwestern North America. *In*: Holdsworth, R. E., Strachan, R. A. & Dewey, J. F. (eds) *Continental Transpressional and Transtensional Tectonics*. Geological Society, London, Special Publications, **135**, 183–202.

Fossen, H. & Tikoff, B. 1998. Extended models of transpression and transtension, and application to tectonic settings. *In*: Holdsworth, R. E., Strachan, R. A. & Dewey, J. F. (eds) *Continental Transpressional and Transtensional Tectonics*. Geological Society, London, Special Publications, **135**, 15–33.

Harland, W. B. 1971. Tectonic transpression in Caledonian Spitzbergen. *Geological Magazine*, **108**, 27–42.

Hubbert, M. K. 1937. Theory of scale models as applied to the study of geologic structures. *Geological Society of America Bulletin*, **48**, 1459–1520.

Hutton, D. H. W. 1988. Granite emplacement mechanisms and tectonic controls: inferences from deformation studies. *Transactions of the Royal Society of Edinburgh: Earth Sciences*, **79**, 245–255.

Jackson, J. & McKenzie, D. 1989. Relation between seismicity and paleomagnetic rotation in zones of distributed continental deformation. *In*: Kissel, C. & Laj, C. (eds) *Paleomagnetic Rotations and Continental Deformation*. Kluwer, Dordrecht, 33–49.

Karlstrom, K. E., Miller, C. F., Kingsbury, J. A. & Wooden, J. L. 1993. Pluton emplacement along an active ductile thrust zone, Piute Mountains, southeastern California: interaction between deformation and solidification processes. *Geological Society of America Bulletin*, **105**, 213–230.

Krabbendam, M. & Dewey, J. F. 1998. Exhumation of UHP rocks by transtension in the Western Gneiss Region, Scandinavian Caledonides. *In*: Holdsworth, R. E., Strachan, R. A. & Dewey, J. F. (eds) *Continental Transpressional and Transtensional Tectonics*. Geological Society, London, Special Publications, **135**, 159–181.

Kukowski, N. & Neugebauer, H. J. 1990. On the ascent and emplacement of granitoid magma bodies—dynamic–thermal numerical models. *Geologische Rundschau*, **79**, 227–239.

Lacroix, S., Sawyer, E. W. & Chown, E. H. 1998. Pluton emplacement within an extensional transfer zone during dextral strike-slip faulting: an example from the late Archaean Abitibi Greenstone Belt. *Journal of Structural Geology*, **20**, 43–59.

LIN, S., JIANG, D. & WILLIAMS, P. F. 1998. Transpression (or transtension) zones of triclinic symmetry: natural example and theoretical modelling. *In*: HOLDSWORTH, R. E., STRACHAN, R. A. & DEWEY, J. F. (EDS) *Continental Transpressional and Transtensional Tectonics*. Geological Society, London, Special Publications, **135**, 41–57.

MARCOS, A. 1973. Las series del Paleozoico Inferior y la estructura herciniana del occidente de Asturias (NW de España). *Trabajos de Geología*, **6**, 1–111.

MARSH, B. D. 1984. Mechanics and energetics of magma formation and ascension. *In*: BOYD, F. R. (ed.) *Explosive Volcanism: Inception, Evolution and Hazards*. National Academy Press, Washington, DC, 67–83.

MARTÍNEZ, F. J., CARRERAS, J., ARBOLEYA, M. L. & DIETSCH, C. 1996. Structural and metamorphic evidence of local extension along the Vivero fault coeval with bulk crustal shortening in the Variscan Chain (NW Spain). *Journal of Structural Geology*, **18**, 61–73.

MARTÍNEZ CATALÁN J. R. 1985. *Estratigrafía y Estructura del Domo de Lugo (Sector Oeste de la zona Asturoccidental–leonesa)*, Lab. Geol. de Lage, La Coruña, Corpus Geologicum Gallaeciae.

——, GONZALEZ-LODEIRO, F., IGLESIAS, M. & DIEZ-BALDA, M. A. 1977. La estructura del Domo de Lugo y del anticlinorio del Ollo de Sapo. *Studia Geológica*, **12**, 109–122.

——, HACAR-RODRÍGUEZ, M. P., VILLAR-ALONSO, P., PÉREZ-ESTAÚN, A. & GONZALEZ-LODEIRO, F. 1992. Lower Paleozoic extensional tectonics in the limit between the West Asturian–Leonese and Central Iberian Zones of the Variscan Fold-Belt in NW Spain. *Geologische Rundschau*, **81**(2), 545–560.

MATTE, P. & RIBEIRO, A. 1975. Forme et orientation de l'ellipsoïde de déformation dans la virgation de Galice. Relations avec le plissement et hypothèses sur la genèse de l'arc ibéro-armoricain. *Comptes Rendus de l'Académie des Sciences, Série D*, **280**, 2825–2827.

MCCAFFREY, K. J. W. 1992. Igneous emplacement in a transpressive shear zone: Ox Mountains igneous complex. *Journal of the Geological Society, London*, **149**, 221–235.

MCKENZIE, D. & JACKSON, J. 1983. The relationship between strain rates, crustal thickening, palaeomagnetism, finite strain and fault movements within a deforming zone. *Earth and Planetary Science Letters*, **65**, 182–202.

—— & —— 1986. A block model of distributed deformation by faulting. *Journal of the Geological Society, London*, **143**, 349–353.

—— & —— 1989. The kinematics and dynamics of distributed deformation. *In*: KISSEL, C. & LAJ, C. (eds) *Paleomagnetic Rotations and Continental Deformation*. Kluwer, Dordrecht, 17–31.

MERLE, O. & VENDEVILLE, B. 1992. Modélisation analogique de chevauchements induits par des intrusions magmatiques. *Comptes Rendus de l'Académie des Sciences*, **315**, 1541–1547.

—— & —— 1995. Experimental modelling of thin-skinned shortening around magmatic intrusions. *Bulletin of Volcanology*, **57**, 33–43.

NALPAS, T. & BRUN, J.-P. 1993. Salt flow and diapirism related to extension at crustal scale. *Tectonophysics*, **228**, 349–362.

PATERSON, S. R. & TOBISCH, O. T. 1992. Rates of processes in magmatic arcs: implications for the timing and nature of pluton emplacement and wall rock deformation. *Journal of Structural Geology*, **14**, 291–300.

PAVLIDES, S., MOUNTRAKIS, D., KILIAS, A. & TRANOS, M. 1990. The role of strike-slip movements in the extensional area of Northern Aegean (Greece). A case of transtensional tectonics. *Annales Tectonicae*, **4**, 196–211.

PÉREZ-ESTAÚN, A. 1978. *La estratigrafía y la estructura de la rama Sur de la zona Asturoccidental Leonesa*. Memorias del Instituto Geológico y Minero de España, **92**.

——, MARTÍNEZ-CATALÁN, J. R. & BASTIDA, F. 1991. Crustal thickening and deformation sequence in the footwall to the suture of the Variscan belt of northwest Spain. *Tectonophysics*, **191**, 243–253.

PITCHER, W. S. 1992. *The Nature and Origin of Granite*. Chapman & Hall, London.

QUESADA, C. 1987. *Mapa Geológico-Minero de Extremadura*. Junta de Extremadura.

QUIRK, D. G., D'LEMOS, R. S., MULLIGAN, S. & RABTI, B. M. R. 1998. Insights into the collection and emplacement of granitic magma based on 3D seismic images normal fault-related salt structures. *Terra Nova*, **10**, 268–273.

RAMBERG, H. 1981. *Gravity, Deformation, and the Earth's Crust in Theory, Experiments and Geological Applications*, Academic Press, New York.

RICHARD, P., LOYO, B. & COBBOLD, P. 1989. Formation simultanée de failles et de plis au-dessus d'un décrochement de socle: modélisation expérimentale. *Comptes Rendus de l'Académie des Sciences*, **309**, 1061–1066.

——, MOCQUET, B. & COBBOLD, P. R. 1991. Experiments on simultaneous faulting and folding above a basement wrench fault. *Tectonophysics*, **188**, 133–141.

RIEDEL, W. 1929. Zur mechanik geologischer Brucherscheinungen. *Zentralblatt für Mineralogie, Geologie und Paläontologie*, **1929 B**, 354–368.

RIES, A. C. & SHACKLETON, R. M. 1971. Catazonal complexes of north-west Spain and north Portugal, remnants of a Hercynian thrust plate. *Nature, Physical Science*, **234**, 65–68.

ROMÁN-BERDIEL, T., BRUN, J. P. & GAPAIS, D. 1995. Analogue models of laccolith formation. *Journal of Structural Geology*, **17**, 1337–1346.

——, —— & ——. 1997. Granite intrusion along strike-slip zones in experiment and nature. *American Journal of Science*, **297**, 651–678.

SCHMIDT, C. J., SMEDES, H. W. & O'NEILL, J. M. 1990. Syncompressional emplacement of the Boulder and Tobacco Root batholiths along old fault zones (Montana, USA) by pull-apart. *Geological Journal*, **25**, 305–318.

SCHREURS, G. & COLLETTA, B. 1998. Analogue modelling of faulting in zones of continental transpression and transtension. *In*: HOLDSWORTH, R. E., STRACHAN, R. A. & DEWEY, J. F. (eds) *Continental Transpressional and Transtensional Tectonics*. Geological Society, London, Special Publications, **135**, 59–79.

TCHALENKO, J. S. 1970. Similarities between shear zones of different magnitudes. *Geological Society of America Bulletin*, **81**, 1625–1640.

TRON, V. & BRUN, J. P. 1991. Experiments on oblique rifting in brittle–ductile systems. *Tectonophysics*, **188**, 71–84.

VENDEVILLE, B. 1987. *Champs de failles et tectonique en extension: modélisation expérimentale*. Mémoires et Documents du CAESS, **15**.

——, COBBOLD, P. R., DAVY, P., BRUN, J. P. & CHOUKROUNE, P. 1987. Physical models of extensional tectonics at various scales. *In*: COWARD, M. P., DEWEY, J. & HANCOCK, P. L. (eds) *Continental Extensional Tectonics*. Geological Society, London, Special Publications, **28**, 95–107.

VIGNERESSE, J. L. 1995. Crustal regime of deformation and ascent of granitic magma. *Tectonophysics*, **249**, 187–202.

WATKEYS, M. K. & SOKOUTIS, D. 1998. Transtension in southeastern Africa associated with Gondwana break-up. *In*: HOLDSWORTH, R. E., STRACHAN, R. A. & DEWEY, J. F. (eds) *Continental Transpressional and Transtensional Tectonics*. Geological Society, London, Special Publications, **135**, 203–214.

WEIJERMARS, R. & SCHMELING, H. 1986. Scaling of newtonian and non-newtonian fluid dynamics without inertia for quantitative modelling of rock flow due to gravity (including the concept of rheological similarity). *Physics of the Earth and Planetary Interiors*, **43**, 316–330.

WICKHAM, S. M. 1987. The segregation and emplacement of granitic magmas. *Journal of the Geological Society, London*, **144**, 281–297.

WILCOX, R. E., HARDING, T. P. & SEELY, D. R. 1973. Basic wrench tectonics. *AAPG Bulletin*, **57**, 74–96.

WITHJACK, M. O. & JAMISON, W. R. 1986. Deformation produced by oblique rifting. *Tectonophysics*, **126**, 99–124.

Indentation of volcanic edifices by the ascending magma

OLIVIER MERLE & FRANCK DONNADIEU

Département des Sciences de la Terre, CNRS-OPGC-CRV, 5 rue Kessler, 63038 Clermont-Ferrand, France (e-mail: merle@opgc.univ-bpclermont.fr)

Abstract: The process by which magma ascends into and deforms a volcanic edifice is studied by analogue modelling. A control experiment is conducted with a wooden piston moving vertically into a sand cone. This reveals a well-defined fault pattern that makes it possible to draw the main compressive stress trajectory within the cone during the ascent of the piston. This makes it possible to show that the deformational process is that of indentation of the cone by the rigid piston. Experiments with an indenter that is viscous, as in nature, show that the motion of the viscous body is controlled by the first fault created in the cone. This fault serves as a structural guide, making the viscous body deviate from the vertical and resulting in deformation of the flank of the cone, which bulges out. Other major shear faults that were observed in the control experiment are then inhibited and do not form. This result emphasizes that the structural evolution of an indentation process within a brittle cone and at low rate depends on the rheology of the indenter.

The 1980 Mount St Helens eruption showed that the ascent of viscous magma into a volcanic edifice triggers severe deformation (Lipman *et al.* 1981; Moore & Albee 1981). The surface fractures and shear faults that developed were associated with the growth of a lateral bulge on the flank of the volcano. The kinematic evolution of surface deformation resulted from the upward push of the highly viscous magma. This deformation ultimately led to catastrophic collapse of the deforming flank and to a Plinian eruption preceded by a dramatic lateral blast.

The goal of this paper is to investigate the process by which the viscous magma can deform a volcanic cone. To a first order of approximation, a strato-volcano can be described as a stack of brittle layers failing according to the Coulomb criterion, whereas the magma is likely to behave as either a Bingham or a non-linear viscous fluid. In the process under consideration, the magma is believed to have very high viscosity, such as that inferred for the Mount St Helens magma (i.e. 10^{11} Pa s, Alidibirov & Dingwell 1996). The ascent of such a fluid can produce failure of a Mohr–Coulomb material. It is shown that this deformational process can be interpreted in terms of indentation of the viscous magma into the brittle edifice.

Indentation process

Numerous works in civil engineering have studied the deformation of a semi-infinite region under the load of a strip or circular footing (e.g. Bottero 1981; Matichard 1981;

From: VENDEVILLE, B., MART, Y. & VIGNERESSE, J.-L. (eds) *Salt, Shale and Igneous Diapirs in and around Europe*. Geological Society, London, Special Publications, **174**, 43–53. 1-86239-066-5/00/$15.00

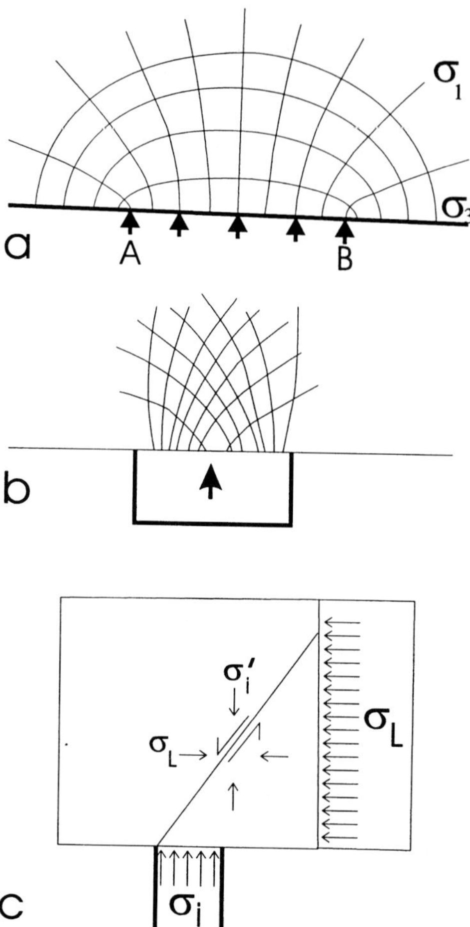

Fig. 1. Basic stress and fault trajectories in the process of indentation. (**a**) Stress trajectories for uniform loading over a strip AB on the surface of a semi-infinite region (after Jaeger & Cook 1979). (**b**) Trajectories of potential failure (after Gerbault *et al.* 1998). (**c**) Oblique strike-slip fault generated from the indenter to the lateral boundary when the principal compressive stress σ_i' overcomes the weak lateral confining stress σ_l (after Davy & Cobbold 1988).

de Borst 1982). The analytical problem is similar to that of a rigid piston pushed into an elastic or plastic half-space and results in the progressive indentation of the deformable area by the piston. Using elasticity theory, indentation produces curved trajectories of the greatest principal stress diverging from the indenter (Jaeger & Cook 1979) (Fig. 1a). According to the Mohr–Coulomb criterion of failure, the acute angle θ between the maximum compressive stress σ_1 and the faults is given by

$$\theta = \frac{\pi}{4} - \frac{\phi}{2}$$

where ϕ is the angle of internal friction of the material. This makes it possible to draw trajectories of potential failures within the half-space. They reveal a symmetric pattern of crescents that develop from the punch (Gerbault *et al.* 1998) (Fig. 1b).

The model of indentation has been used in plate tectonics to explain lateral escape of a continent in the direction of unconfined boundaries as for the Himalayas (Tapponnier & Molnar 1976) or the Austrian Alps (Ratschbacher *et al.* 1989). Lateral escape following indentation is one of the most popularized large-scale tectonic models of continent–continent collision proposed in the past 20 years. It has been studied both numerically (e.g. England & McKenzie 1982; Vilotte *et al.* 1982) and experimentally (e.g. Davy & Cobbold 1988; Ratschbacher *et al.* 1991). In all studies, the indenter is considered to be rigid (non-deformable indenter) whereas the rheology of the indented continent has generally been considered as viscous, ideally plastic or non-linear viscous. Indentation within a continent having a weak lateral confinement generates an oblique large-scale strike-slip fault, straight or curved depending on the geometry of the system, along which lateral escape of part of the continent takes place. The oblique wrench zone initiates from the inner corner of the indenter and propagates laterally to the opposite unconfined margin (Fig. 1c).

At small scale, salt diapirs in sedimentary basins also intrude the overlying cover to reach the surface. This process corresponds to an indenter moving towards a horizontal free surface. Both experimental and numerical studies show that active diapirism produces a central graben above the diapir crest (e.g. Davison *et al.* 1993; Schultz-Ela *et al.* 1993). The central graben is flanked by upward and outward flaps, which may be separated from the regional overburden by curved reverse faults. This deformation is the distinctive feature of an indentation beneath a free surface compared with indentation of an infinite half-space.

However, as shown from finite-element modelling (fig. 14g of Schultz-Ela *et al.* 1993), the main compressive stress axis in the overburden is vertical above the central part of the indenter and diverges outward from both corners of the indenter. This stress pattern is consistent with outward dipping normal faults and inward dipping reverse faults initiated at the two corners of the indenter as observed in some physical models (e.g. fig. 9 in Davison *et al.* 1993). The normal faults may cross each other before reaching the upper free surface in a way similar to those depicted in Fig. 1b. This shows that indentation of an infinite half-space or indentation beneath a free surface present remarkable similarities in fault and stress patterns.

The ascent of magma into a volcanic edifice may be viewed as a similar process of indentation where the forceful intrusion of the magma plays the role of the indenter. Main differences lie in the cone-like geometry of the volcano with unconfined boundaries in all directions, and the highly viscous rheology of the indenter, which is no longer rigid. Such an indentation process produces complex deformation, which is difficult to model numerically. To understand better the deformation, we have chosen to use an experimental approach, which proves to be a powerful method.

Experimental approach

Three sets of basic 2D and 3D experiments have been conducted to assess the role of the rheology of the indenter in the deformation of the cone. Two different structural evolutions are then clearly identified when considering rigid or viscous indenters.

In 2D experiments, the model is built within a box of 20 cm length by 15 cm height by 2.5 cm width so that the deformation is of plane strain type. The analogue volcano is composed of alternating coloured sand layers or sand–flour mix, the dip of those on either side of the vertical being equal to the angle of repose of the sand, that is about

33°. Indentation into the sand is achieved through an aperture of 1 cm width situated at the base and centre of the model from which the indenter slowly moves upward. To avoid boundary effects the vertical walls were coated with translucid oil. In 3D experiments, a cone of alternating sand layers or sand–flour mix is poured onto the base of the model so that the aperture is located under the apex of the cone.

The scaling procedure has been explained in detail by Donnadieu & Merle (1998). These experiments are scaled to simulate the evolution of silicic stratovolcanoes such as Mount St Helens. Similarity conditions show that experiments match the evolution of all edifices with similar properties, that is, edifices with a high-viscosity magma (10^8–10^{12} Pa s) and a time deformation of about 1 month or a few months.

Two-dimensional rigid indenter

A rigid wooden indenter has been used in the first set of experiments. As sand or sand–flour mix are the only analogue materials involved in the indentation process, these experiments are rate independent and yield the same results regardless of the velocity of the rigid indenter.

Symmetric deformation occurs in the analogue volcano once the wooden indenter moves upward. It is achieved through two curved major shear faults (MSF), which appear in succession and initiate at each corner of the indenter, propagate upward to cross each other in the upper part of the model along an axis of vertical symmetry and then merge at the two opposite flanks of the model (Fig. 2). Clear offsets along the layering indicate normal faulting along the two MSFs resulting in the inflation and outward movements of both flanks whereas the summit area undergoes subsidence.

This simple deformation is rapidly inhibited, as the two conjugate MSFs cannot operate long together for compatibility reasons. From that moment, the triangle limited by the two MSFs and located ahead of the rigid indenter remains undeformed and represents the effective indenter during the following deformation. Further upward movement of the wooden indenter results in the formation of new pairs of shear faults ahead of the previous ones. The deformation then becomes increasingly complicated with time, involving rotations of several blocks bounded by shear faults (Fig. 2c). It should be noted that the complex deformation above the effective indenter is flanked by upward and outward flaps, which may be separated from the undeformed lateral part by a reverse fault (Fig. 2c, right side) in a way similar to the deformation described for salt diapirs (Schultz Ela *et al.* 1993).

Two-dimensional viscous indenter

In these experiments, the deformation is rate dependent and the time of the experiments calculated from similarity conditions is about 1 day, corresponding to 1 month in nature. The deformation produced in the analogue volcano is completely different when using silicone putty as an indenter instead of the wooden indenter. The first stage of the deformation shows that the silicone enlarges symmetrically when it comes out of the aperture, resulting in a drop shape for the viscous indenter. This first stage produces a symmetric deformation ahead of the intrusion corresponding to an outward upturning of sand layers on both sides of a vertical line passing through the summit. The end of the first stage is associated with a slight decrease

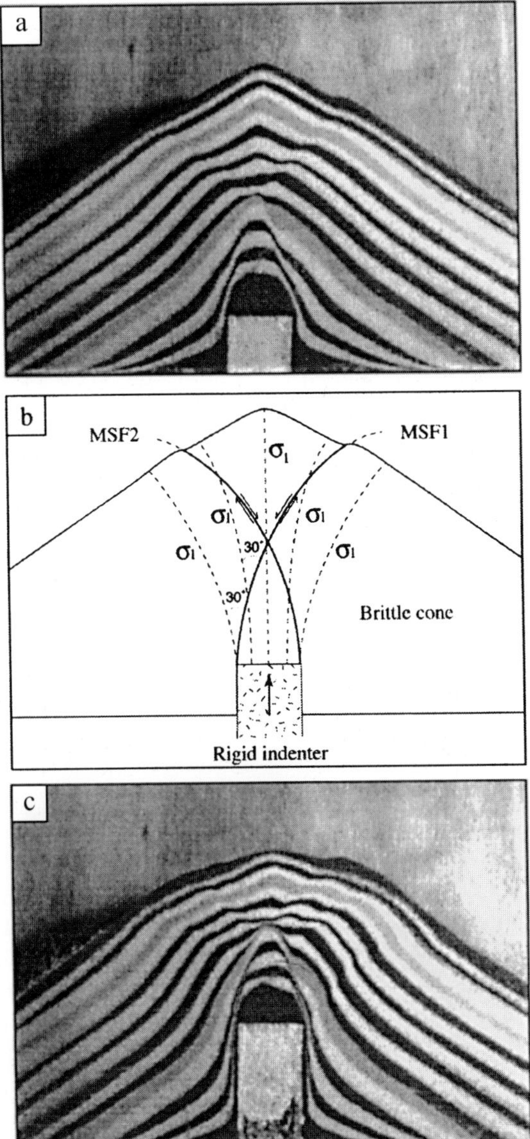

Fig. 2. Two successive stages of indentation of a sand cone by a rigid wooden indenter. Trajectories of the main compressive stress axis σ_1 for the first stage as deduced from the trace of the two MSFs. The reverse fault on the right side of the cone at the second stage should be noted (see text).

of the intrusion rate while the viscous indenter enlarges at its base to take a more rectangular shape (still symmetric on either side of a vertical line).

Then, the pressure from below leads to failure of the brittle cone. Only one MSF initiates from a corner of the viscous indenter and the silicone deviates from the vertical to rise obliquely along the MSF (Fig. 3). The conjugate MSF never forms and the oblique motion of the silicone results in one inflated side whereas the opposite

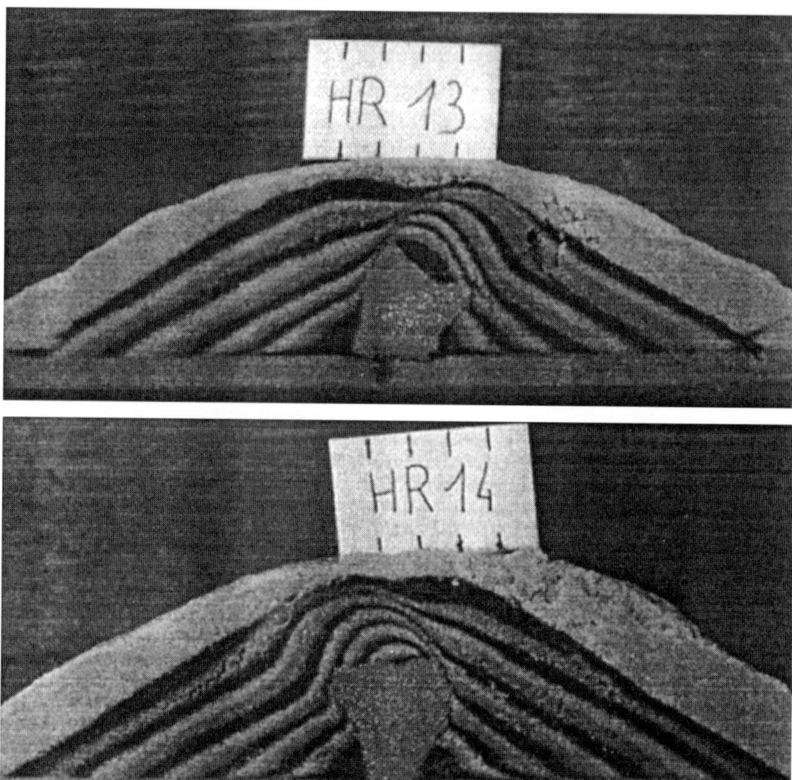

Fig. 3. Viscous indenter: initiation of a unique MSF, which acts as a structural guide for the viscous intrusion, which rises obliquely along it. In both experiments, the uppermost white layer of sand is added after deformation for technical convenience.

side of the cone remains undeformed. Normal faults at the summit area accommodate the extension produced by the MSF and a graben structure develops upslope above the inflated flank (Fig. 3). The growth of the lateral bulge is accompanied by steepening of the sand ahead of the ascending silicone. This flap is bounded by a reverse fault starting from the other corner of the indenter. The silicone slightly intrudes the two mechanical discontinuities, the MSF and the reverse fault, so that the shape of the indenter may exhibit two tips, at its corners (Fig 3, top). This simple deformation pattern is observed throughout the experiment until the silicone ultimately pierces the sand envelope and flows downward along the surface slope.

A clearly different structural evolution is thus established, from a complex symmetric evolution with a rigid indenter to a simple asymmetric evolution with a viscous indenter. The formation of the first MSF at the start of the viscous indenter experiment corresponds to an important mechanical discontinuity that influences and induces the further structural evolution of the model.

Three-dimensional viscous indenter

Three-dimensional experiments have been presented in some detail in a previous publication (Donnadieu & Merle 1998). They dealt with viscous intrusion only and

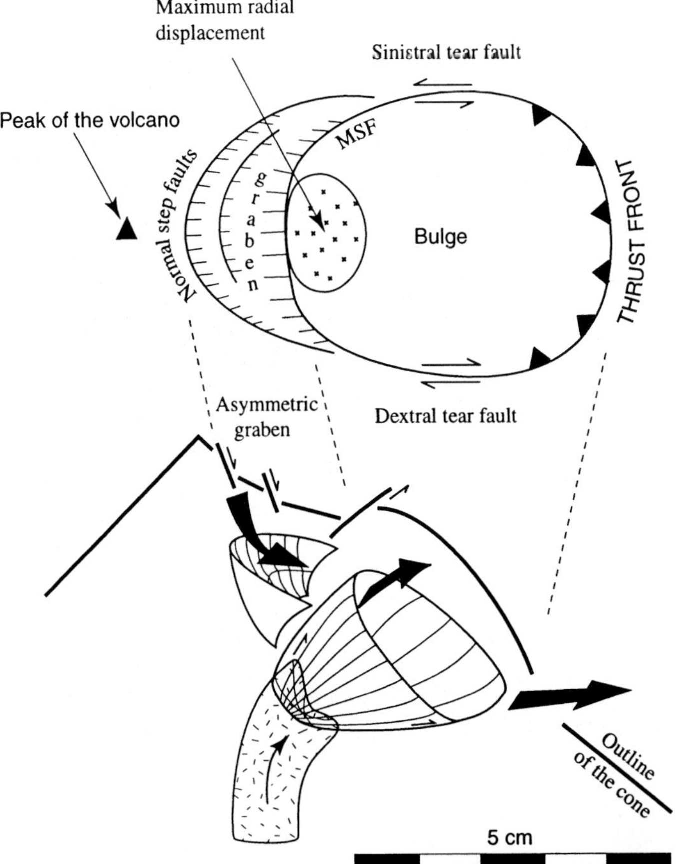

Fig. 4. Top: schematic representation of the structures observed at the surface of the cone during the ascent of the silicone. Bottom: the conical shape of the MSF and the basal thrust fault as deduced from serial cross-sections (full description in text).

yielded results matching viscous intrusion in two dimensions, especially the asymmetric evolution of the deformation, and will not be reported here (see Donnadieu & Merle (1998) for further details).

Observations at the surface and serial cross-sections at different stages of the deformation give insight into the 3D shape of the MSF (Fig. 4). The asymmetric evolution of the deformation is clearly depicted from the displacement field. A slightly radial pattern of displacement restricted to a portion of the cone corresponds to a moving lateral bulge obliquely pushed by the ascending silicone (Fig. 5). The upward limit of this bulge where maximum displacement is observed coincides with the MSF outcrop. The MSF initially displays a rather linear trace at the surface. It then curves to take a horseshoe shape with tips developing downward as the inner part is uplifted. The outward motion of the bulge is further accommodated by the curving of the MSF into tear faults bounding the bulge laterally. Normal faults accommodate the summit extension upslope, forming an asymmetric graben. The upper part of the bulge is intensely fractured and highly prone to crumbling down. The two transcurrent branches of the MSF ultimately join each other, together with the creation

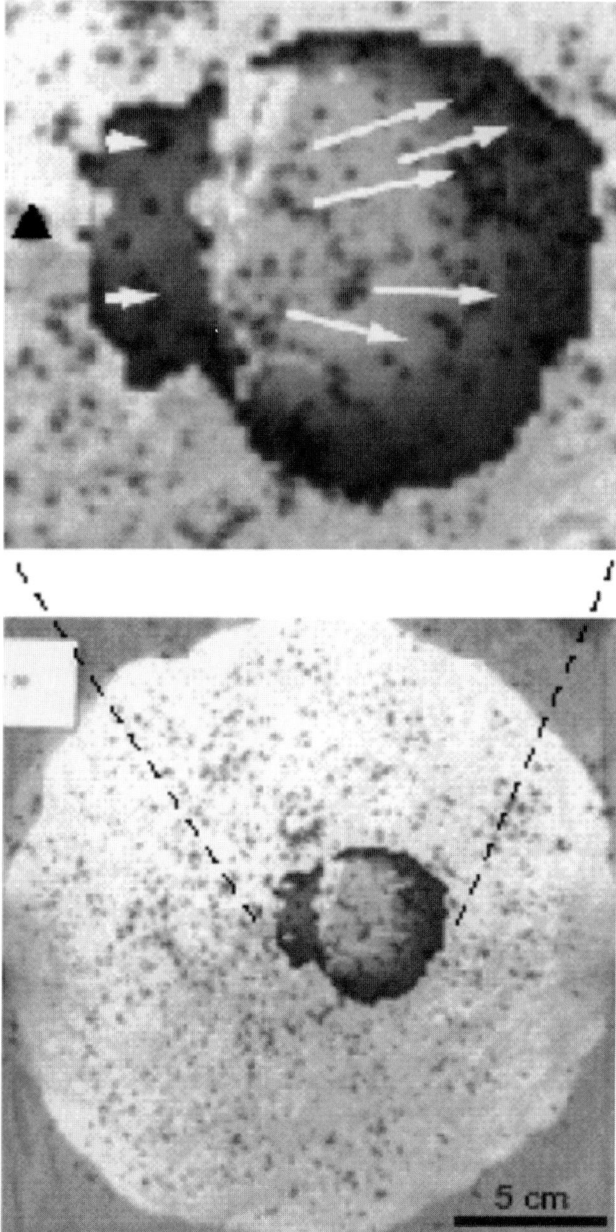

Fig. 5. Top views of the deformation in 3D experiments. Translation vectors (arrows) display radial pattern of displacement associated with the moving lateral bulge (top). Lengths of vectors are proportional to the amount of displacement.

of a basal thrust fault. This reverse fault is gently dipping inward and makes the cone-like shape of the MSF especially clear (Fig. 4). Thus, the initial MSF evolves to form a conical shape that terminates at depth around the head of the intruding silicone. Serial cross-sections make it possible to restore the 3D shape of the unique MSF (Fig. 4).

It is not possible to predict the orientation towards which the bulge will grow during experiments. Numerous models show that the bulge growth orientation, that is, the orientation of the MSF at the beginning of the experiment, is totally random when using homogeneous sand cones. However, the orientation of the MSF may result from slight heterogeneities within the analogue volcano, which cannot be a perfect cone when formed by pouring sand on the basal surface. Complementary experiments reveal that for a cone composed of sand on one side and sand–flour mix on the other side the MSF always initiates within the cohesionless side. This indicates that the orientation of the MSF is controlled by the initial distribution of mechanical strengths within the cone.

Interpretation

Considering the angle of internal friction of the granular material used in experiments, the two MSFs formed at the beginning of experiments with a rigid indenter make it possible to infer the main compressive stress trajectory within the cone. This reveals a fan-like trajectory, the main compressive stress axis diverging from the indenter to the flanks of the analogue volcano (Fig. 2). Such a trajectory is similar to that calculated by Jaeger & Cook (1979) for uniform loading over a strip on the surface of a semi-infinite region (Fig. 1a) or to that calculated by Schultz-Ela *et al.* (1993) for the initial phase of active diapirism beneath a free surface. This means that the cone-like geometry in our experiments does not alter the general stress pattern calculated from numerical methods for the indentation of a rigid piston into a deformable material (e.g. Jaeger & Cook 1979; Gerbault *et al.* 1998). However, it is worth noting that the free surface of the model exerts an influence on the trajectories of the principal stresses. When approaching the upper free surface, stress trajectories are likely to deviate from those calculated for a semi-infinite region as the free surface of the triangular cone must be parallel to a principal stress (no shear stress is possible).

Using a viscous indenter results in the formation of a unique MSF, which remains active until the end of the experiment. It is clear and we believe significant that the formation of the second MSF is inhibited once a viscous indenter is used in experiments. The rheology of the indenter is the key factor controlling the structural evolution of the system. A rigid indenter moves vertically upward regardless of the fault system developed within the cone. Conversely, the translation path of a viscous and deformable indenter is controlled by the structural discontinuities encountered during motion. The strength of the indenter is not high enough to cross the upper limit of the cohesionless zone existing along the brittle fault. The MSF is then a major mechanical discontinuity that causes a deviation in the upward motion of the viscous indenter in its direction. It follows that the indenter moves obliquely and that the stress orientation depicted in Fig. 2 and triggering the formation of the second MSF is no longer valid, the further structural evolution being entirely controlled by the orientation of the first MSF.

Our experiments can be compared with experiments on indentation at continent scale in which a unique major wrench fault initiates from a corner of the indenter to the lateral unconfined boundary (Davy & Cobbold 1988) (Fig. 1c). In these experiments, it is the weak lateral confinement at one margin that prevents development of the conjugate wrench fault and permits lateral escape, the orientation of the single wrench fault being controlled by the orientation of the unconfined boundary. In our experiments, unconfined boundaries in all directions make the orientation of the unique MSF unpredictable. The asymmetric evolution of the deformation results from the rheology of the indenter once the MSF is formed.

Concluding remarks

Scaled models of forcible intrusion of magma into a volcanic edifice show an asymmetric evolution of the deformation. The process by which the magma penetrates into the brittle edifice can be related to indentation tectonics, as has been done for some other geological processes. Lateral escape of the magma, which deviates from the vertical to bulge out a flank of the edifice, may be interpreted as the mechanical response of the viscous indenter to the brittle MSF formed in the cone. It is likely that the first MSF created serves as a structural guide for the viscous body motion, which is unable to cross such mechanical discontinuities.

Experiments with the viscous indenter were scaled to match volcanic edifices similar to Mount St Helens (Donnadieu & Merle 1998). Two-dimensional models clearly indicate that the rheology of the indenter is one of the key factors controlling the growth of a lateral bulge. It appears that the effect of the rigid indenter is strongly due to the inexorable rise of the undeformed corners and that both corners are constrained to uplift the overburden equally. Conversely, a viscous indenter can suffer internal strain according to mechanical discontinuities initiated within the brittle cone at the very beginning of the upward motion of the magma.

It may be interesting to speculate about the role played by the velocity of the ascending magma. The rheology of the magma can range from linear viscous magma to Bingham material, that is, having a yield stress making it more difficult to be deformed. The magma can also exhibit different rheological behaviours at high strain rate, being either nearly unable to deform internally or even showing failures similar to brittle material. Thus, it is possible that very high velocity may diminish the guide effect played by the MSF, permitting the vertical motion of the magma and resulting in a more or less symmetrical evolution with the formation of the other MSF. Further experimental studies are needed to quantify the roles played by the viscosity and the velocity of the indenter in the general indentation process described in this paper.

The authors thank M. de Saint-Blanquat and D. D. Schultz-Ela for helpful reviews.

References

ALIDIBIROV, M. & DINGWELL, B. D. 1996. Magma fragmentation by rapid decompression. *Nature*, **380**, 146–148.

BOTTERO, A. 1981. *Contribution B l'étude du tassement et de la force portante des fondations superficielles reposant sur un multicouche*. Thèse d'Etat, Université de Grenoble.

DAVISON, I., INSLEY, M., HARPER, M., WESTON, P., BLUNDELL, D., MCCLAY, K. & QUALLINGTON, A. 1993. Physical modelling of overburden deformation around salt diapirs. *Tectonophysics*, **228**, 255–274.

DAVY, PH. & COBBOLD, P. R. 1988. Indentation tectonics in nature and experiment. I. Experiments scaled for gravity. *Bulletin of the Geological Institution of the University of Uppsala*, **14**, 129–141.

DE BORST, R. 1982. Calculation of collapse loads using higher order elements. *In*: WERMEER, P. A. & LUGER, H. J. (eds) *Deformation and Failure of Granular Materials*, Balkema, Rotterdam, 503-513.

DONNADIEU, F. & MERLE, O. 1998. Experiments on the indentation process during cryptodome intrusion: new insights into Mt St Helens deformation. *Geology*, **26**, 79–82.

ENGLAND, P. C. & MCKENZIE, D. 1982. A thin viscous sheet model for continental deformation. *Geophysical Journal of the Royal Astronomical Society*, **70**, 295–321.

GERBAULT, M., POLIAKOV, A. N. B. & DAIGNIERES, M. 1998. Prediction of faulting from the theories of elasticity and plasticity: what are the limits? *Journal of Structural Geology*, **20**, 301–320.

JAEGER, J. C. & COOK, N. G. W. 1979. *Fundamental of Rock Mechanics*. 3rd edn. Chapman & Hall, London.

LIPMAN, P. W., MOORE, J. G. & SWANSON, D. A. 1981. Bulging of the north flank before the May 18 eruption—geodetic data. *In*: LIPMAN, P. W. AND MULLINEAUX, D. R. (eds) *The 1980 Eruptions of Mount St Helens*. US Geological Survey Professional Paper, **1250**, 143–155.

MATICHARD, Y. 1981. *Sol bicouche renforcé par géotextile. Application aux chaussées provisoires*. Thèse Docteur d'Ingénieur, Université de Grenoble.

MOORE, J. G. & ALBEE, W. C. 1981. Topographic and structural changes, March–July 1980—Photogrammetric data. *In*: LIPMAN, P. W. AND MULLINEAUX, D. R. (eds) *The 1980 eruptions of Mount St Helens*. US Geological Survey Professional Paper, **1250**, 123–134.

RATSCHBACHER, L., FRISCH, W., NEUBAUER, F., SCHMID, S. & NEUGEBAUER, J. 1989. Extension in compressional orogenic belts: the Eastern Alps. *Geology*, **17**, 404–407.

——, MERLE, O., DAVY, PH. & COBBOLD, P. R. 1991. Lateral extrusion in the Eastern Alps, Part I: Boundary conditions and experiments scaled for gravity. *Tectonics*, **10**, 245–256.

SCHULTZ-ELA, D. D., JACKSON, M. P. A. & VENDEVILLE, B. C. 1993. Mechanics of active salt diapirism. *Tectonophysics*, **228**, 275–312.

TAPPONNIER, P. & MOLNAR, P. 1976. Slip line field theory and large scale continental tectonics. *Nature*, **264**, 319–324.

VILOTTE, J. P.; DAIGNIERES, M. & MADARIAGA, R. 1982. Numerical modelling of intraplate deformation: simple mechanical models of continental collision. *Journal of Geophysical Research*, **87**, 10709–10728.

Pluton emplacement in the Northern Tyrrhenian area, Italy

F. ROSSETTI[1], C. FACCENNA[1], V. ACOCELLA[2], R. FUNICIELLO[1], L. JOLIVET[3] & F. SALVINI[1]

[1] *Dipartimento di Scienze Geologiche, Università 'Roma Tre', Largo S.L. Murialdo, 1, 00146 Rome, Italy (e-mail: rossetti@uniroma3.it)*

[2] *Dipartimento di Scienze della Terra, Università di Siena, Via del Laterino 8, 53100 Siena, Italy*

[3] *Département de Géotéctonique, Université Pierre et Marie Curie, 4 Place Jussieu, 75252, Paris cedex 05, France*

Abstract: The post-orogenic extensional processes that affected the inner sector of the Northern Apennine orogenic wedge (i.e. the Northern Tyrrhenian region) were accompanied by the emplacement of chiefly anatectic intrusive rocks of Late Miocene to Mid-Pleistocene age. In this paper, we compare geological and structural data from Messinian–Pliocene monzogranitic intrusions located both offshore (Monte Capanne, Porto Azzurro, Montecristo and Giglio) and onshore (Gavorrano and Botro ai Marmi) in the Northern Tyrrhenian region to constrain modes of pluton emplacement. Offshore, eastward non-coaxial extensional shear zones active both in ductile and brittle conditions accompanied the emplacement of the monzogranitic intrusions, and accommodated extension oriented E–W to WNW–ESE. Onshore, N–S dextral strike-slip faulting was active both during and after the late stage of emplacement of both Botro ai Marmi and Gavorrano plutons, and controlled their rise in coincidence with releasing bends. In our interpretation, the N–S, Late Miocene–Pliocene strike-slip faulting constitutes a secondary shear feature in a context of generalized post-orogenic extension, accommodating in the brittle upper crust the non-coaxial ductile extension in the lower crust. In this framework, N–S strike-slip faults localized the rise of early anatectic magma, generated during regional post-orogenic extension and residing at the base of the extending crust.

Thermal relaxation processes accompanying the late extensional attenuation of previously thickened orogenic crust are usually associated with widespread magmatic activity (e.g. Gans *et al.* 1989; Turner *et al.* 1992; Lister & Baldwin 1993; Platt & England 1994). In particular, these post-orogenic extensional processes usually show a close spatial and temporal relationship with igneous activity and granitic pluton emplacement at higher levels in the continental crust, as in the Basin and Range province of the western USA (e.g. Wernicke *et al.* 1987; Gans *et al.* 1989; Fletcher *et al.* 1995) and in the Cyclades region of the Aegean Sea (Lister *et al.* 1984; Gautier *et al.* 1993).

Studying pluton emplacement mechanics may provide a record of the chemical, thermal and strain conditions involved in the crustal-scale material transfer processes needed for magma generation and flow during regional deformation (e.g. Turner *et al.*

From: VENDEVILLE, B., MART, Y. & VIGNERESSE, J.-L. (eds) *Salt, Shale and Igneous Diapirs in and around Europe*. Geological Society, London, Special Publications, **174**, 55–77. 1-86239-066-5/00/$15.00

1992, Paterson & Fowler 1993; Hanson & Glazner 1995; Vigneresse 1995). This understanding may thus have important implications for interpreting the origin and evolution of the developing deformation structures in sites of crustal extension.

A well-documented example of extension-related magmatic activity derives from the central Mediterranean region, where magmatism accompanied and followed crustal thinning during the formation of the Tyrrhenian and Aegean basins (Jolivet *et al.* 1998, and references therein). In this framework, the post-orogenic (Late Miocene–Quaternary) formation of the Northern Tyrrhenian basin in the inner sector of the Northern Apennine chain (Fig. 1) provides a good geological test site to study

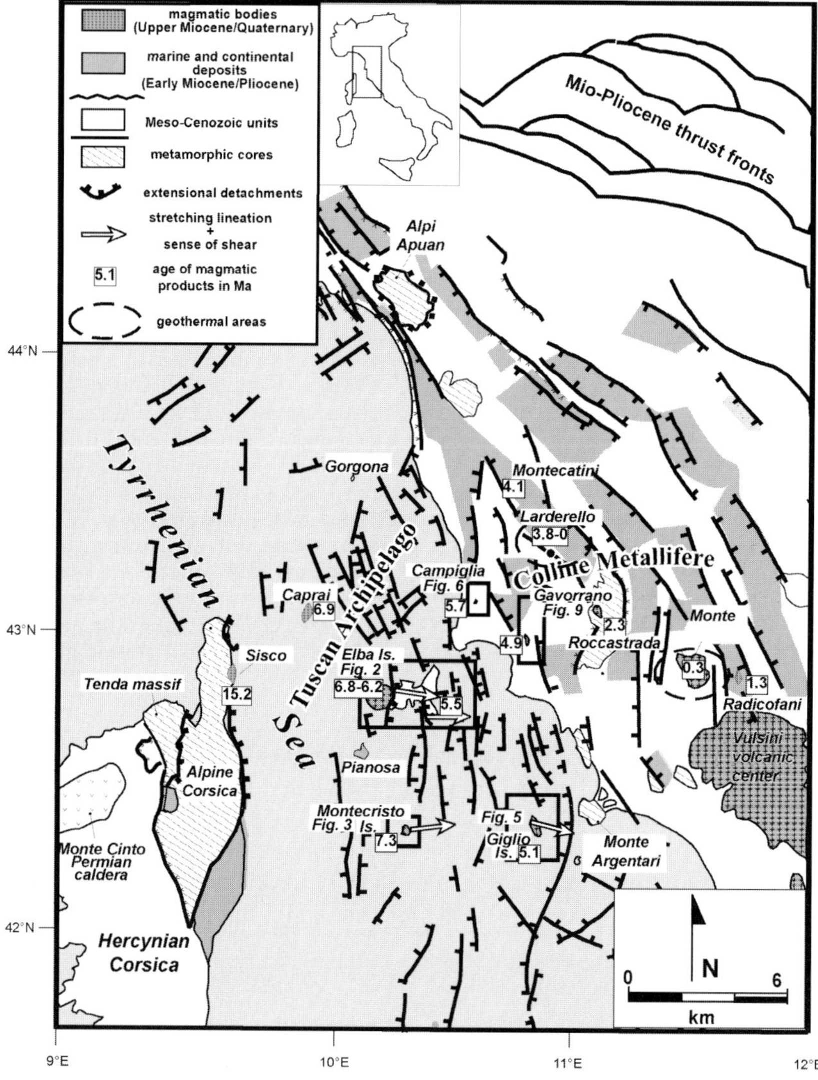

Fig. 1. Simplified tectonic map of the Northern Tyrrhenian region (modified after Jolivet *et al.* (1998) and references therein) showing the study areas (boxes) (age of magmatism after Serri *et al.* (1993) and references therein).

relationships between extensional processes and the related activity of the Tuscan Magmatic Province (Marinelli 1967; Innocenti *et al.* 1992). In fact, despite the fact that the mean source for the Tuscan magmatic suite is considered the result of crustal anatexis in an overall extensional regime (Serri *et al.* 1993; Innocenti *et al.* 1997), the kinematics of pluton emplacement in the region is still poorly detailed.

To address such a problem, in the present study we discuss structural and field data on the kinematics of pluton emplacement in various Tuscan intrusive stocks from the Northern Tyrrhenian region, namely the Montecristo, Monte Capanne and Porto Azzurro (Elba Island), and Giglio intrusions located offshore in the Tuscan Archipelago, and the Campiglia Marittima and Gavorrano intrusions located onshore in the Tuscan Colline Metallifere (Fig. 1). These intrusions were chosen because they have similar radiometric cooling ages (from *c.* 7 to *c.* 5 Ma; Fig. 1), similar monzogranitic composition (Poli *et al.* 1989; Innocenti *et al.* 1992), but have different crustal depths of crystallization, deeper offshore (12–15 km) in the Tuscan Archipelago (Westerman *et al.* 1993; Innocenti *et al.* 1997), and shallower (2–4 km) in the Colline Metallifere region (e.g. Franceschini 1994).

Taking into account the overall extensional regime of the area, we present evidence for different structural settings controlling the rise of magma to the surface. The ascent and emplacement of the offshore plutons of Montecristo, Elba and Giglio were controlled by top-to-the-east extensional shear zones located at the brittle–ductile transition; whereas the onshore plutons of Campiglia Marittima and Gavorrano rose within the crust in zones of localized extension along N–S trending right-lateral strike-slip fault zones. We interpret such N–S-trending strike-slip shear zones as secondary shear features that occurred during generalized post-orogenic extension, when the upper crust accommodated an eastward-directed ductile non-coaxial flow occurring in the lower crust.

Geological framework

The Northern Tyrrhenian domain (Fig. 1) developed as a consequence of post-orogenic extensional processes occurring at the back of the growing and eastward migrating Apennine fold-and-thrust belt (e.g. Patacca *et al.* 1990; Jolivet *et al.* 1998). Extensional tectonics strongly modified the original Oligocene architecture of the thrust belt (Carmignani *et al.* 1994; Jolivet *et al.* 1998). The following tectono-stratigraphic units constituted the original thrust–nappe complex, from top to bottom (Carmignani *et al.* 1994, and references therein): the Ligurian units consist of Jurassic low-grade metabasic and mafic rocks, their Upper Jurassic–Lower Cretaceous pelagic sedimentary cover, and Cretaceous–Paleogene flysch deposits; the Tuscan units, locally reduced by extensional tectonics, comprise the sedimentary units of the Tuscan Nappe (Upper Triassic–Cretaceous shallow-marine to pelagic carbonaceous–carbonaceous siliceous and Upper Cretaceous–Tertiary pelagic terrigenous formations) and its metamorphic basement (Tuscan Metamorphic Complex).

Extensional tectonics, active since Early Miocene time in Alpine Corsica, progressively migrated eastwards, reworking the early thickening phase (Jolivet *et al.* 1998, and references therein) and leading to crustal attenuation (Nicolich 1989; Ponziani *et al.* 1994). Extension formed eastward, shallow-dipping extensional shear zones at depth (Jolivet *et al.* 1998) and N–S to NW–SE oriented sedimentary basins (e.g.

Bartole 1995). Extensional basins were filled by lacustrine–marine (Upper Tortonian–Messinian), marine (Lower–Middle Pliocene) and fluvial-lacustrine (Quaternary) sequences (Bossio *et al.* 1993; Martini & Sagri 1993). Magmatic activity, consisting of the volcano-plutonic associations of the Tortonian–Pleistocene Tuscan Magmatic Province (Marinelli 1967; Innocenti *et al.* 1992), accompanied and followed extension with a similar eastward space–time migration (Serri *et al.* 1993). From about 7 Ma onwards, magmatic activity, mainly represented by acidic intrusive stocks, was widely diffused both offshore and onshore (Fig. 1). Deep boreholes in the geothermal areas of southern Tuscany encountered either plutonic bodies or their related thermometamorphic units (Franceschini 1994, and references therein), whereas seismic reflection data showed several buried intrusions offshore (Zitellini *et al.* 1986). This magmatic activity is at present responsible for the high, positive heat flow anomaly that characterizes the geothermal areas of the Tuscan–Latium peri-Tyrrhenian region (Mongelli & Zito 1991) (Fig. 1). In the geothermal areas, the present elevation of the marine Pliocene deposits suggests that the whole region underwent uplift of several hundreds of metres during Mid–Late Pliocene time (Bossio *et al.* 1993, and references therein). This circumstance, coupled with the correspondence between positive heat flow anomaly and gravity minima in the geothermal region, has been interpreted as the result of pluton emplacement at shallower crustal levels during Plio-Pleistocene times (Marinelli *et al.* 1993; Baldi *et al.* 1994; Barberi *et al.* 1994). All this evidence supports the hypothesis that the outcropping igneous bodies represent apophyses of an extensive deep-seated intrusion (Baldi *et al.* 1994; Franceschini 1994).

Structural study

The structural study presented here comprises a compilation of previous works on the Tuscan Archipelago intrusive rocks (Elba Island: Daniel & Jolivet 1995; Montecristo Island: Jolivet *et al.* 1998; Giglio Island: Rossetti *et al.* 1999) and new investigations related to the Campiglia Marittima and Gavorrano areas in the Tuscan Colline Metallifere mining region, and partly to Montecristo Island (Fig. 1).

Tuscan Archipelago

Elba Island. Elba Island is made up of a nappe complex intruded by two diachronous plutons (Fig. 2). These plutons are the older Monte Capanne (granodiorite to monzogranite) and the younger Porto Azzurro (monzogranite) intrusions, dated at 6.8–6.2 Ma (K/Ar and Rb/Sr methods: Ferrara & Tonarini 1985; U/Pb method on zircon: Juteau *et al.* 1984) and 5.5 Ma (K/Ar and Rb/Sr methods: Saupé *et al.* 1982, Ferrara & Tonarini, 1985), respectively. Thermobarometric data in the aureole of Porto Azzurro indicates that the intrusion crossed the 600°C isotherm at a pressure of 2 kbar (Duranti *et al.* 1992). Daniel & Jolivet (1995) have inferred that the syn-emplacement deformation of Monte Capanne and Porto Azzurro intrusive rocks was related to a non-coaxial flow accommodated by a top-to-the-east shear along shallow-dipping extensional shear zones, active in the ductile and brittle domains of the continental crust (Fig. 2). The overall asymmetry of the extensional deformation is well documented around the Monte Capanne intrusion. The eastern side is characterized by shallow east-dipping foliations and E–W stretching lineations,

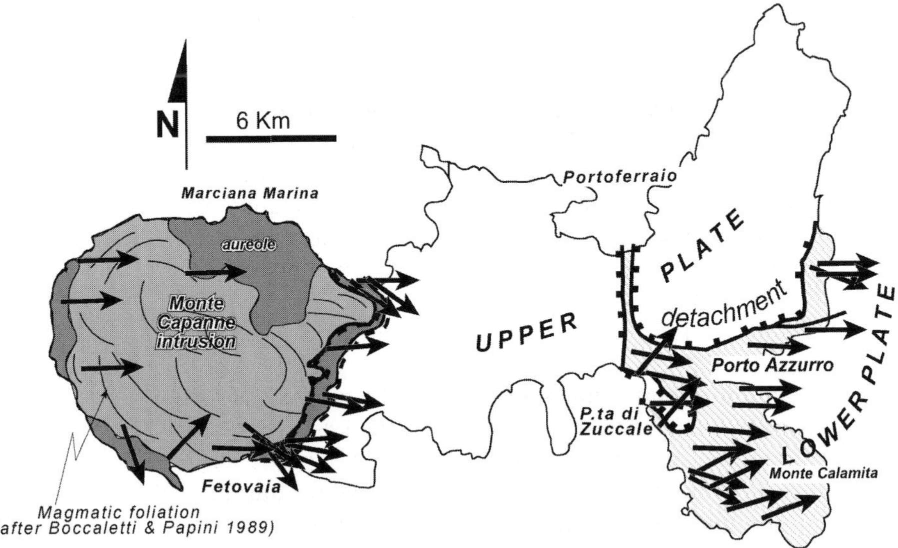

Fig. 2. Simplified structural map of Elba Island showing attitude of top-to-the-east asymmetric extension associated with the emplacement of the Monte Capanne and Porto Azzurro intrusions; arrows indicate stretching lineations and sense of shear (modified after Daniel & Jolivet 1995).

within both the pluton itself and in its aureole. Kinematic indicators, coeval with crystallization of high-temperature minerals of the aureole, show a consistent top-to-the-east sense of shear (Daniel & Jolivet 1995). Conversely, the western margin shows a steep foliation that wraps around the pluton. There, finite strain recorded shortening perpendicular to the foliation. Deformation on the eastern margin of the island shows a progressive transition to more brittle conditions, expressed by the shallow-dipping Zuccale detachment fault, which controls the contacts between upper-plate Tuscan non-metamorphic units and lower-plate metamorphic ones (Keller & Pialli 1990). Ductile deformation is also observed below the Zuccale detachment and can be related to the emplacement of the younger Porto Azzurro intrusion (Daniel & Jolivet 1995). Here, a consistent E–W trending stretching lineation (Fig. 2) is associated with top-to-the-east kinematic indicators. The deformation evolved with time to brittle style and finally localized along the Zuccale fault with the same top-to-the-east sense of shear.

Montecristo Island. Montecristo Island (Fig. 3) comprises mainly intrusive rocks having an overall monzogranitic composition (Innocenti *et al.* 1997) and Rb/Sr cooling ages between 7.0 and 7.3 Ma (Ferrara & Tonarini 1985; Innocenti *et al.* 1997). Using the ternary plot of normative composition (quartz, albite and orthoclase, recalculated to 100%), a minimum pressure of 5 kbar was proposed for the crystallization of the Montecristo monzogranite (Innocenti *et al.* 1997). The bulk of the intrusion is very poorly deformed, except in the southern edge of the island (Fig. 3), where both ductile (solid-state) and brittle deformation textures are present (Innocenti *et al.* 1997; Jolivet *et al.* 1998). There, shear deformation is attested by shallow west-dipping NNW–SSE trending shear zones and low-grade mylonites. Stretching lineations on

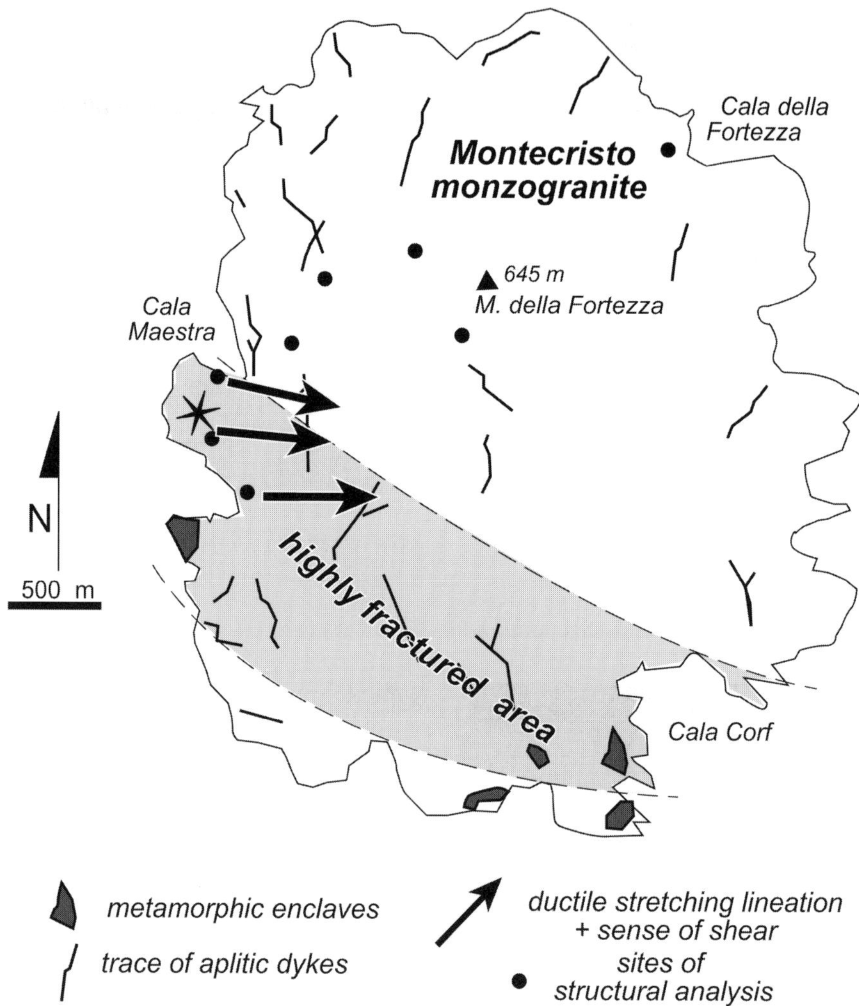

Fig. 3. Simplified geological and structural map of Montecristo Island (modified after Innocenti *et al.* (1997) and references therein). Asterisk indicates location of the structural features shown in Fig. 4a.

C-surfaces, provided mostly by tourmaline and quartz, trend roughly E–W. The sense of shear as provided by the S–C relationships is consistently top-to-the-east, as also suggested by dragging of late dykes along the C-surfaces (Fig. 4a). Moreover, a late episode of NW–SE to N–S conjugate normal faults, in association with high-angle joints and dykes, is recorded in the monzogranite, accommodating a late, roughly E–W extension direction (Fig. 4a).

These data suggest that, during emplacement and cooling, the Montecristo monzogranite was sheared along an eastward asymmetric fault zone. Shearing continued after solidification of the pluton, forming low-grade mylonites and cataclasites along the boundary of the intrusion. The final ascent of the pluton was accompanied by normal faulting, which segmented the original plutonic boundaries and controlled the late emplacement of dyke arrays.

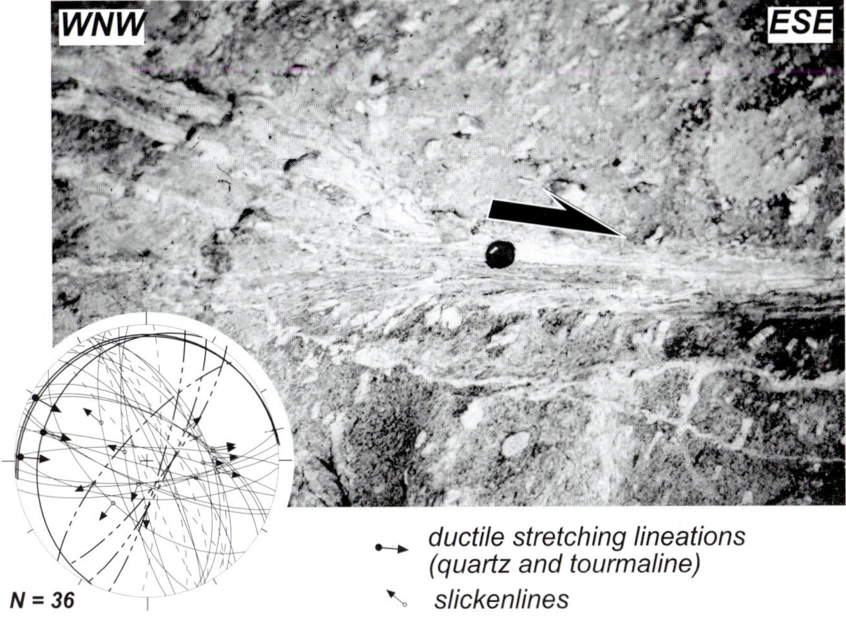

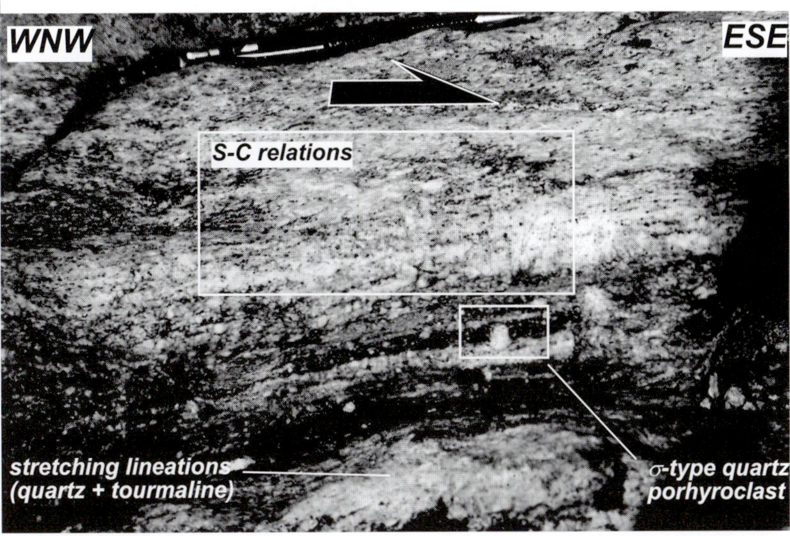

Fig. 4. Examples of solid-state tectonic shear features recorded in the Tuscan Archipelago monzogranites. (**a**) Ductile, top-to-the-east shear zone at Montecristo Island (exposure parallel to the stretching lineation and normal to foliation; see Fig. 3 for location); representative stereoplot (Schmidt equal-area projection, lower hemisphere) of the shear deformation within the monzogranite (bold lines represent the main top-to-the-east ductile shear planes (continuous) with their conjugates (dotted); fine lines represent faults (continuous) and fractures (dashed) associated with the late ENE–WSW to E–W extensional event). (**b**) S–C tectonites at Giglio Island (P.ta del Fenaio in Fig. 5) showing consistent top-to-the-east shear sense indicators (exposure parallel to the stretching lineation and normal to foliation).

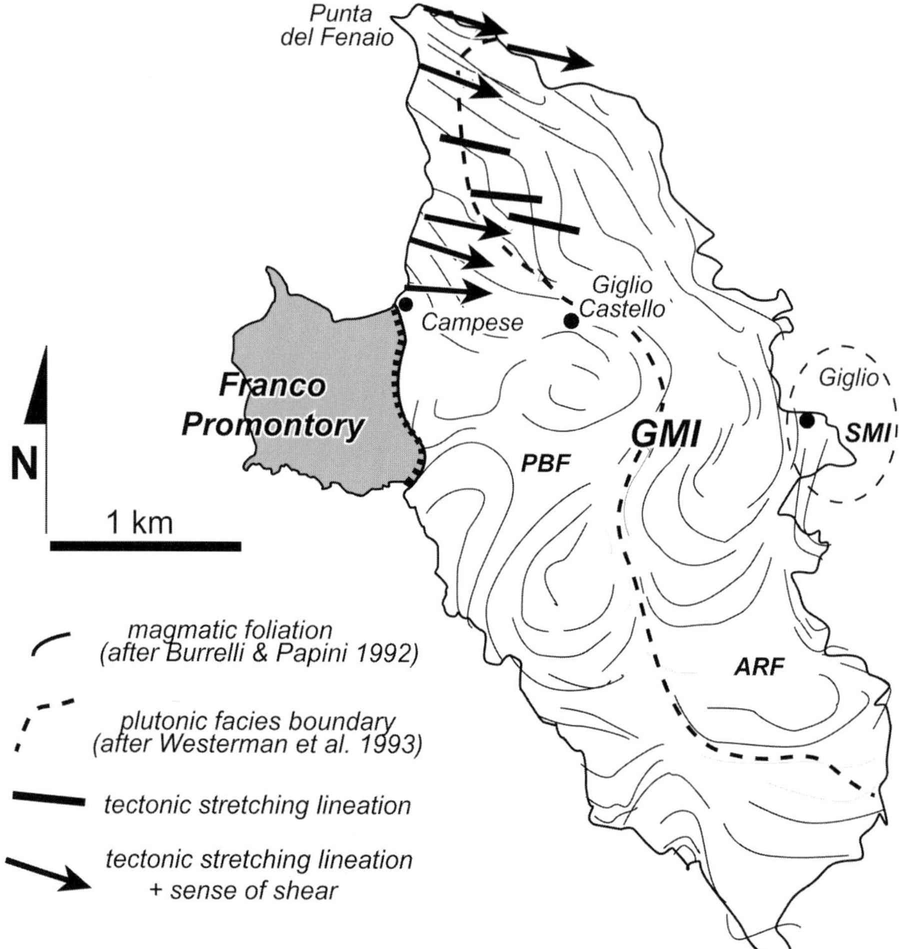

Fig. 5. Geological map of Giglio Island (modified after Lazzarotto *et al.* 1964). The attitude of tectonic shear deformation within the Giglio Monzogranite Intrusion (GMI) is also shown (modified after Rossetti *et al.* 1999). PBF, Pietrabona facies; ARF, Arenella facies; SMI, Scole Monzogranite Intrusion.

Giglio Island. Giglio Island comprises S-type peraluminous and sub-alkaline monzogranitic intrusions with an average Rb/Sr radiometric cooling age of 5.0 Ma (Westerman *et al.* 1993) and, subordinately, metamorphic and sedimentary rocks of the Franco Promontory (Lazzarotto *et al.* 1964; Rossetti *et al.* 1999) (Fig. 5). On a ternary plot of normative quartz, albite and orthoclase, all of the intrusive rocks plot around a minimum crystallization pressure of *c.* 4 kbar (Westerman *et al.* 1993). Intrusive contacts with country rocks are exposed only at Punta del Fenaio and near the village of Giglio Porto (Barrese *et al.* 1987) (Fig. 5). A NNW–SSE trending, westward-dipping extensional fault zone marks the contact between the intrusive rocks and the Franco Promontory. At this location, the country rocks lack any evidence of thermal metamorphism (Lazzarotto *et al.* 1964; Westerman *et al.* 1993).

The main intrusive body (Giglio Monzogranite Intrusion, GMI) is characterized by an outer zone that is strongly foliated and locally layered (Pietrabona facies, PBF) and an inner zone that is highly porphyritic and homogeneously textured (Arenella facies, ARF) (Westerman *et al.* 1993). The latter facies has preserved the magmatic, concentric structures and it may be considered the core of the pluton (Burrelli & Papini 1992) (Fig. 5). Transition to solid-state structures is mostly localized in the northwestern part of the island, an area that may represent the original plutonic rim. In particular, solid-state deformation increases at Punta del Fenaio (Fig. 5), where shear strain localization occurs with formation of S–C foliations and low-grade mylonites, overprinting the primary igneous structure (Fig. 4b). C- surfaces, defined by zones rich in micas and fine-grained recrystallized quartz, show NNE–SSW strikes and shallow dips. Stretching lineations on C-surfaces, mostly marked by quartz and tourmaline crystals, trend N120°E and the sense of shear is top-to-the-east, as indicated by S–C relationships and other kinematic indicators (Rossetti *et al.* 1999) (Figs 4b & 5). At Campese (Fig. 5), ductile shear deformation is localized on low-angle east-dipping aplitic dykes. Stretching directions strike N120°E and the sense of shear, as deduced from the offset of pre-existing dykes and S–C relationships, is always top-to-the-east (Rossetti *et al.* 1999). Locally, late cataclasites and ultra-cataclasites to pseudotachylytes developed along the C-surfaces accompanying a late, eastward-directed, extensional shear. This is consistent with shear deformation continuing with fall in temperature.

The attitude of the tectonic deformation thus indicates that the top-to-the-east extensional shear affected the GMI in its late stage of cooling. In addition, the presence of an eastward shallow-dipping extensional fault, located at the top of the GMI and controlling basin sedimentation on the eastern side of the island, is indicated by seismic survey results (Bartole 1995). This suggests that the rising GMI was emplaced along an east-dipping extensional shear zone, recording a progressive transition from ductile to brittle shearing (Rossetti *et al.* 1999). A late NW–SE normal faulting episode accompanied the late cold plutonic rise of the GMI, possibly during Mid–Late Pliocene time. This faulting episode is responsible for the present-day contacts between the Franco Promontory and the Giglio intrusive rocks.

The Colline Metallifere region

Campiglia Marittima area. The Campiglia Ridge, located on the westernmost part of the Colline Metallifere area (Fig. 1), is made of rocks of the Tuscan Nappe sequence and is bounded to the west and east by N–S and NW–SE trending faults, respectively (Giannini 1955; Costantini *et al.* 1993) (Fig. 6a). The Botro ai Marmi intrusion, cropping out on the western edge of the ridge (Fig. 6a), has an overall quartz monzonitic composition (Barberi *et al.* 1967; Poli *et al.* 1989) and an average K/Ar cooling age of 5.7 Ma (Barberi *et al.* 1967; Borsi *et al.* 1967). The intrusion is surrounded by a rather static thermal aureole, affecting Lower Jurassic Tuscan carbonates and interrupted by the westward N–S bounding fault (Giannini 1955; Barberi *et al.* 1967) (Fig. 6a). This aureole extends N–S for *c.* 5 km and is 300 m thick (Rodolico 1945). Contact metamorphism assemblages indicate a temperature of about 500°C and a pressure lower than 1 kbar (Barberi *et al.* 1967). Magmatic activity in the area was also recorded by the San Vincenzo Rhyolites with Ar/Ar ages of 4.7 Ma (Feldstein *et al.* 1994),

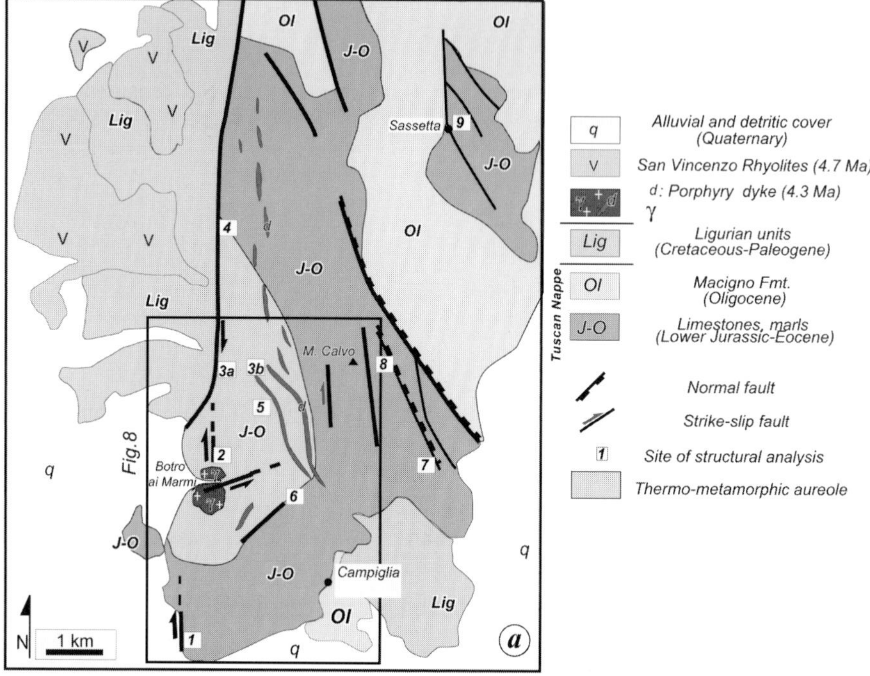

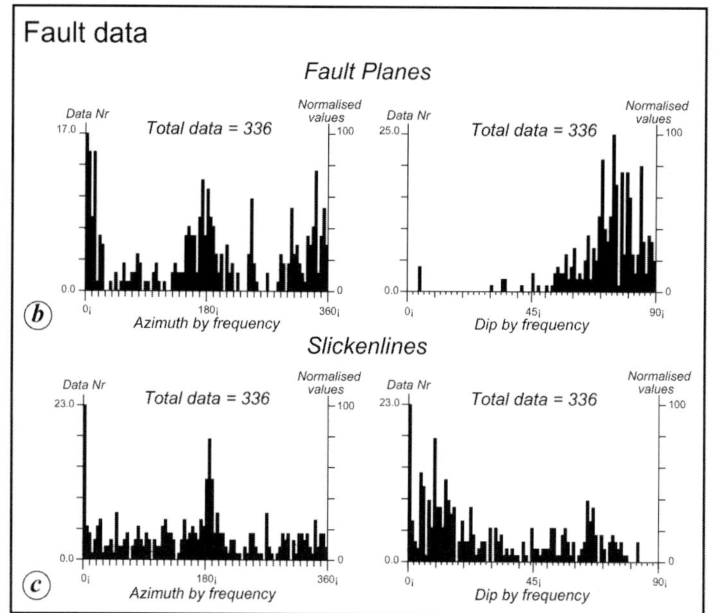

Fig. 6. (a) Geological and structural map of the Campiglia Ridge (modified after Giannini 1955; Costantini *et al.* 1993). (b) Frequency distribution analysis of collected faults (right-hand rule): fault strikes cluster around N–S trends (0–180° in the left histogram), with dips clustering around high dip values (70–80° in the right histogram). (c) Frequency distribution analysis of fault slickenlines (right-hand rule) showing a maximum concentration along a N–S trend (left histogram) with near-horizontal dips (right histogram), indicating prevalent strike-slip motions.

and by a N–S to NW–SE trending porphyry dyke swarm with K/Ar ages of 4.3 Ma (Barberi *et al.* 1967; Borsi *et al.* 1967) (Fig. 6a).

A N–S right-lateral strike-slip fault zone, consisting of steeply dipping fault segments (Figs 6 and 7a), marks the western margin of the ridge. Slickensides on fault planes are mainly sub-horizontal (Figs 6c and 7a). NW–SE sub-vertical cleavage planes are reoriented progressively N–S and reactivated as right-lateral strike-slip faults approaching the main fault zone (Fig. 7b). Cataclastic breccias and hydrothermal sulphide mineralizations (mostly pyrite) are often present along the strike-slip fault planes (Figs 7a and c). Subordinate, high-angle, ENE–WSW trending left-lateral strike-slip faults and sub-vertical N40°E trending tensional joints are also present. N–S dextral faults are occasionally reactivated as normal faults mainly in the northern sector, with a displacement of about a thousand metres (Giannini 1955). The eastern border of the Campiglia Ridge is represented by dextral NNW–SSE faults that were partly reactivated as normal faults. We also found sub-vertical N–S dextral faults, NNW–SSE sub-vertical cleavage planes and subordinate N–S normal faults inside the ridge. Mafic dykes cut across the N–S strike-slip fault zones and associated cleavage fractures with a simple dilatational displacement, suggesting that strike-slip faults pre-dated dyke intrusion.

The recognized fracture pattern within the ridge can adequately fit into a general kinematic framework of N–S, right-lateral simple shear movement (see, e.g. Sylvester 1988). The strike-slip related structures were mostly reactivated during a roughly NE–SW extension, either by reactivating the existing fractures or be generating new ones.

The main structural feature within the aureole is a second-phase foliation that, mostly related to flattening-type deformation, wraps around the intrusion (Fig. 8). At the contact between the pluton and the host-rock, the aureole foliation is parallel to the magmatic layering and early dykes are folded along the main foliation. Aureole foliation reworked a pre-existing sub-vertical cleavage, possibly related to the regional (strike-slip) deformation (Fig. 7d), and is crosscut by N–S trending, right-lateral strike-slip fault segments and associated cleavage fractures (Fig. 7b and e). Foliation in the thermal aureole is folded to form a broad NE–SW to N–S trending antiformal structure that culminates where the pluton crops out. This major antiform plunges in periclinal closure through rim synforms (Fig. 8). Bedding attitudes outside the thermal aureole are parallel to the main foliation in the aureole (Fig. 8). Mineral stretching lineations are not well developed in the aureole, but when present (mostly calcite fibres in marbles) have trend N90° to 110°E (Fig. 8). No clear shear sense indicators associated to these lineations were recognized.

The bulk of the intrusion is weakly solid-state deformed, except for localized semi-brittle shear zones, developed during the late stage of cooling of the intrusion. The shear zones consist of: a NNE–SSW extensional shear zone that crosscuts the contact with the host-rocks on the western margin of the pluton, with a top-to-the-west sense of shear; a conjugate set of strike-slip shear zones, N–S right-lateral and ENE–WSW left-lateral inside the pluton (Fig. 8).

The data collected within the thermal aureole and the pluton suggest an interplay between the near-field deformation associated with the magma flow during ascent and the far-field tectonic regime (strike-slip) recognized in the area (see, e.g. Brun & Pons 1981; Castro 1987; Hutton 1998*a*; Paterson *et al.* 1989; Guglielmo 1993). The following points are indicative of deformation induced by the upward movement of magma

(a)

(b)

(c)

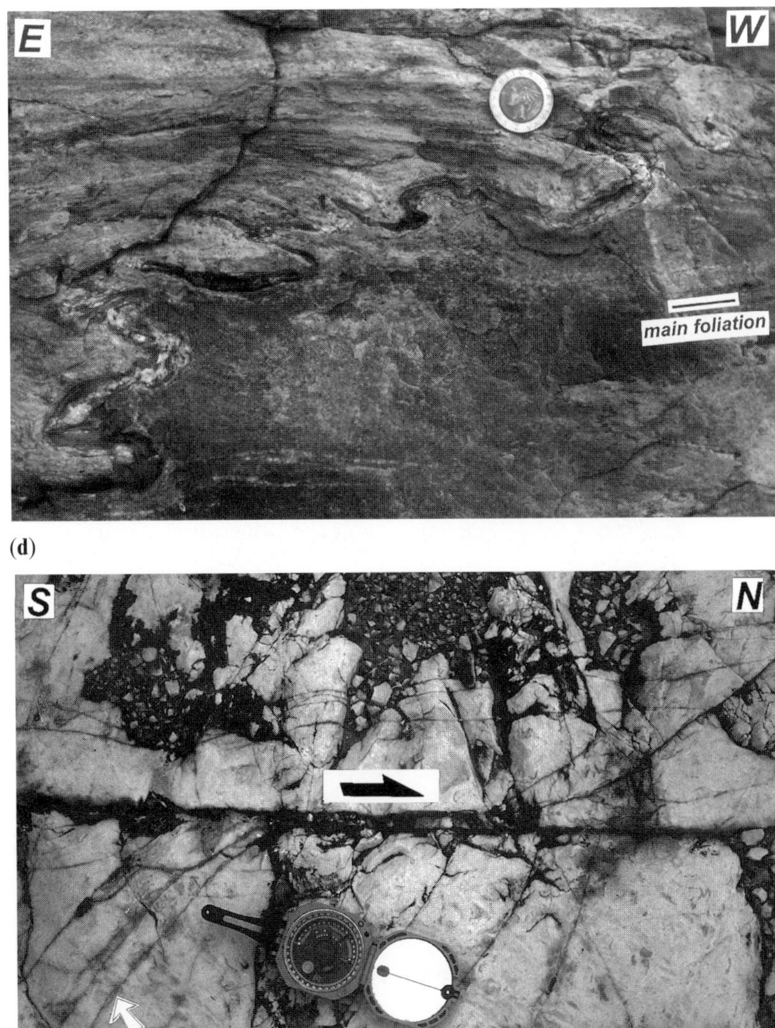

Fig. 7. Structural features recognized in the Campiglia Marittima area. (a) Aspect of the cataclastic deformation associated with a N–S trending dextral fault plane exposed within the thermal aureole of the Botro ai Marmi intrusion (westward bounding fault; site 3a in Fig. 6a, looking north); the stereoplot (Schmidt equal-area projection, lower hemisphere) is representative of the N–S, right-lateral strike-slip shear recognized within the Campiglia Ridge (fault planes are represented as great circles and slickenlines as dots; arrows indicate the sense of shear). (b) Attitude of the strike-slip related fracture cleavage (westward bounding fault, site 3a in Fig. 6a, looking north). (c) Slickensides on the fault plane shown in (a); sulphide mineralization (pyrite) is present along the slickenlines, provided by calcite fibres. (d) Second-phase foliation in the thermal aureole of the Botro ai Marmi intrusion, reworking a pre-existing cleavage, possibly connected to regional strike-slip faulting (southern edge of the intrusion). (e) Post-emplacement deformation within the Botro ai Marmi thermal aureole: NW–SE fracture cleavage (white arrow) is dragged along a N–S trending dextral fault crosscutting the second-phase foliation (plan view; site 5 in Fig. 6a).

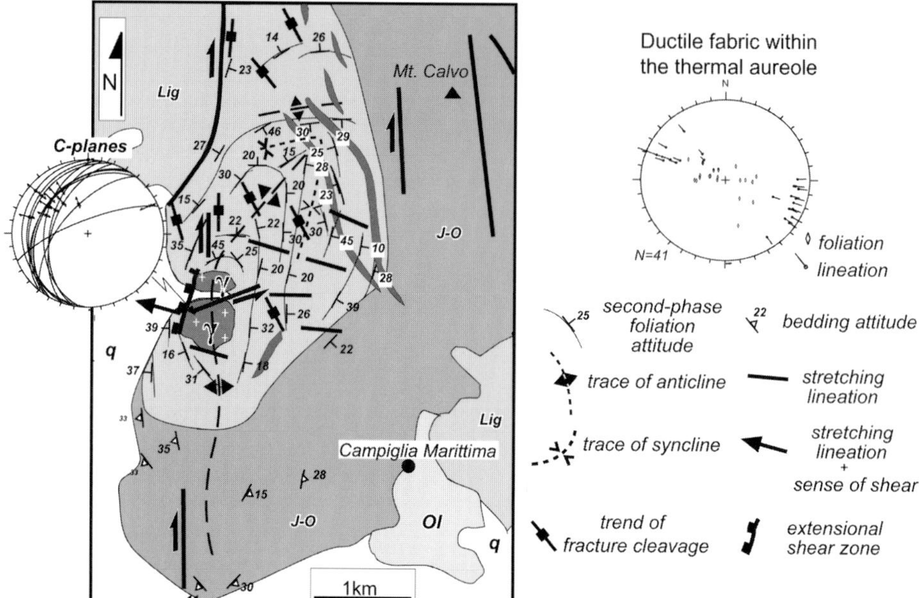

Fig. 8. Structural map of the thermal aureole around the Botro ai Marmi monzogranite. Patterns of the S–L fabric within the aureole and attitude of the shear deformation within the monzogranite are shown with the related stereoplots (Schmidt equal-area projection, lower hemisphere). The antiform marked by the main foliation within the thermal aureole culminates where the pluton crops out at the releasing bend along the westward bounding N–S dextral fault.

through the surrounding country rocks: the nearly concentric attitude of foliation both in the intrusion (mostly magmatic with little or no solid-state deformation) and in the thermal aureole, folded parallel to the margins of the intrusion (Paterson *et al.* 1989; England 1992); the lack of a continuous transition from magmatic to solid-state flow within the pluton (Paterson *et al.* 1989); the late folding of the aureole associated with the final ascent of the monzogranite resembles the rise of a laterally expanding, solid-state diapir across stiffer wall-rocks as described in the experimental work of Weinberg & Podladchikov (1995). The following points are indicative of the far-field tectonic deformation: occurrence in the contact rocks of constant WNW–ESE trending mineral stretching lineations, suggesting that regional, non-flattening-type deformation processes accompanied the development of ductile fabric in the aureole (Paterson & Vernon 1995); the prior existence of a cleavage oblique to the main foliation in the thermal aureole; the presence of a continuous fracture–cleavage in the aureole maintaining a NNW–SSE strike parallel to the strike of the regional cleavage in the wall-rock (Paterson *et al.* 1989); the presence of semi-brittle NNE–SSW extensional shear zones localized at the western border of the intrusion and brittle N–S right-lateral shear crosscutting the entire dome, both accommodating a common WNW–ESE extension.

Time relationships between the major N–S strike-slip faulting recognized in the whole ridge and the Botro ai Marmi monzogranite emplacement are envisaged by the following structural data: strike-slip related cleavage, which constitutes the main structural feature in the wall-rocks outside the thermal aureole, is folded along the main foliation in the aureole, suggesting that strike-slip tectonics was active before pluton emplacement; the orientation of the ductile linear fabric within the thermal aureole, as well as the shear zones inside the pluton, are consistent with the major N–S right-lateral faulting, suggesting that strike-slip shear was active during pluton emplacement; the post-emplacement deformation in the aureole is represented by N–S right-lateral strike-slip faulting and associated structures. In addition, the fact that the culmination of the intrusion coincides with the location of the releasing bend of the westward, N–S trending dextral fault zone suggests an active role of the strike-slip shear on pluton emplacement.

On these bases, we interpret the intrusion as having been emplaced in an overall strike-slip regime, along the westward N–S right-lateral strike-slip fault zone, where a jog initiated a pull-apart into which the granitic magma passively rose and cooled. The persistence of strike-slip right-lateral shear localized extension within the releasing bend and provided the space required for the Botro ai Marmi monzogranite to rise diapirically.

Gavorrano area. The Pliocene Gavorrano intrusion crops out in fault contact on the northern termination of the N–S trending Gavorrano Ridge, where the complete Tuscan Nappe sequence is exposed (Fig. 9a). The intrusion has a monzo–syenogranitic composition (Marinelli 1961; Poli *et al.* 1989) with K/Ar and Rb/Sr cooling ages of 4.9 Ma and 4.4 Ma (Borsi *et al.* 1967; Ferrara & Tonarini 1985), respectively. The northern edge of the intrusion is bounded by an ENE–WSW trending fault (Arisi Rota & Vighi 1971). In the Capanne Vecchie mining district, located north of the intrusion, the Tuscan Nappe sequence is strongly reduced, bordering to the south the Fosso dei Noni basinal area (Giannini *et al.* 1971; Elter *et al.* 1994) (Fig. 9a). Continental Late Tortonian and marine Lower Pliocene sequences, with a major Messinian lacuna, fill this basin (Pascucci & Sandrelli, pers. comm.). The rapid deposition of the Pliocene sequence (up to 600 m Ma^{-1}), suggests an increase in basin subsidence at that time (Pascucci & Sandrelli, pers. comm.).

The exposed portion of the Gavorrano intrusion displays a flat-lying magmatic foliation defined by biotite and K-feldspar aggregates (Boccaletti & Conticini 1983), with no evidence of syn-tectonic deformation. In addition, the lack of thermal aureole indicates that NW–SE trending normal faults bounding the monzogranite (Marinelli 1961; Arisi Rota & Vighi 1971) removed and cut-out the original contacts (Fig. 9a).

The main tectonic feature of the Gavorrano area is a N–S trending steeply dipping strike-slip fault zone, constituted of segments arranged in a right-stepped en echelon array (Fig. 9a & b). The fault zone extends from the Caldana village to the south to the Capanne Vecchie area to the north, marking contacts among the various units (Fig. 9a). Shear fibres, striations and grooves on fault surfaces have pitch values indicating from strike- to oblique-slip right-lateral fault movement (Figs 9c and 10). NNW–SSE secondary synthetic and ENE–WSW antithetic faults accompany the N–S right-lateral shear. Cleavage planes are steeply dipping and trend mainly N120°–140°E, and, approaching the main fault zone, they became parallel to the N–S general trend and are reactivated in a right-lateral shear sense. To the north, in the Fosso dei Noni area, stepover between two right-lateral, en echelon, N–S faults occurs by means of a N40°E extensional accommodation zone bounding the basin (Fig. 9a). The increasing subsidence of the basin during Early Pliocene time

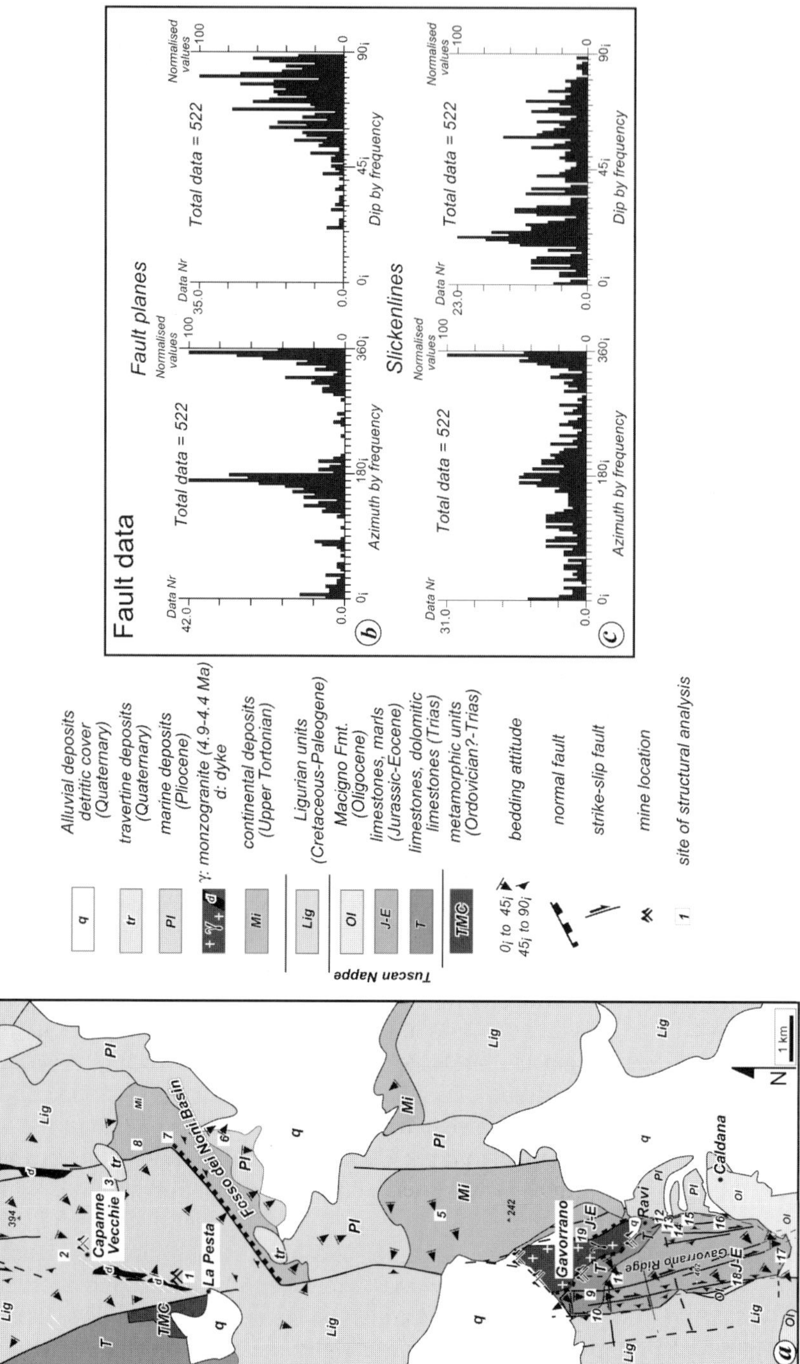
Fault data
Fault planes
Slickenlines
Total data = 522
Azimuth by frequency
Dip by frequency
Data Nr
Normalised values
Alluvial deposits detritic cover (Quaternary)
travertine deposits (Quaternary)
marine deposits (Pliocene)
γ: monzogranite (4.9-4.4 Ma) d: dyke
continental deposits (Upper Tortonian)
Ligurian units (Cretaceous-Paleogene)
Macigno Fmt. (Oligocene)
limestones, marls (Jurassic-Eocene)
limestones, dolomitic limestones (Trias)
metamorphic units (Ordovician?-Trias)
Tuscan Nappe
bedding attitude
normal fault
strike-slip fault
mine location
site of structural analysis
Capanne Vecchie
La Pesta
Fosso del Noni Basin
Gavorrano
Ravi
Caldana
Gavorrano Ridge
1 km

Fig. 10. N–S trending dextral fault planes (arrows) developed in the Lower Jurassic limestones of the Tuscan Nappe, Gavorrano area (site 11 in Fig. 9a, looking south). These fault planes describe negative flower-structures (height of the cliff 50 m); the stereoplot (Schmidt equal-area projection, lower hemisphere) shows the main fault systems (fault planes are represented as great circles and slickenlines as dots; arrows indicate the sense of shear).

and the associated distribution of the clastic deposition, indicating a WSW provenance (Pascucci & Sandrelli, pers. comm.), may be explained by the pull-apart geometry along N–S fault segments undergoing right-lateral strike-slip shear (see e.g. Sylvester 1988).

Commonly, slickensides on fault planes show early strike-slip superimposed by oblique or dip-slip extensional motions (Fig. 9c), corresponding to a late NE–SW extension. In addition, intense hydrothermal mineralization on fault planes and the

Fig. 9. (a) Structural map of the Gavorrano–Capanne Vecchie area (modified after Bertini *et al.* 1969; Giannini *et al.* 1971; Elter *et al.* 1994). (b) Frequency distribution analysis of collected faults (right-hand rule): fault strikes cluster around N–S trends (0–180° in the left histogram), with dips clustering around high dip values (70–90° in the right histogram). (c) Frequency distribution analysis of fault slickenlines (right-hand rule) shows two maximum concentrations along a N–S trend (left histogram), with two populations of dips (around 20° and 60–70° in the right histogram), indicating both strike-slip and dip-slip motions.

presence of N–S to NNE–SSW trending dykes in the northern sector (Fig. 9a) indicate that the N–S strike-slip fault zone acted as a preferential pathway for late circulation of magmatic fluids.

Our data thus suggest that the strike-slip regime in the Gavorrano area was coeval with the Early Pliocene sedimentation and subsidence of the Fosso dei Noni basin, and therefore with pluton emplacement. As a consequence, ascent and location of the Gavorrano monzogranite might have been controlled by the N–S right-lateral shear, possibly along a releasing bend of the main strike-slip fault zone bounding the Gavorrano Ridge to the west (Fig. 9a). A late-stage NW–SE extension, compatible with the orientation of the normal faults bounding the monzogranite, accompanied the late rise of the Gavorrano intrusion.

A general hypothesis for emplacement of the Tuscan granitoids

We have shown that coexisting extensional and strike-slip tectonics controlled the emplacement of the Messinian–Pliocene Tuscan intrusive rocks in the Northern Tyrrhenian area. Previous studies on various areas have shown that extensional and strike-slip tectonics are effective mechanisms for creating the necessary space for granitic plutons emplacement in the crust. In particular, pluton emplacement is controlled by zones of decompression either at the footwall of major extensional shear zones (see, e.g. Lister *et al.* 1984; Hutton *et al.* 1990; Gautier *et al.* 1993; Fletcher *et al.* 1995) or at sites of local extension along major strike-slip fault systems, (see, e.g. Guineberteau *et al.* 1987; Hutton 1988*b*; Morand 1992; Tikoff & Teyssier 1992). However, any model for pluton emplacement in the Northern Tyrrhenian region has to consider the following points: regional extension connected with the Northern Apennine post-orogenic crustal relaxation processes caused thinning of the previously thickened crust and diffuse crustal melting; eastward, non-coaxial extensional shear at mid-crust level controlled pluton emplacement offshore (Elba, Montecristo and Giglio); active right-lateral strike-slip faulting during pluton emplacement localized the emplacement of intrusions onshore: strike-slip faulting controlled the intrusions' location and ascent in sectors of localized extension on releasing bends along the main faults bordering both the Gavorrano and Campiglia ridges; a common E–W to WNW–ESE sense of extension is recorded both offshore and onshore during pluton emplacement; the depth of pluton emplacement was shallower onshore than offshore.

Extension contemporary with horizontal constriction inducing strike-slip faulting is widely documented during the extensional collapse of the thickened crust undergoing non-coaxial shear (Fletcher & Bartley 1994; Mancketelow & Pavlis 1994). In particular, strike-slip faulting may accomplish simple shear along crustal-scale ductile extensional shear zones to maintain strain compatibility with upper-plate rocks (Fletcher & Bartley 1994). In an extending crust, a switch between maximum (σ_1) and intermediate (σ_2) stress axes is required to obtain this configuration: the vertical load must equal the intermediate stress, which is horizontal and perpendicular to the extension direction (σ_3). From this point of view, the N–S trending dextral strike-slip faulting episode recognized in the Colline Metallifere area may be interpreted as a second-order brittle shear process accommodating the regional E–W to WNW–ESE ductile eastward asymmetric flow, stretching apart the previously thickened Northern Tyrrhenian crust. Within this framework, we propose a model in which

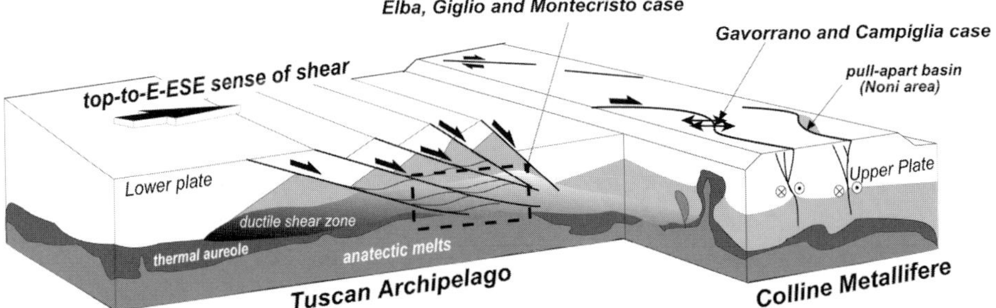

Fig. 11. Block diagram showing modes of pluton emplacement in the Tyrrhenian area during Messinian–Early Pliocene time(not to scale; location of structures is only indicative). Ductile eastward non-coaxial extension in the lower plate is responsible for the emplacement of offshore intrusions, whereas N–S dextral strike-slip faulting in the upper plate controlled the rise of onshore ones. Strike-slip faulting is interpreted as an upper-plate shear process accommodating the eastward ductile stretching of lower-plate rocks.

the strike-slip shear zones provided a pathway for the emplacement at upper-crustal levels of an early anatectic magma, generated by a regional post-orogenic extensional tectonics and residing at the base of the extending crust (Fig. 11). The inherited steeply dipping, mainly N–S to NW–SE oriented, strike-slip discontinuities were then reactivated during an NE–SW extension. Associated NW–SE normal faults drove the final intrusives ascent path and fragmented the original intrusions' contacts, as observed in the Montecristo, Giglio and Gavorrano areas. The uplift of the Campiglia and Gavorrano ridges is possibly connected to this final brittle-faulting event.

The pre-existing strike-slip deep sets of fracture also constitute a preferred pathway for dyke intrusion and late magmatic products and hydrothermal circulation in the region. The revealed connection between non-coaxial extensional shear at depth and strike-slip shear zones at shallow crustal levels may then provide an innovative frame of reference for deep exploration strategies in geothermal research on the region.

V. Pascucci and F. Sandrelli of the Siena University, coworkers in the Gavorrano area, are gratefully acknowledged for providing the stratigraphic data for the Noni area. M. Pietrini and A. Venuti partly participated in fieldwork in the Campiglia Marittima and Gavorrano areas. Comments, suggestions and critical reviews by A. Castro, C. Fernández and B. Vendeville are gratefully acknowledged. This paper benefited from discussions with A. Lazzarotto, M. Mattei, V. Pascucci, F. Sandrelli and F. Storti. Structural data analysis was performed by the use of DAISY software (Salvini 1998).

References

Arisi Rota, F. & Vighi, L. 1971. Le mineralizzazioni a pirite e solfuri misti della Toscana meridionale. *Rendiconti della Società Italiana di Mineralogia e Petrografia*, **27**, 368–423.

Baldi, P., Bellani, S., Ceccarelli, A., Fiordelisi, A., Squarci, P. & Taffi, L. 1994. Correlazioni tra le anomalie termiche ed altri elementi geofisici e strutturali della Toscana Meridionale. *Studi Geologici Camerti, Special Volume*, **1994**(1), 139–149.

Barberi, F., Buonasorte, G., Cioni, R. *et al.* 1994. Plio-Pleistocene geological evolution of the geothermal area of Tuscany and Latium. *Memorie Descrittive della Carta Geologica d'Italia*, **69**, 77–134.

——, INNOCENTI, F. & MAZZUOLI, R. 1967. Contributo alla conoscenza chimico-petrografica e magmatologica delle rocce intrusive, vulcaniche e filoniane del campigliese (Toscana). *Memorie della Società Geologica Italiana*, **6**, 643–681.

BARRESE, E., DELLA VENTURA, G., DI SABATINO, B., GIAMPAOLO, C. & DI LISA, A. 1987. The thermometamorphic contact aureole around the isola del Giglio Granodiorite (Tuscany, Italy): petrography and petrogenetical considerations. *Geologica Romana*, **26**, 349–357.

BARTOLE, R. 1995. The North Tyrrhenian–Northern Apennines post-collisional system: constraints for a geodynamic model. *Terranova*, **7**, 7–30.

BERTINI, M., CENTAMORE, E., JACOBACCI, A., & NAPPI, G. 1969. *Foglio 127, Piombino*. Carta Geologica 1:100.000. Servizio Geologico d'Italia.

BOCCALETTI, M. & CONTICINI, F. 1983. Ricerche meso e microstrutturali sui corpi ignei della toscana—1. L'intrusione di Gavorrano. *Memorie della Società Geologica Italiana*, **26**, 553–562.

—— & PAPINI, P. 1989. Ricerche meso e microstrutturali sui corpi ignei neogenici della Toscana 2: l'intrusione del Monte Capanne (Isola d'Elba). *Bollettino della Società Geologica Italiana*, **108**, 699–707.

BORSI, S., FERRARA, G. & TONGIORGI, E. 1967. Determinazione con il metodo del K/Ar delle età delle rocce magmatiche della Toscana. *Bollettino della Società Geologica Italiana*, **86**, 403–410.

BOSSIO, A., COSTANTINI, A. *et al.* 1993. Rassegna delle conoscenze sulla stratigrafia del Neautoctono Toscano. *Memorie della Società Geologica Italiana*, **49**, 17–98.

BRUN, J. P. & PONS, J. 1981. Strain patterns of pluton emplacement in a crust undergoing non-coaxial deformation, Sierra Morena, Southern Spain. *Journal of Structural Geology*, **3**, 219–230.

BURRELLI, F. & PAPINI, P. 1992. Structural analysis of Giglio island pluton. *In*: Carmignani, L. & Sassi, P. (eds) *Contribution to the Geology of Italy with Special Regard to the Palaeozoic Basements. A Volume Dedicated to Tommaso Cocozza*. IGCP 276, Newsletters, **15**, 449–452.

CARMIGNANI, L., DECANDIA, F. A., FANTOZZI, P. L., LAZZAROTTO, A., LIOTTA, D. & MECCHERI, M. 1994. Tertiary extensional tectonics in Tuscany (Northern Apennines, Italy). *Tectonophysics*, **238**, 295–315.

CASTRO, A. 1987. On granitoid emplacement and related structures. A review. *Geologische Rundschau*, **76**, 101–124.

COSTANTINI, A., LAZZAROTTO, A., MACCANTELLI, M., MAZZANTI, R., SANDRELLI, F., TAVARNELLI, E. & ELTER, F. M. 1993. Geologia della Provincia di Livorno a Sud del Fiume Cecina. *Quaderni del Museo di Storia Naturale Livorno*, **13**, 1–164.

DANIEL, J. M. & JOLIVET, L. 1995. Detachment faults and pluton emplacement: Elba Island (Tyrrhenian Sea). *Bulletin de la Societé Géologique de France*, **4**, 341–354.

DURANTI, S., PALMERI, R., PERTUSATI, P. C. & RICCI, C. A. 1992. Geological evolution and metamorphic petrology of the basal sequence of the eastern Elba (Complex II). *Acta Vulcanologica*, **2**, 213–229.

ELTER, F. M., PASCUCCI, V. & SANDRELLI, F. 1994. Geologia dell'area a sud di Massa Marittima (GR): studi preliminari. *Studi Geologici Camerti, Special Volume*, **1994**(1), 57–64.

ENGLAND, R. W. 1992. The identifications of granitic diapirs. *Journal of the Geological Society, London*, **147**, 931–933.

FELDSTEIN, S. N., HALLIDAY, A. N., DAVIES, G. R. & HALL, C. M. 1994. Isotope and chemical microsampling: constraints on the history of an S-type rhyolite, San Vincenzo, Tuscany, Italy. *Geochimica et Cosmochimica Acta*, **58**, 943–958.

FERRARA, G. & TONARINI, S. (1985). Radiometric geochronology in Tuscany: results and problems. *Rendiconti della Società Italiana di Mineralogia e Petrografia*, **40**, 111–124.

FLETCHER, J. M. & BARTLEY, J. M. 1994. Constrictional strain in a non-coaxial shear zone: implications for fold and rock fabric development, central Mojave metamorphic core complex, California. *Journal of Structural Geology*, **16**, 555–570.

——, ——, MARTIN, M. W., GLAZNER, A. F. & WALKER, J. D. 1995. Large-magnitude continental extension. An example from the Central Mojave metamorphic core complexes. *Geological Society of American Bulletin*, **107**, 1468–1483.

FRANCESCHINI, F. 1994. 'Larderello Plutono-Metamorphic core complex': metamorfismo regionale ercinicodi bassa pressione o metamorfismo di contatto Plio-Quaternario? *Studi Geologici Camerti, Special Volume,* **1994**(1) 113–128.

GANS, P. B., MAHOOD, G. A. & SCHERMER, E. 1989. *Synextensional Magmatism in the Basin and Range Province: a Case Study from the Eastern Great Basin.* Geological Society of America, Special Papers, **233**.

GAUTIER, P., BRUN, J. P. & JOLIVET, L. 1993. Structure and kinematics of upper Cenozoic extensional detachment on Naxos and Paros (Cyclades islands, Greece). *Tectonics*, **12**, 1180–1194.

GIANNINI, E. 1955. Geologia dei Monti di Campiglia Marittima (Livorno). *Bollettino della Società Geologica Italiana*, **84**, 1–77.

——, LAZZAROTTO, A. & SIGNORINI, R. 1971. Lineamenti di stratigrafia e tettonica nella Toscana meridionale. *Rendiconti della Società Italiana di Mineralogia e Petrografia*, **27**, 33–168.

GUGLIELMO, G. 1993. Interference between pluton expansion and non-coaxial tectonic deformation: three-dimensional computer model and field implications. *Journal of Structural Geology*, **15**, 593–608.

GUINEBERTEAU, B., BOUCHEZ, J. L. & VIGNERESSE, J. L. 1987. The Mortagne granite pluton (France) emplaced by pull-apart along a shear zone: structural and gravimetric arguments and regional implications. *Geological Society of the American Bulletin*, **99**, 763–770.

HANSON, R. B. & GLAZNER, A. F. 1995. Thermal requirement for extensional emplacement of granitoids. *Geology*, **23**, 213–216.

HUTTON, D. H. W. 1988*a*. Granite emplacement and tectonic control: inferences from deformation studies. *Transactions of the Royal Society of Edinburgh: Earth Sciences*, **79**, 245–255.

—— 1988*b*. Igneous emplacement in a shear zone termination: the biotite granite at Strontian, Scotland. *Journal of the Geological Society, London*, **139**, 615–631.

——, DEMPSTER, T. J., BROWN, P. E. & BECKER, S. D. 1990. A new mechanism of granite emplacement: intrusion in active extensional shear zones. *Nature*, **343**, 452–455.

INNOCENTI, F., SERRI, G., FERRARA, G., MANETTI, P. & TONARINI, S. 1992. Genesis and classification of the rocks of the Tuscan Magmatic Province: Thirty years after the Marinelli's model. *Acta Vulcanologica*, **2**, 247–265.

——, WESTERMAN, D., ROCCHI, S. & TONARINI, S. 1997. The Montecristo monzogranite (Northern Tyrrhenian Sea, Italy) a collisional pluton in an extensional setting. *Geological Journal*, **32**, 131–151.

JOLIVET, L., FACCENNA, C., GOFFÉ, B. *et al.* 1998. Mid-crustal shear zone in post-orogenic extension: example from the Tyrrhenian Sea. *Journal of Geophysical Research*, **103**, 12123–12160

JUTEAU, M., MICHARD, A., ZIMMERMAN, J. L. & ALBARÈDE, F. 1984. Isotopic heterogeneities in the granitic intrusion of Monte Capanne (Elba Island, Italy) and dating concepts. *Journal of Petrology*, **25**, 532–545.

KELLER, J. V. & PIALLI, G. 1990. Tectonics of the Island of Elba: a reappraisal. *Bollettino della Società Geologica Italiana*, **109**, 413-425.

LAZZAROTTO, A., MAZZANTI, R. & MAZZONCINI, F. 1964. Geologia del Promontorio dell'Argentario (Grosseto) e del Promontorio del Franco (Isola del Giglio-Grosseto). *Bollettino della Società Geologica Italiana*, **83**, 1–124.

LISTER, G. S. & BALDWIN, S. 1993. Plutonism, and the origin of metamorphic core complexes. *Geology*, **21**, 607–610.

——, BANGA, G. & FEENSTRA, A. 1984. Metamorphic core complexes of cordilleran type in the Cyclades, Aegean Sea, Greece. *Geology*, **12**, 221–225.

MANCKTELOW, N. S. & PAVLIS, T. 1994. Fault–fold relationships in low-angle detachment systems. *Tectonics*, **13**, 668–685.

MARINELLI, G. 1961. L'intrusione terziaria di Gavorrano. *Atti della Società Toscana di Scienze Naturali*, **66**, 117–194.

—— 1967. Genèse des magmas du vulcanisme Plio-Quaternaire des Apennines. *Geologische Rundschau*, **57**, 127–141.

——, BARBERI, F. & CIONI, R. 1993. Sollevamenti neogenici e intrusioni acide della Toscana e del Lazio settentrionale. *Memorie della Società Geologica Italiana*, **49**, 279–288.

MARTINI, I. P. & SAGRI, M. 1993. Tectono-sedimentary characteristics of Late Miocene–Quaternary extensional basins of the Northern Apennines, Italy. *Earth-Science Reviews*, **34**, 197–233.

MONGELLI, F. & ZITO, G. 1991. Flusso di calore nella regione Toscana. *Studi Geologici Camerti, Special Volume*, **1991**(1), 91–98.

MORAND, V. J. 1992. Pluton emplacement in a strike-slip fault zone: the Doctor Flat Pluton, Victoria, Australia. *Journal of Structural Geology*, **14**, 205–213.

NICOLICH, R. 1989. Crustal structures from seismic studies in the frame of the European Geotraverse (southern segment) and CROP project. *In*: BORIANI, A., BONAFEDE, M., PICCARDO, G. B. & VAI, G. B. (eds) *The Lithosphere in Italy*. Accademia dei Lincei, Rome, 41–61.

PATACCA, E., SARTORI, E. & SCANDONE, P. 1990. Tyrrhenian basin and Apenninic arcs: kinematic relations since Late Tortonian times. *Memorie della Società Geologica Italiana*, **45**, 425–451.

PATERSON, S. R. & FOWLER, T. K., JR 1993. Re-examining pluton emplacement mechanism. *Journal of Structural Geology*, **15**, 191–206.

—— & VERNON, R. H. 1995. Bursting the bubble of ballooning plutons: a return to nested diapirs emplaced by multiple processes. *Geological Society of America Bulletin*, **107**, 1356–1380.

——, —— & TOBISCH, O. T. 1989. A review of criteria for identification of magmatic and tectonic foliations in granitoids. *Journal of Structural Geology*, **11**, 349–363.

PLATT, J. P. & ENGLAND, P. C. 1994. Convective removal of lithosphere beneath mountains belts: thermal and mechanical consequences. *American Journal of Science*, **293**, 307–336.

POLI, G., MANETTI, P. & TOMMASINI, S. (1989). A petrological review on Miocene–Pliocene intrusive rocks from Southern Tuscany and Tyrrhenian Sea (Italy). *Periodico di Mineralogia*, **58**, 109–126.

PONZIANI, F., DE FRANCO, R., MINELLI, G., BIELLA, G., FEDERICO, C. & PIALLI, G. 1994. Caratteristiche della crosta dell'Appennino settentrionale in base alla revisione dati dei profili N–C–S e B–C–A della campagna DSS 1978. *Studi Geologici Camerti, Special Volume*, **1994**(1), 151–162.

RODOLICO, F. 1945. Ragguagli sul granito del Campigliese. *Atti della Società Toscana di Scienze Naturali*, **52**, 125–132.

ROSSETTI, F., FACCENNA, C., JOLIVET, L., TECCE, F., FUNICIELLO, R. & BRUNET, C. 1999. Syn- and post-orogenic extension: the case study of Giglio island (Northern Tyrrhenian Sea, Italy). *Tectonophysics*, **304**, 73–92.

SALVINI, F. 1998. *DAISY 2.1 Software*. Rome.

SAUPÉ, P., MARIGNAC, C., MOINE, B., SONET, J. & ZIMMERMANN, J. L. 1982. Datation par les méthodes K/Ar et Rb/Sr de quelques roches de la partie orientale de l'Tle d'Elbe (Province de Livourne, Italie). *Bulletin de Minéralogie*, **105**, 236–245.

SERRI, G. F., INNOCENTI, F. & MANETTI, P. 1993. Geochemical and petrological evidence of the subduction of delaminated Adriatic continental lithosphere in the genesis of the Neogene–Quaternary magmatism of Central Italy. *Tectonophysics*, **223**, 117–147.

SYLVESTER, A. G. 1988. Strike-slip faults. *Geological Society of America Bulletin*, **100**, 1666–1703.

TIKOFF, B. & TEYSSIER, C. 1992. Crustal-scale, en-echelon 'P-shear' tensional bridges: a possible solution to the batholithic room problem. *Geology*, **20**, 927–930.

TURNER, S., SANDIFORD, M. & FODEN, J. 1992. Some geodynamic and compositional constraints on 'postorogenic' magmatism. *Geology*, **20**, 931–934.

VIGNERESSE, J. L. 1995. Crustal regime of deformation and ascent of granitic magma. *Tectonophysics*, **249**, 187–202.

WEINBERG, R. F. & PODLADCHIKOV, Y. Y. 1995. The rise of solid-state diapirs. *Journal of Structural Geology*, **17**, 1183–1195.

WERNICKE, B. P., CHRISTIANSEN, R. L., ENGLAND, P. C. & SONDER, L. J. 1987. Tectonomagmatic evolution of Cenozoic extension in the North America Cordillera. *In*: COWARD, M. P., DEWY, J. & HANCOCK, P. L. (eds) *Continental Extensional Tectonics*. Geological Society, London, Special Publications, **28**, 203–221.

WESTERMAN, D., INNOCENTI, F., TONARINI, S. & FERRARA, G. 1993. The Pliocene intrusions of island of Giglio. *Memorie della Società Geologica Italiana*, **49**, 345–363.

ZITELLINI, F., TRICARDI, F., MARANI, M. & FABBRI, A. 1986. Neogene tectonics of the Northern Tyrrhenian Sea. *Giornale di Geologia*, **48**, 25–40.

Seismic tomography of the Dead Sea region: thinned crust, anomalous velocities and possible magmatic diapirism

NITZAN RABINOWITZ[1] & YOSSI MART[2]

[1]*Geophysical Institute of Israel, 1, HaMashbir Street, Holon, Israel*

[2]*Leon Recanati Center for Marine Studies, University of Haifa, Haifa 31905, Israel*

Abstract: Analysis of first arrivals of P waves from 113 earthquakes in the Dead Sea region and the calculation of a tomographic model indicates anomalous distribution of seismic velocities in the crust and the upper mantle. The seismic tomography model was applied to the lower and intermediate crust and the uppermost mantle there, at depths greater than the detection limits of seismic reflection surveys. At these depths the inversion shows two deep layers, at depths of 10–22 km and 22–32 km. The deeper layer shows average seismic velocity of 7.7 km s^{-1}, and the shallower one 6.5 km s^{-1}. The model suggests therefore that the Moho under the central Dead Sea is found at depth of 22 km, and the seismic velocity in the upper lithospheric mantle is anomalously low. The velocity of the layer overlying the Moho suggests a modelled layer of composite mineralogy, but high velocity anomalies were encountered in the 10-22 km layer under the boundary faults of the Dead Sea Rift, and a low velocity anomaly under the central Dead Sea is plausible. The probable interpretation of these anomalies is that the mantle under the central Dead Sea is shallow and anomalously hot, and that the lower and intermediate crust under the axial zone of the central and northern Dead Sea is also anomalously hot. Furthermore, it seems probable that magmatic diapirs ascend along the boundary faults of the rift into the intermediate crust. Structural similarity of the upper mantle and the lower crust between the Dead Sea and the northern Red Sea suggests an analogous tectonic regime for these two regions.

The Levant Rift system is a seismically active segment of the tectonic boundary between the Arabian and the African plates. (The Levant Rift and the Dead Sea Rift are synonyms. The present paper discusses the geophysics of the Dead Sea, namely, the hypersaline lake that is the namesake of the entire regional rift. To avoid confusion between the local and the regional phenomena, we refer to the regional structure as the 'Levant Rift'.) It extends from the northern Red Sea to southern Anatolia, and comprises a linear series of deep, faulted basins, some submerged under water and some exposed, separated by structural thresholds (Picard 1987). The length of the rift system is *c.* 1000 km, and its width varies from about 10 km in the structural basins to more than 35 km in the thresholds. The Dead Sea basin, the namesake of the entire feature, is a large structural depression within the Levant Rift system, bounded by large N–S trending faults on both flanks. It is 130 km long and 15 km wide, and the famous hypersaline Dead Sea occupies its

From: VENDEVILLE, B., MART, Y. & VIGNERESSE, J.-L. (eds) *Salt, Shale and Igneous Diapirs in and around Europe*. Geological Society, London, Special Publications, **174**, 79–92. 1-86239-066-5/00/$15.00

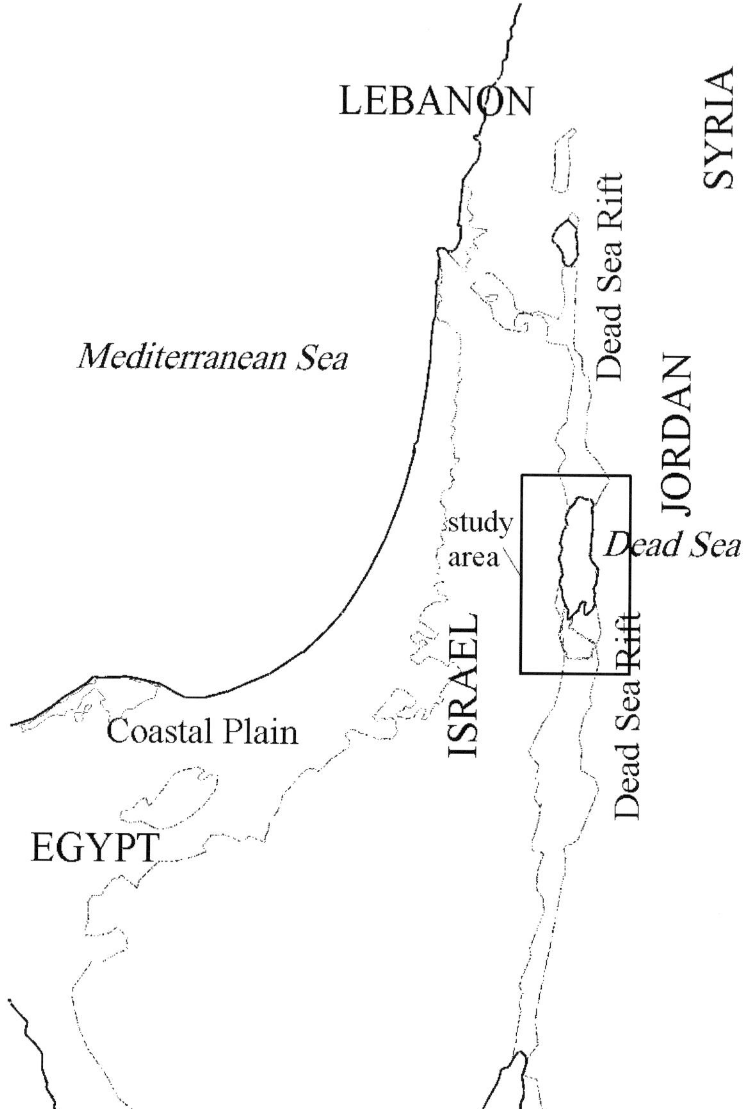

Fig. 1. The southern section of the Levant Rift system and the Dead Sea basin. The area between the coastal plain and the Rift is mountainous.

centre (Fig. 1). The water level of the lake is 410 m below mean sea level, and its bottom is at depth of nearly 800 m below mean sea level. In contrast, the topographic elevation of the western margin is *c*. 800 m, and the mountains to the east reach elevations of more than 1000 m, so that the present topographic relief in the Dead Sea area exceeds 1500 m.

The Levant Rift system is seismically active; the average recurrence rate of damaging earthquakes along it is *c*. 100 years (Poirier *et al.* 1980), and small earthquakes there are very numerous (Van Eck & Hofstetter 1989). Earthquake monitoring by the

Geophysical Institute of Israel and the Jordanian Seismological Service has been carried out in the region of the Dead Sea for years, and 113 earthquakes were selected for inversion of their P-wave direct arrivals into a tomographic model. The initial regional crustal stratification of the model was derived from the findings of a seismic refraction survey (Ginzburg *et al.* 1979), which showed that the crust under the Rift is considerably thinner than that under the western flank. Deep penetrating seismic reflection profiles showed a 7 km thick lithological sequence of a low seismic velocity, which was presumed to be underconsolidated Neogene–Quaternary deposits (Kashai & Krocker 1987). Such thickness of young depositional material suggests active subsidence. Concurrently, there is ample evidence that the flanks of the Dead Sea Rift are rising (Mart & Horowitz 1981; Zak & Freund 1981; Begin & Zilberman 1997). An earlier tomographic model of the region suggested the occurrence of magmatic diapirs in the lower and intermediate crust (Rabinowitz *et al.* 1996).

Data reduction and model calculation

Seismic tomography

Tomography of the Earth's crust uses the travel time of numerous earthquakes, measured at numerous seismograph stations, to show the distribution and the variability of seismic velocities in the crust and mantle. Tomography uses mathematical inversion of many ray paths from different directions, which cross the region of interest, to produce a 3D model (Romanowicz 1991). To study a particular area, a set of sources and a set of recorders are used to produce differential travel times of the P waves, which emphasize local heterogeneities. Very similarly to medical tomography, the investigated area is broken up into a large number of small blocks, to which an initial velocity is attributed based on previous studies, or on correlations with similar terrains elsewhere. With many rays passing through each block, mathematical methods, such as least squares, are used to adjust the initial velocities to match the measured travel times (Doyle 1995). Optimization presents a 3D model of the distribution of seismic velocities in the study area. As seismic velocities introduce constraints on estimates of the density and the viscosity of Earth materials, the tomographic inversion can be used for tectonic analysis purposes.

The horizontal variations of seismic velocities in a studied area can result from changes in the chemical–mineralogical or thermal characteristics, or a combination of both. The basic stratification in the Earth's crust and mantle is the product of such variations, where pressure and density increase with depth, crystallization becomes more compact, and the ratio of heavy elements in magma increases. Consequently, seismic velocity increases with depth. However, as temperature increases with depth as well, the density of a stable crust or mantle component would decrease if it becomes hotter, irrespective of depth. Such a thermal anomaly could take place, for example, in a zone of magmatic upwelling, such as hot spots and mid-ocean ridges, where ascending hot magma affects adjacent rocks. When mantle material becomes gravimetrically metastable as a result of heating, it flows upwards diapirically, and could affect the seismic velocities of its surroundings in two adverse ways. On one hand, the seismic velocity of the ascending mantle material is likely to be higher than that of its surroundings, and thus form a fast tomographic anomaly,

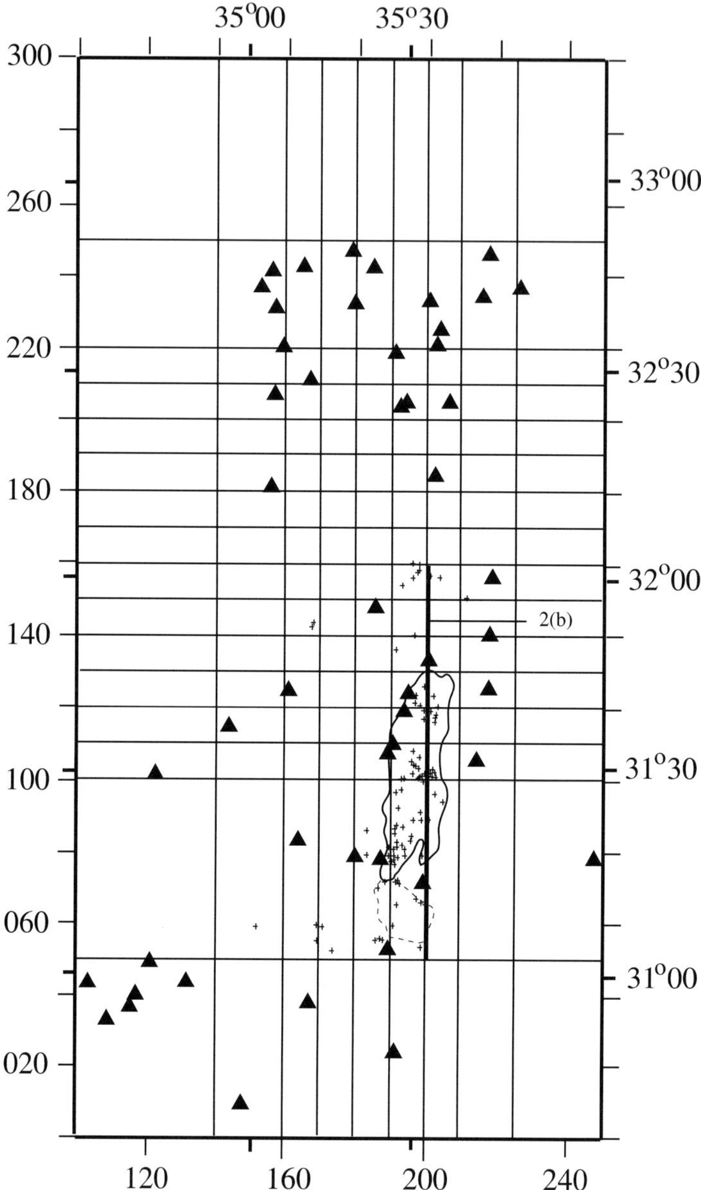

Fig. 2. (**a**) Map of earthquake epicentres (dots) that occurred in the Dead Sea region in January 1991–July 1993, and the seismograph stations (▲) that recorded them. Both geographic (heavy ticks) and Israeli (fine ticks) coordinate grids are presented. Heavy segment of coordinate 200 Israeli grid shows the location of (**b**).

which is essentially mineralogical. On the other hand, the penetrating material heats the crustal rocks, which become less dense, and their seismic velocity could drop, so a slow tomographic anomaly is formed, which is essentially thermal. Tectonic considerations should be applied to differentiate between thermal and chemical–mineralogical anomalies of seismic velocities in tomographic inversions.

Dataset

Analysis of direct P-wave arrivals of local earthquakes in the Dead Sea region was carried out using seismograph stations on both sides of the Rift (Fig. 2). These P readings were reported as such by the annual bulletins of the Israeli Seismic Network (ISN) and by the reports of the Jordan Seismological Observatory. These datasets were used to calculate a tomographic model of the distribution of seismic velocities in the crust. Seismograph stations on both sides of the Rift recorded 113 earthquakes, which occurred in the Dead Sea region in the period of January 1991–July 1993 and were selected for the present inversion. The inclusion of seismograph stations from both the Jordanian and the Israel sides of the Levant Rift provides a good overall coverage for the northern basin of the Dead Sea. Simultaneous inversion of this dataset was carried out to model the velocity structure in the underlying lower crust and uppermost mantle, to a depth of 32 km. The inversion procedure was based on Thurber's (1981) approach, utilizing the concept of approximate ray tracing (ART) for effective 3D travel-time calculations, and the method of parameter separation for handling the inversion of large matrices.

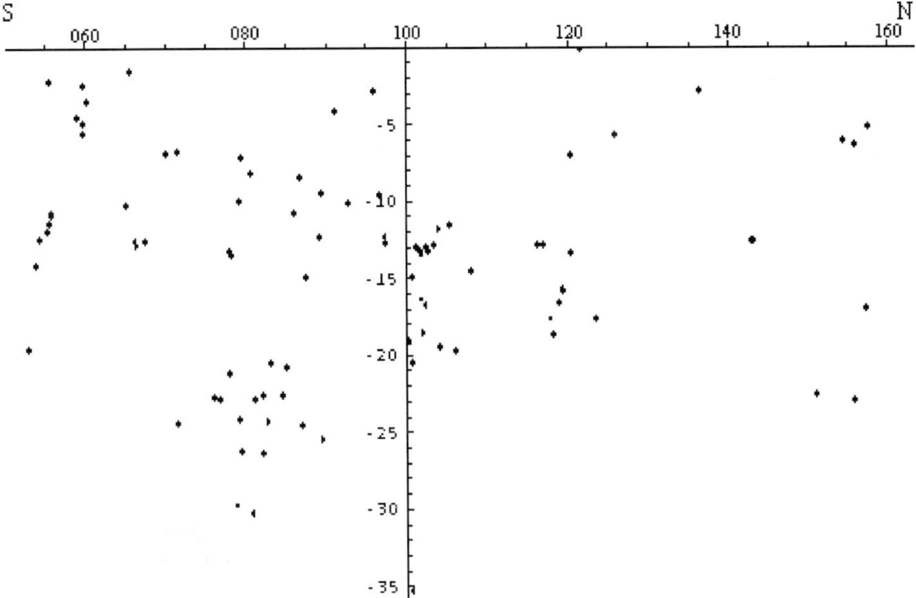

Fig. 2 (b) Composite N–S depth section of the Dead Sea region showing the depth distribution of projected earthquake foci used in the present inversion. Foci were projected, following VELEST correction, on coordinate 200, Israeli grid.

Problem formulation

Thurber (1981) showed that the arrival time t of a seismic wave generated by an earthquake is a nonlinear function of the station coordinates (s), the hypocentral parameters (h) and the velocity field (m):

$$t = \mathrm{f}(s, h, m) \tag{1}$$

Applying a first-order Taylor series expansion to (1), the coupled velocity model parameter relationship can be written in matrix form as

$$\boldsymbol{t} = \boldsymbol{Hh} + \boldsymbol{Mm} + \boldsymbol{e} = \boldsymbol{Ad} + \boldsymbol{e} \tag{2}$$

where $\boldsymbol{t}$ is the vector of travel time residual, $\boldsymbol{H}$ is the matrix of partial derivative of travel time with respect to hypocentral parameters, $\boldsymbol{h}$ is the vector of hypocentral parameter adjustments, $\boldsymbol{M}$ is the matrix of partial derivatives of travel time with respect to model parameters, m is the vector of model parameter adjustments, $\boldsymbol{e}$ is the vector of travel time errors, $\boldsymbol{A}$ is the matrix of all partial derivatives, and $\boldsymbol{d}$ is the vector of hypocentral and model parameter adjustments. The solution is based on iterative refinement of a linearized approximation of the vectors of hypocentral and model parameter adjustments ($\boldsymbol{h}$ and $\boldsymbol{m}$ in equation (2)) to the nonlinear function f (see Thurber 1981, 1983; Kissling *et al.* 1994).

The resolution matrix $\boldsymbol{R}$ of the inversion is given by the formula

$$\boldsymbol{R} = (\boldsymbol{A}^{\mathrm{T}}\boldsymbol{A} + \lambda^2\boldsymbol{I})^{-1}\boldsymbol{A}^{\mathrm{T}}\boldsymbol{A} \tag{3}$$

where λ is the damping parameter that should be selected to represent the variance of the data, i.e. errors in arrival time readings, and $\boldsymbol{I}$ is the identity matrix.

The size of the diagonal elements of $\boldsymbol{R}$ reflects the resolution of the model parameters, where unity being perfect resolution. If any diagonal element or $\boldsymbol{R}$ is near zero, the corresponding model parameter is unresolved (see Wiggins (1972) for further discussion). 'Areas of acceptable resolution' were defined (Kissling, pers. comm.) as blocks whose resolution values exceed 0.4, and confine our interpretation to these areas.

Computation procedure

The inversion scheme was based on the commonly accepted two-stage procedure:

- Stage I: travel times were jointly inverted to obtain a 1D tomographic solution, together with revised hypocentre coordinates and station corrections. This procedure is referred to as the '1D minimum model' (Kissling *et al.* 1994).
- Stage II: 3D tomographic inversion was carried out using the minimum 1D model, obtained in stage I (Thurber 1981; Evans & Eberhart-Phillips 1994).

As in many other nonlinear problems, the chances for successful estimation of the true model depend on the selection of the starting model in the neighbourhood of the true model, one for which the linearization assumptions hold. Kissling *et al.* (1994) showed how systematic error in the starting model will not only influence the 3D results, but will also distort the resulting error estimates. The determination of the minimum 1D model is actually a trial-and-error process. It is very likely that more than one such model will be found, and the decision to adopt one particular model should take into consideration additional *a priori* information.

Table 1. *1D model of seismic velocity, computed by* VELEST *procedure (Kissling* et al. *1995), which was used as the initial model of the inversion*

Layer (km depth)	Velocity ($km\,s^{-1}$)
0–2	2.5
2–10	5.6
10–22	6.5
22–32	7.0
>32	–

To estimate the minimum 1D model for the Dead Sea dataset, we used the VELEST program (Kissling *et al.* 1995), and the starting model was set to that of Ginzburg *et al.* (1979). The VELEST program produced a 1D velocity model (Table 1). One important consequence of this procedure was encountered in the layout of the focal depth re-estimation. A relatively high number of earthquakes was placed by the VELEST algorithm at focal depths deeper than 25 km (18 earthquakes, in comparison with six earthquakes as reported in the bulletins). The occurrence of deep events underneath the Dead Sea is of major importance regarding the capability of the 3D tomographic procedure to resolve sub-crustal velocity structures in this area.

To produce the inversion of stage II we have divided the medium under the Dead Sea region into unevenly spaced blocks, based on the composite distribution of seismometers and earthquakes. Figure 2 depicts the horizontal demarcation of the blocks of uniform velocity. These blocks were sliced into five layers, and average seismic velocity was calculated for each layer. The computed depth–velocity distribution is presented in Table 1. We then prepared a tomographic model of the velocity stratification of the crust and the upper mantle under the Dead Sea basin and its proximal margins following verified procedures (Aki & Richards 1980; Thurber 1983; Eberhart-Phillips 1986; Kissling 1988).

Geological setting

The Levant Rift comprises a series of basins s*et al.*ong a regional topographic valley, the floor of which has been subsiding and its flanks rising at least since late Miocene time (Begin & Zilberman 1997). The internal basins, which show a general trend of northward decrease in size, are separated by axial thresholds (Fig. 1). The basins are sediment filled and their margins are steeply faulted, whereas the thresholds are structurally shallow, their marginal faults are small and dispersed, and their rifted configuration is diffuse (Mart 1991).

The geological set-up of the Dead Sea region is built of three distinctly different geological provinces, the two facing margins and the rift between them. The rift floor is built mainly of thick Neogene–Quaternary clastic fill, interbedded with lacustrine carbonates. A prominent early Pliocene evaporitic layer, which also occurs within the alluvial fill, is the source of a series of salt diapirs that ascended into the rift floor and along its boundary faults (Neev & Hall 1979). Some of these salt diapirs crop out along the rift floor and its margins, and others were detected in the subsurface (Ten Brink & Ben-Avraham 1989). Seismic reflection surveys suggest further that the combined thickness of the Miocene and Plio-Quaternary sequences exceeds 7 km in the southern section of the Dead Sea basin, whereas their thickness in the northern basin is merely 3–4 km (Kashai & Krocker 1987). This considerable accumulation probably reflects axial subsidence, which was enhanced by normal displacement along the boundary faults (Mart and Rabinowitz 1986; Garfunkel and Ben-Avraham 1996). Furthermore, the subsidence of the rift floor occurred concurrently with the uplift of its margins, which caused the steep escarpments and the high rate of erosion along the margins (Mart & Horowitz 1981; Begin & Zilberman 1997). Similar phenomena of axial subsidence and marginal uplift were discerned in the southern Suez Rift, and were considered to be a common feature in most rifts (Jackson *et al.* 1988). The uplift of the margins of the Dead Sea basin has been asymmetric, and the eastern flank was raised more than the western one, and was were deeply eroded (Garfunkel 1981; Wdowinsky & Zilberman 1997). Consequently, the western margin is built mainly of mid- and late Cretaceous carbonates, whereas Lower Cretaceous sandstones, with some significant occurrences of Precambrian series, are exposed along the eastern margin. Basaltic outcrops of Plio-Pleistocene age, which occur in the plateau southeast of the Dead Sea, do not have an equivalent in the west (Steinitz & Bartov 1991).

The tomographic model

The calculated tomographic model divides the crust and the upper mantle into five distinct layers (Table 1). The upper-crustal layers have been studied extensively by seismic reflection and boreholes, delineating stratification of thick layers of unconsolidated sediments and salt, and the Mesozoic strata underlying the pre-Miocene sedimentary sequence. The verifiable inversion of the 2–10 km layer in the Dead Sea region shows 11% lower than average velocity of 5.6 km s^{-1} in the axial zone and 11% higher than the average velocity in the margins. Lateral lithological changes of the shallow section of the crust, which was encountered in exploration wells drilled in the region (Kashai & Krocker 1987), suggest that the tomographic model generally conforms with other datasets at these depths. The deeper inversion layers, at depths of 10-22, and 22–32 km, have been poorly known. Analyses using gravimetry (Ten Brink *et al.* 1993), seismic refraction (Ginzburg *et al.* 1979), and preliminary tomography (Rabinowitz *et al.* 1996) have reached conflicting results. The gravimetric model suggested that the crustal thickness under the Dead Sea Rift and its margins is uniform, the seismic refraction survey suggested that the Moho is found at depth of *c.* 27 km, and the early tomographic model presumed magmatic diapirs, which ascended into the intermediate crust. These findings rendered the structural analysis and composition of the lower crust and upper mantle rather dubious (Hofstetter *et al.* 1998; Rabinowitz & Mart 1998).

The resolution of the present model shows that the available dataset opens for analysis only a relatively small section of the 22–32 km layer, probably the uppermost mantle under the central Dead Sea. In spite of its size, the analysis of this segment is verifiable (Fig. 3). The results of the present inversion set the Moho Discontinuity under the central Dead Sea at depth of 22 km. The average seismic velocity at the 22–32 km depth interval is 7.4–7.7 km s^{-1}, whereas the average velocity of the overlying lower crust is 6.5 km s^{-1} in the same place. The velocity suggests that this layer comprises the upper lithospheric mantle. It should be noted that the tomographic model outside the boundary of the reliable resolution zone (Fig. 3) is probably unreproducible. The velocity measured in this layer is low for a normal upper mantle, but it is clearly higher than that of lower crust elsewhere. The shallow depth of the Moho and the low velocity of the upper part of the lithospheric mantle under the central Dead Sea suggest doming of anomalously hot magma under this section of the Levant Rift. Similar velocity structures were encountered under a mid-ocean ridge (Purdy 1987)

Average seismic velocity of 6.5 km s^{-1} was calculated for the 10–22 km depth interval in the crust, where several anomalies of high and low seismic velocities can be discerned in the area of acceptable resolution (Fig. 4). The reliable resolution of the tomographic model in this layer covers the northern basin of the Dead Sea and its flanks. Four high velocity anomalies can be discerned along the margins of the Dead Sea basin, two along the eastern boundary fault, and two along the western fault. The anomalies are 4–6% higher than the average velocity, reaching seismic velocities of *c.* 6.8 km s^{-1}. Whereas high velocity anomalies occur under the marginal faults of the Dead Sea Rift, a low velocity anomaly is located under the axial zone, deviating *c.* 1% from the average velocity. The average seismic velocity of 6.5 km/s suggests that the crustal layer at 10–22 km depth probably comprises more than one lithological unit. Ginzburg *et al.* (1979) measured seismic velocities of 6.7 and 6.0 km s^{-1} to depths of 28 km, and they did not determine the stratification between the lower and the intermediate crust. In view of the velocity structure of the underlying layer, the high velocity anomalies under the boundary faults of the Dead Sea Rift could indicate penetration of material of higher velocity from the mantle into the crust. The large low velocity anomaly at the 10–22 km layer that underlies the northern Negev west of the southern Dead Sea basin, could probably be derived from increasing thickness of the sequence of sedimentary rocks there (Fig. 4).

The verifiable inversion of the 2–10 km layer in the Dead Sea region shows 11% lower than average velocity of 5.6 km s^{-1} in the axial zone and 11% higher than the average velocity in the margins. Lateral lithological changes of the shallow section of the crust, which were encountered in exploration wells drilled in the region (Kashai & Krocker 1987), suggest that the tomographic model, which generally conforms with other datasets at these depths, is inferior to other exploration tools at that range.

Discussion

The tomographic model of the Dead Sea region illuminates some aspects in the structure of the lower crust and the upper mantle under a divergent tectonic plate boundary. Earlier geophysical studies in this region utilized two useful methods to explore the deep crustal structure in the Dead Sea region, seismic refraction and gravimetry. Seismic reflection surveys did not penetrate deeper than 10–11 km (Kashai & Krocker 1987; Ginzburg & Ben-Avraham 1997). Magnetic surveys did not encounter large anomalies in the region (Folkman & Yuval 1976; Rybakov *et al.* 1997). Presuming that the temperature of the magma under the rift is probably higher than Curie temperature, this finding is hardly surprising (Rabinowitz & Mart 1998). Seismic refraction surveys measured crustal thickness of *c.* 28 km under the Rift, and more than 40 km under its western margin (Ginzburg *et al.* 1979). The refraction data also showed that seismic velocities measured at the upper mantle under the Rift axis were slightly slower than those measured at equivalent crustal units under the western flank. Early gravimetric measurements indicated that the gravity acceleration at the Dead Sea was slower than along its western margin, therefore a crustal thermal anomaly under the rift axis was suggested (Ginzburg & Makris 1979). However, subsequent gravimetric measurements from both flanks of the Dead Sea were used to compile a different crustal model, which suggested uniformity between the crustal thickness under the Rift and its margins (Ten Brink *et al.* 1993). This latter model attributed the low gravimetric anomaly of the Rift to thick layers of salt and loosely consolidated sediments in the rift. Ten Brink *et al.* (1993) supported their interpretation with seismic reflection profiles that encountered 7 km thickness of such strata in the southern basin of the Dead Sea (Kashai & Krocker 1987).

The present tomographic model, which continues earlier tomographic analyses (Rabinowitz *et al.* 1996; Mart & Rabinowitz 1997), is basically conformable with the crustal model of Ginzburg & Makris (1979) in showing crustal thinning and a thermal anomaly under the Dead Sea Rift. Our model suggests that the Moho occurs at depth of 22 km under the northern basin of the Dead Sea, and the seismic velocity in the upper mantle there is *c.* 7.7 km s^{-1}. Such thickness and velocity are anomalously low for stable lithosphere, but compatible with zones of ascending and upwelling magma. Indeed, not very many earthquakes had their foci at depths greater than 22 km, therefore the resolution analysis showed that the velocity of the 22–32 km layer is reliable only in a small area in the central Dead Sea. However, that area, where the tomographic model is statistically significant, opens a verifiable window to the structure of the upper mantle (Fig. 3). Therefore it is suggested that the occurrence of that small anomaly under the central Dead Sea is tectonically meaningful. The occurrence of the Moho at shallow depth and the low seismic velocity in the upper mantle probably indicate the upwelling of magma of higher temperature under the axis of the Dead Sea basin. Magmatic ascent under the Dead Sea was predicted by Schubert & Garfunkel (1984), who presumed that the Rift system is a leaky transform fault, and that the Dead Sea segment is a pull-apart basin. However, Rabinowitz *et al.* (1996), who also endorsed the magmatic upwelling model, presumed that the Dead Sea Rift is an incipient oblique spreading centre.

The interpretation of the velocity anomalies in the lower crust is somewhat ambiguous. Descending cold blocks, having lower–intermediate crust mineralogical composition, could have caused the low velocity anomalies along the boundary faults of the Rift (Fig. 4). Alternately, the ascent and penetration of hot block from the upper mantle or the lower crust could have formed these anomalies. The density of such a block would be reduced because of its elevated temperatures, but its seismic velocity would still be higher than that of the surrounding material. In view of the large thermal anomaly in the upper mantle, we prefer the latter interpretation. This interpretation suggests the ascent and penetration of magmatic diapirs into

Fig. 3 (opposite). Velocities distribution in the 22–32 km layer of the inversion model. Broad red line marks the boundary of the verifiable area, where resolution exceeds 0.4. The seismic velocity in the plausible area is *c.* 7.7 km s^{-1}, suggesting that this model layer represents the upper mantle under the Dead Sea. Colour bar scale is percentage point deviation from the average velocity. ○, location of epicentres; ▲, location of seismographs.

Fig. 4 (opposite). Distribution of seismic velocities in the 10–22 km crustal layer shows fast anomalies under the boundary faults of the Dead Sea Rift, whereas the northern basin in underlain by a slow velocity anomaly. Average seismic velocity is 6.5 km s^{-1}, suggesting a composite modelled layer. The broad red line delineates the area of acceptable resolution, and broad pink lines indicate the approximate location of the boundary faults of the Rift. Colour bar scale is percentage point deviation from the average velocity. ○, location of epicentres; ▲, location of seismographs.

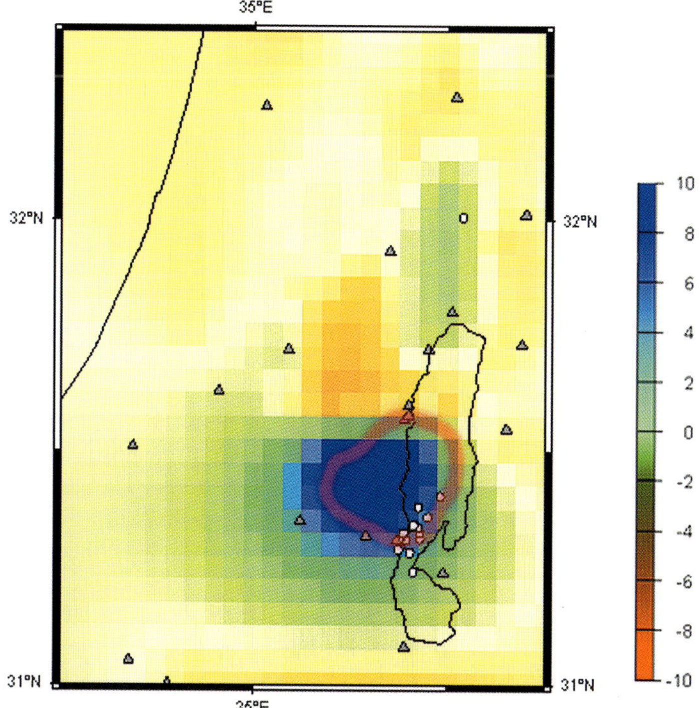
35°E
32°N
32°N
31°N
31°N
35°E
10
8
6
4
2
0
-2
-4
-6
-8
-10

Fig. 3

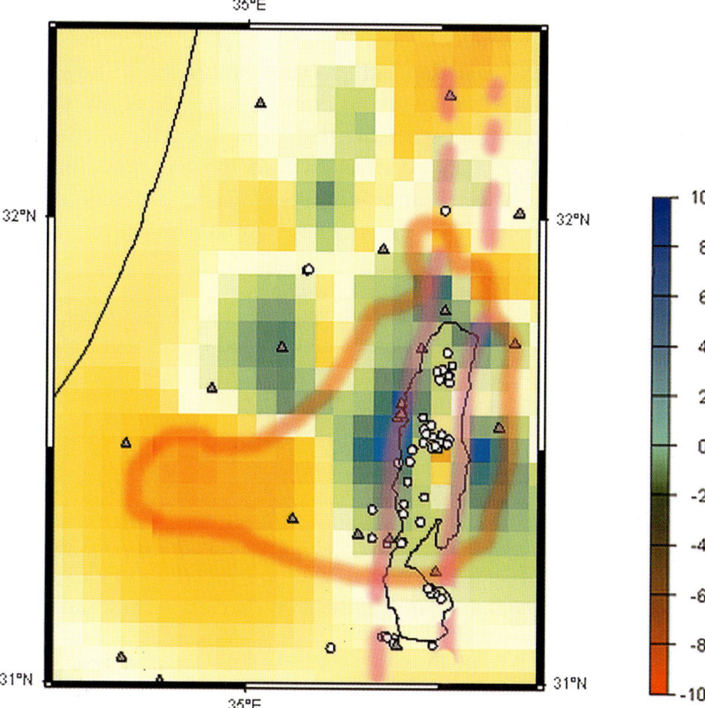
35°E
32°N
32°N
31°N
31°N
35°E
10
8
6
4
2
0
-2
-4
-6
-8
-10

Fig. 4

the 10–22 km layer of the crust. There is no evidence that these diapirs affect the upper crust or the surface, and there seems to be no association between them and the salt diapirs in the upper crust. The temperature of the magmatic diapirs would be above their Curie temperature, therefore they were not likely to produce magnetic anomalies (Folkman & Yuval 1976; Rybakov *et al.* 1997). Presuming some chemical interaction between the ascending hot upper mantle and the overlying crust, it seems plausible that the shallow Moho and the thermal anomalies are probably the products of magmatic underplating under the axial zone of the Dead Sea Rift. Furthermore, the diapiric features discerned in the lower–intermediate crust could have been formed as a result of metastable crustal stratification that resulted from the magmatic ascent.

Conclusions

Inversion of first arrivals of P waves from earthquakes in the Dead Sea region led to the calculation of a tomographic model of the upper lithospheric mantle and the lower crust. Moho was encountered at depth of 22 km under the central segment of the Dead Sea, and the average seismic velocity there is *c.* 7.7 km s^{-1}.

Magmatic ascent in the lithospheric mantle and thinning of the crust were also encountered under the northern Red Sea. Bonatti (1985) showed that crustal accretion has produced a linear array of magmatic diapirs along the axial zone of the Red Sea. The diapirs in the central Red Sea have reached the surface, and accreted kernels of oceanic basalts, but the diapirs of the northern Red Sea are noticeably smaller, and commonly they do not crop out and do not produce basaltic accretion at the sea floor. The crustal structure of the Dead Sea, as reconstructed in the tomographic model, is similar to the model Bonatti (1985) suggested for the northern Red Sea. Consequently, the presented tomographic model of the crust and the upper mantle underneath the Dead Sea Rift, and its interpretation of the variable seismic velocities as products of a thermal event, implies upwelling in the lithospheric mantle. The model is compatible with the distribution of the earthquakes in that region, because most earthquakes occurred within the rift, and their focal depth rarely exceeds 30 km. The model does not resolve the problem of whether the Levant Rift is a transform fault or an incipient spreading centre, but it certainly keeps the debate vividly unresolved.

We are deeply indebted to E. Kissling for his help in computing the inversion model and evaluating its resolution. Two anonymous reviewers contributed significantly to improvement of the early version of this paper.

References

Aki, K. & Richards, P. 1980. *Quantitative Seismology: Theory and Methods*, Vols I and II. Freeman, New York.

Begin, Z. B. & Zilberman, E. 1997. *Main stages and rate of the relief development in Israel*. Geological Survey of Israel, Report GSI/24/97, Jerusalem [in Hebrew, abstract in English].

Bonatti, E. 1985. Punctiform initiation of seafloor spreading in the Red Sea during transition from a continental to an oceanic rift. *Nature*, **316**, 33–37.

Doyle, H. 1995. *Seismology*. Wiley, Chichester.

Eberhart-Phillips, D. 1986. Three dimensional velocity structure in northern California Coast Ranges from inversion of local earthquake arrival times. *Bulletin of the Seismological Society of America*, **76**, 1025–1052.

EVANS, J. R & EBERHART-PHILLIPS, D. 1994. *Users manual guide for simulps12 for imaging* v_p *and* v_p/v_s*: a derivative of the 'Thurber' tomographic inversion* SIMUL3 *for local earthquakes and explosions.*

FOLKMAN, Y. & YUVAL, Z. (compilers) 1976. *Israel—Aeromagnetic Map*. Survey of Israel, Tel Aviv.

GARFUNKEL, Z. 1981. Internal structure of the Dead Sea leaky transform (rift) in relation to plate kinematics, *Tectonophysics*, **80**, 81–108.

—— & BEN-AVRAHAM, Z. 1996. The structure of the Dead Sea basin. *Tectonophysics*, **266**, 155–176.

GINZBURG, A. & BEN-AVRAHAM, Z. 1997. A seismic refraction study of the north basin of the Dead Sea, Israel. *Geophysical Research Letters*, **24**, 2063–2066.

—— & MAKRIS, J. 1979. Gravity and density distribution in the Dead Sea rift and adjoining areas, *Tectonophysics*, **54**, T17–T26.

——, ——, FUCHS, K., PRODEHL, C., KAMINSKI, W. & AMITAI, U. 1979. A seismic survey of the crust and upper mantle of the Jordan—Dead Sea rift and their transition toward the Mediterranean Sea. *Journal of Geophysical Research*, **84**, 1569–1582.

HOFSTETTER, A., RYBAKOV, M. & TEN BRINK, U. 1998. Comment on Magmatic diapirs in the intermediate crust under the Dead Sea by Rabinowitz *et al. Tectonics*, **17**, 819–821.

JACKSON, J. A., WHITE, N. J., GARFUNKEL, Z. & ANDERSON, H. 1988. Relations between normal-fault geometry, tilting and vertical motions in extensional terrains: an example from the southern Gulf of Suez. *Journal of Structural Geology*, **10**, 155–170.

KASHAI, E. & CROCKER, P.F. 1987. Structural geometry and evolution of the Dead Sea–Jordan rift system as deduced from new subsurface data. *Tectonophysics*, **141**, 33–60.

KISSLING, E. 1988. Geotomography with local earthquake data. *Reviews of Geophysics*, **26**, 659–698.

——, ELLSWORTH, W. L., EBERHART-PHILLIPS, D. & KRADOLFER, U. 1994. Initial reference models in local earthquake tomography. *Journal of Geophysical Research*, **99**, 19635–19646.

——, SOLARINO, S. & CATTANEO, M. 1995. Improved seismic velocity reference model from local earthquake data in northwestern Italy. *Terra Nova*, **7**, 528–534.

MART, Y. 1991. The Dead Sea Rift: from continental rift to incipient ocean. *Tectonophysics*, **197**, 155–179.

—— & HOROWITZ, A. 1981. The tectonics of the Timna region in southern Israel and the evolution of the Dead Sea Rift. *Tectonophysics*, **79**, 165–199.

—— & RABINOWITZ, P. D. 1986. The northern Red Sea and the Dead Sea rift. *Tectonophysics*, **124**, 85–113.

—— & —— 1997. Thermal anomaly in the upper mantle and lower crust in the Dead Sea region. *In*: JACOB, A. W. B., DELVAUX, D. & KHAN M. A. (eds), *Lithospheric Structure, Evolution and Sedimentation in Continental Rifts. Proceedings of IGCP 400: Continental Rifts*. Dublin, May 1997, 66–70.

NEEV, D. & HALL, J. K. 1979. Geophysical investigations in the Dead Sea. *Sedimentary Geology*, **23**, 209–238.

PICARD, L. 1987. The Elat (Aqaba)–Dead Sea–Jordan subgraben system. *Tectonophysics*, **141**, 23–32.

POIRIER, J. P., ROMANOWICZ, B. A. & TAHER, M. A. 1980. Large historical earthquakes and seismic risk in northwest Syria. *Nature*, **285**, 217–220.

PURDY, G. M. 1987. New observations of the shallow seismic structure of young oceanic crust. *Journal of Geophysical Research*, **92**, 9351–9362.

RABINOWITZ, N. & MART, Y. 1998. Reply to comment, Magmatic diapirs in the intermediate crust under the Dead Sea. *Tectonics*, **17**, 821–822.

——, STEINBERG, J. & MART, Y. 1996. New evidence of magmatic diapirs in the intermediate crust under the Dead Sea, Israel. *Tectonics*, **15**, 237–243.

ROMANOWICZ, B. 1991. Seismic tomography of the Earth's mantle. *Annual Review of Earth and Planetary Sciences*, **19**, 77–99.

RYBAKOV, M., GOLDSHMIDT, V. & ROTSTEIN, Y. 1997. New regional gravity and magnetic maps of the Levant. *Geophysical Research Letters*, **24**, 33–36.

SCHUBERT, G. & GARFUNKEL, Z. 1984. Mantle upwelling in the Dead Sea and Salton trough—Gulf of California, leaky transforms. *Annales Geophysicae*, **2**, 633–648.

STEINITZ, G. & BARTOV, G. 1991. The Miocene–Pleistocene history of the Dead Sea segment of the rift in light of K–Ar ages of basalts. *Israel Journal of Earth Sciences*, **40**, 199–208.

TEN BRINK, U. S. & BEN-AVRAHAM, Z. 1989. The anatomy of a pull-apart basin: seismic reflection observations of the Dead Sea basin. *Tectonics*, **8**, 333–350.

——, BEN-AVRAHAM, Z., BELL, R. E. *et al.* 1993. Structure of the Dead Sea pull-apart basin from gravity analyses. *Journal of Geophysical Research*, **98**, 21877–21894.

THURBER, C. H. 1981. *Earth structure and earthquake locations in the Coyote Lake Area, Central California*. PhD thesis, Massachusetts Institute of Technology, Cambridge, MA.

—— 1983. Earthquake location and three dimensional crustal structure in the Coyote Lake area, central California. *Journal of Geophysical Research*, **88**, 8226–8236.

VAN ECK, T. & HOFSTETTER, R. 1989. Microearthquake activity in the Dead Sea region. *Geophysical Journal International*, **99**, 605–620.

WDOWINSKY, S. & ZILBERMAN, E. 1997. Systematic analysis of the large-scale topography and structure across the Dead Sea rift. *Tectonics*, **16**, 409–424.

WIGGINS, R. A. 1972. The general linear inversion problem: implication of surface waves and free oscillation for Earth structure. *Reviews of Geophysics and Space Physics*, **1**, 251–285.

ZAK, I. & FREUND, R. 1981. Asymmetry and basin migration in the Dead Sea rift. *Tectonophysics*, **80**, 27–38.

Salt extrusion at Kuh-e-Jahani, Iran, from June 1994 to November 1997

CHRISTOPHER J. TALBOT[1], SERGEI MEDVEDEV[1], MEHDI ALAVI[2], HASSAN SHAHRIVAR[2] & ESMAEL HEIDARI[2]

[1] *Hans Ramberg Tectonic Laboratory, Department of Earth Sciences, Uppsala University, SE-752 36 Uppsala, Sweden (e-mail: Christopher.Talbot@geo.uu.se*

[2] *Institute for Earth Sciences, Geological Survey of Iran, PO Box 131851-1494, Tehran, Iran*

Abstract: Kuh-e-Jahani is one of the largest extrusions of salt currently active in the Zagros mountains. Salt rises from about 4 km below sea level to nearly 1.5 km, above, where, unable to support its own weight, it spreads over the surrounding scenery in a process responsible for present and past allochthonous salt sheets elsewhere. We report vertical movements and apparent horizontal displacements of 43 markers dispersed over this mountain of salt for 4.5 years in three consecutive intervals, the first of 18 months and two others of 12 months. The geometry and inferred flow rate of the salt changed between surveys emphasizing that the gravity spreading is not steady. Our field readings of the dimensions and velocities of the salt at Kuh-e-Jahani are used to tune a simple numerical model and constrain the viscosity of the salt to between 10^{16} and 10^{17} Pa s^{-1}, its rate of surface dissolution to 2–3 cm a^{-1}, and its rate of rise out of its orifice at 2–3 m a^{-1} for *c.* 55 ka. These results imply that vigorous extrusion of salt at Kuh-e-Jahani is probably close to evacuating its deep source and that this mountain will soon begin to waste as salt dissolution overtakes extrusion. This progress report is warranted because our results have significant implications for sophisticated engineering in salt.

Something like 60% of the world's hydrocarbon reserves occur in traps associated with salt structures, which can also be used as mines and for caverns intended for the storage of energy or the isolation of wastes. Improving our understanding of the rheology and dynamics of natural salt structures is therefore important to sophisticated engineering projects as well as to science. The potential of the unique natural salt laboratory in the Zagros for better understanding of these problems is the focus of the long-term Zagros Halokinetics research programme. This programme is run by geologists from the Hans Ramberg Tectonic Laboratory of the Department of Earth Sciences at Uppsala University, and the Institute for Earth Sciences of the Geological Survey of Iran (GSI) under the auspices of GSI's general director (A. Zadeh-Heravi until 1997, and Deputy Minister M. T. Korehei since then).

Since 1993, GSI teams of geologists and drivers (and C.J.T.) have spent about a month a year visiting most of the salt diapirs emergent in the Zagros and choosing 10 for monitoring. By the end of 1997, the positions of about 270 markers on the

From: VENDEVILLE, B., MART, Y. & VIGNERESSE, J.-L. (eds) *Salt, Shale and Igneous Diapirs in and around Europe*. Geological Society, London, Special Publications, **174,** 93–110. 1-86239-066-5/00/$15.00

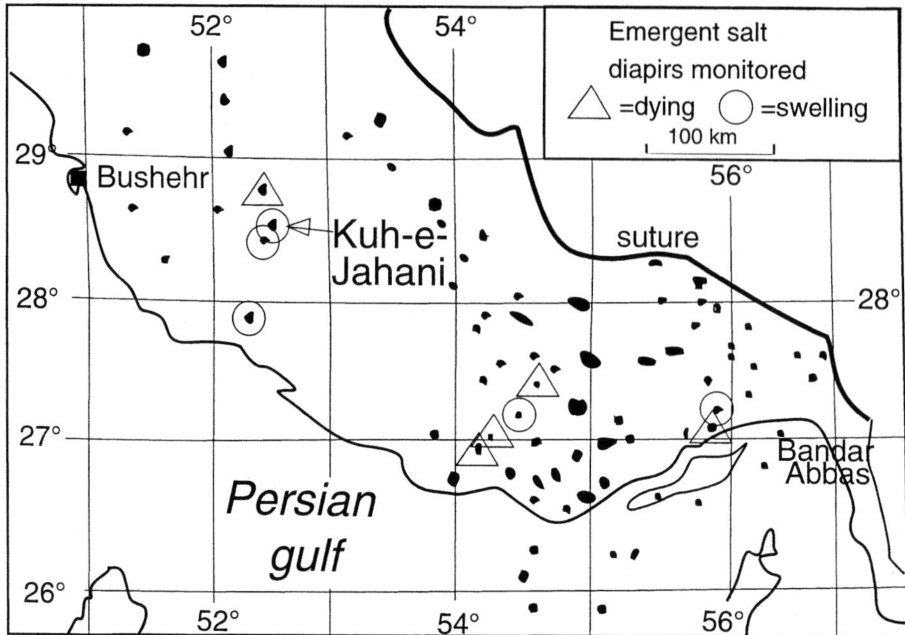

Fig. 1. Extrusions of Hormuz salt (black) shown on a map of southern Iran, with the 10 diapirs chosen for monitoring being distinguished into five dying diapirs that appear to be dissolving faster than they flow and five swelling diapirs that appear to be flowing faster than they dissolve (adapted from the 1:250 000 geological map of the SE Zagros, Oil Service Company of Iran, 1977).

salt had been measured up to four times in different years from 15 theodolite stations, all but one on bedrock. Our measurements to date suggest that five of the 10 monitored mountains are dissolving faster than they flow and that the other five are flowing faster than they dissolve (Fig. 1).

The results for the single mountain we report here are sufficiently spread in space and time for a general picture of the rates of flow and dissolution to be emerging. We already have other markers on this mountain and hope to improve our future surveys on all of them. However, we consider this progress report justified because the velocities of salt flow we document have important practical implications. We start with a brief outline of the Hormuz salt and reconstruct a likely evolutionary history for its three main generations of extrusions. We then describe the survey technique used for our measurements, which are then used to dimensionalize a non-dimensional numerical model of the extrusion process. We end by contrasting our results with impressions of past salt flow rates given in the literature and suggest reasons for the significant discrepancies.

Hormuz salt and its history

The Hormuz salt, estimated to have a total initial thickness of at least 1 km (Kent 1958), is of two types. The Neoproterozoic component is a cyclic sequence of multi-coloured halite interbedded with anhydrite and foetid dolomites deposited in broad half-grabens along the passive margin of Proto-Tethys between what is now the head of the Persian Gulf and Australia (Talbot & Alavi 1996). Apart from dispersed anhydrite and volcanic inclusions, the Cambrian component is generally much purer halite and was deposited above red beds in local grabens trending across Proto-Tethys (Talbot & Alavi 1996).

The Hormuz salt near the Persian Gulf was buried by Palaeozoic clastic sediments and Triassic–Miocene platform carbonates but did not move until late Jurassic to early Cretaceous time, by which time it was at 2.5–3 km depth (Talbot & Alavi 1996). The northward subduction of Neo-Tethys beneath central Iran then began to riffle fault blocks along the northern margin of Gondwana like piano keys. This reactivated N and NNE trending faults in the basement above which the salt began to migrate into the deep conformable pillows that still survive in Arabia south of the Zagros front (Kent 1970). Neo-Tethys began to close like a zip fastener pulled eastward in Lower Cretaceous time. During subsequent convergence of Arabia and Asia, serial thrusts and folds of the Zagros mountains have been propagating episodically southwestward, driving the Persian Gulf in front of them as their foreland basin.

West of the N–S Kazeroun line (Fig. 1), Zagros structures are long and verge consistently southwestward in the NW Zagros thrust–fold belt, which is narrow with a steep (4°) front facing SSW above a high-friction basal décollement (Talbot & Alavi 1996). East of the Kazeroun line, Zagros structures are shorter and more upright in the SE Zagros fold–diapir festoon, which is twice as wide because it is propagating southward twice as fast with a gentle (<1°) front over a basal décollement lubricated by Hormuz salt (Talbot & Alavi 1996).

Lenses of black, red and green clasts of insoluble Hormuz inclusions distinctive among the honey-coloured cover rocks (e.g. Talbot 1998, fig. 4a) around several of the diapirs indicate that extrusion of Hormuz salt has often been recurrent and began locally as early as Early Cretaceous time (Kent 1958; Player 1969). Most episodes of extrusions can be distinguished into three generations, which pre-date, are contemporaneous with, and/or post-date the local arrival of the still-advancing Zagros front (Talbot & Alavi 1996). Some of the major faults in the basement reactivate before the Zagros front arrives, so that some of the salt pillows develop into diapirs that supply the first-generation ('pre-Zagros') extrusions (Talbot & Alavi 1996). Most first-generation diapirs probably reactivate as second-generation ('syn-Zagros') diapirs that herald the local arrival of the Zagros front by the deflation of successive compartments in the deep salt pillows (from Eocene to Miocene onshore). Such pillow deflation is signalled by dramatic increases in thickness of beds of conglomerates in deep salt-withdrawal basins immediately SSW of a growth (thrust-) anticline as clasts derived from a local salt extrusion (re)appear in the local stratigraphy. Sufficient Hormuz salt remains rucked-up in the deep cores of Zagros anticlines to supply a third generation of ('post-Zagros') diapirs, which now extrude from extensional offsets along strike-slip faults behind the Zagros front (Talbot & Alavi 1996). These rise among wasting masses of second-generation salt extrusions and lenses of dirty anhydritic mylonites that mark secondary welds where first-generation diapirs have lost their salt (like that repeated by thrusting at the south end of the profile in Fig. 3).

About 130 diapirs of Hormuz salt now emerge in the SE Zagros fold–diapir festoon. The surface expressions of all three generations can be rearranged to allow us to infer different stages in the evolution of submarine or subaerial salt extrusions of any age in similar environments elsewhere (Talbot & Alavi 1996). Gleaming mountains of salt near the Gulf coast generally diminish in height inland as the salt is increasing hidden by accumulations of soils of insoluble Hormuz components on the way to becoming breccia pipes, which are then flattened into the NW–SE tectonic grain of the high Zagros (Kent 1970). Ala (1974) attributed this general inland increase in maturity of Zagros diapirs to the increase of rainfall into the high Zagros; however, as the first appearance of Hormuz components in the local stratigraphy generally decreases in age southward, it also reflects the southwesterly advance of the Zagros front along N–S chains of pre-Zagros salt pillows (Talbot & Alavi 1996).

On land, the salt extrusions that reach highest (2.2 km) above sea level are vigorous third-generation extrusions. However, none are higher than at least one peak of the surrounding country rocks, suggesting that they are driven largely by gravity (see below). Of all the current extrusions of Hormuz salt, the only examples that rise significantly higher than their surrounding country rocks are out in the Gulf (Talbot 1998). There, tectonic forces combine with gravity to drive second-generation diapirs above the floor of the foreland basin to form the salt islands marking the current front of the Zagros mountains.

Third-generation extrusions in the Zagros first surface in partial ring-dykes of Cambrian salt that rise less than 100 m above their surroundings (Fig. 2b). These soon rise and spread to simple hemispherical salt domes that reach *c*. 500 m above their orifice (Fig. 2c). Extrusive domes develop to salt fountains, in which a summit

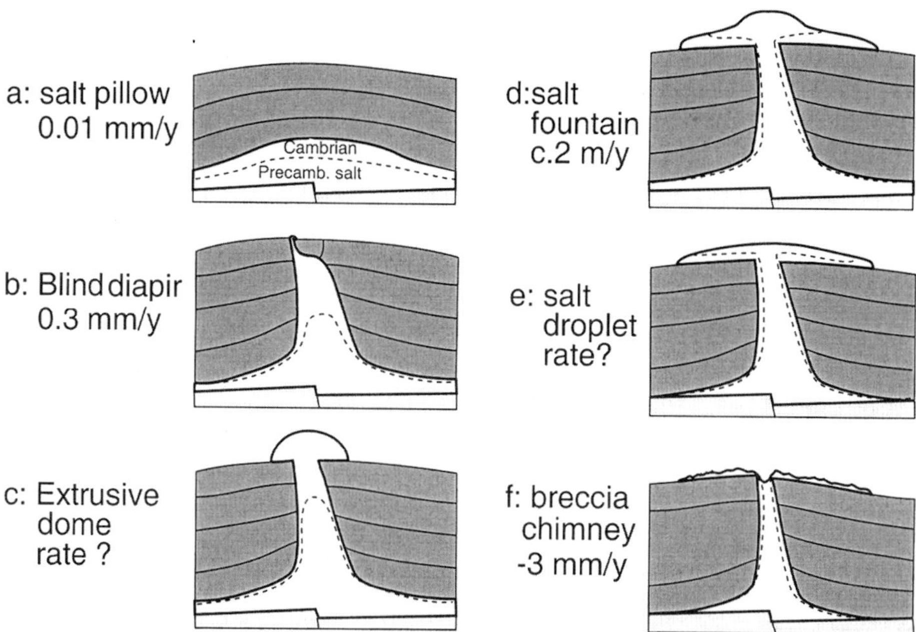

Fig. 2. Schematic representation of stages in the development of diapirs of Hormuz salt near the Persian Gulf, with characteristic rise rates indicated where known.

dome reaching about 1 km above its orifice is at least partially surrounded by a sheet of allochthonous salt spreading downslope over the surrounding scenery (Fig. 2d). On scales of kilometres, the profiles of vigorous salt fountains show a reasonable fit to the mathematical profile of a steady-state extrusion of viscous fluid from a narrow slot (Talbot & Jarvis 1984). However, they all display better fits to the expanding profiles of experimental axisymmetric extrusions of viscous fluids spreading to distances along horizontal surfaces during successive growth stages (Talbot 1993, 1998). Many first- and second-generation diapirs shrink after being thrust from their deep salt source, but third-generation diapirs appear to exhaust their deep source compartmentalized by Zagros thrusts and/or folds. Following source isolation or exhaustion, summit domes sink into salt sheets (Fig. 2e), which, unless they are buried by superimposed roof rocks (Kent 1958), retain the profiles of viscous droplets spreading over a no-slip surface as the salt degrades beneath increasing accumulations of insoluble Hormuz soils (Talbot 1998).

Kuh-e-Jahani

At 28°37′N, 52°25′E, Kuh-e-Jahani (Farsi for 'Mountain of the Universe') is one of the largest third-generation extrusions of Hormuz salt in the current Zagros (Fig. 3a). Jahani rises where a Zagros syncline is offset about 5 km by right-lateral oblique slip along the Mangarak transfer fault zone (Fig. 1). Triassic rocks crop out in the core of the Surmeh anticline to the southeast, but only Cretaceous rocks crop out in the equivalent anticline to the northwest (Fig. 3a). Jahani has the profile of a viscous salt fountain spreading downslope along a gentle valley eroded along the fault offsetting a gentle antiform (with a superimposed rim syncline?) of Fars (Miocene) rocks in the north, and over gravel plains in the south (Fig. 3b).

The summit of Kuh-e-Jahani is 1485 m above sea level and 1000 m above the plain and, by inference, its invisible bedrock orifice in the fault valley (Fig. 3). Four bands of red soils are concentric within the distal salt margins (Kent 1958). All four red bands are derived from broken sandstones, marls and carbonates. We interpret these bands as thrust repetitions of the same Cambrian formation. Salt more distal than the outermost red band is generally uniform beneath buff-coloured anhydritic soils containing blocks of greenstone and is typical of Cambrian salt in the region (Talbot & Alavi 1996). Salt within the outermost red band is rich in blocks of dark foetid carbonates and typical of the cyclic Neoproterozoic Hormuz succession.

The first-order structure of Jahani (Fig. 3) suggests that Cambrian salt extruded first and was overridden by Neoproterozoic salt, which decoupled over the inverted base of the red bed and was extruded over the surface as two or three ductile nappes. Most of the halite is coarse grained (1–3 cm) with a slight gneissic foliation but foliation-parallel lenses of coarser grained halite (e.g. 10 cm) a few decimetres thick occur locally (for photographs, see Talbot & Alavi 1996, fig. 8b and e). On a scale of a kilometre or so, the foliation generally parallels the nearest boundary to the flowing salt whether this is a top boundary free to air or a side or bottom boundary, which is non-slip against country rocks or gravels. On scales of <100 m, the foliation is truncated by steep erosion surfaces (Talbot 1998). Topographic contours at 250 m intervals in Figs 3a, 4 and 5 illustrate that solution etching of salt topography is on scales too small to disrupt the smooth salt profile shown in Figs 3b, 6 and 7. Salt exposed in (usually dry) stream channels displays very few joints and is probably

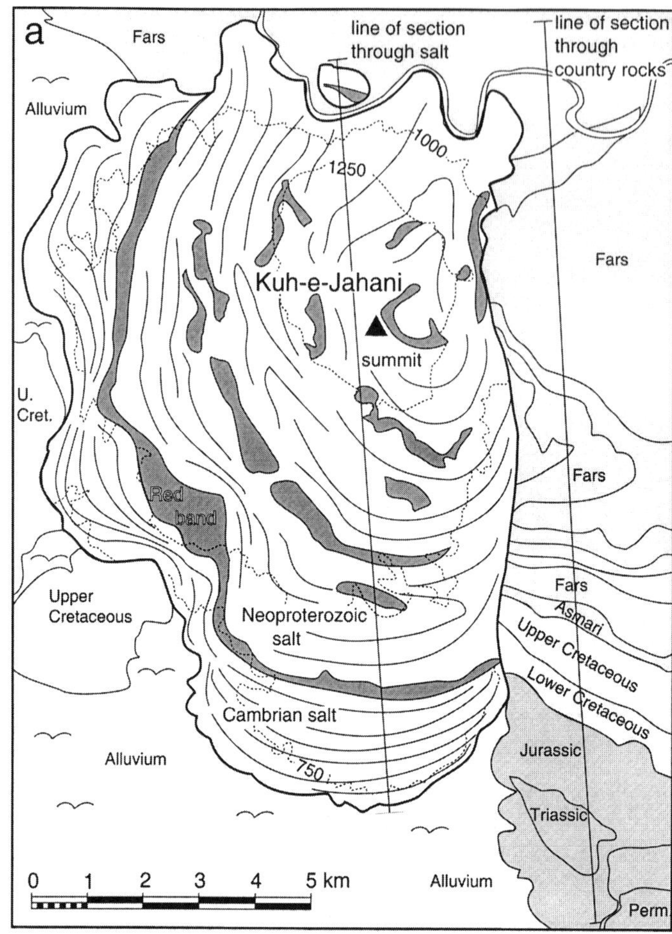

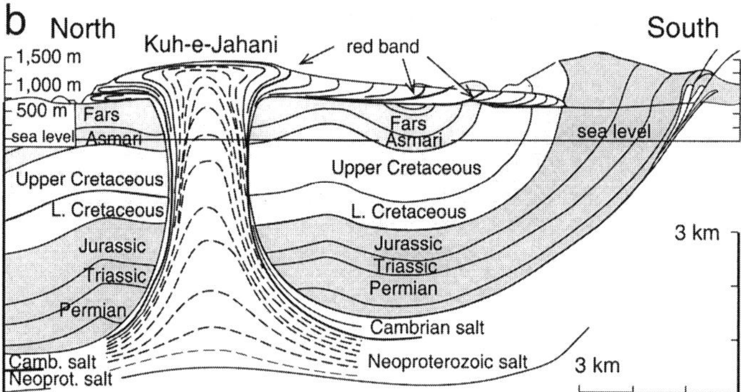

Fig. 3. (**a**) Geological map of Kuh-e-Jahani and its surroundings. Red bands (dark grey) and trend lines of layering in the salt (clear) are from air photographs and field mapping. Topographic contours at 250 m intervals are shown on the salt. (**b**) North–south profiles of the salt and country rock to the east (indicated in (**a**)), are well constrained by surface geology; trend lines shown in deep salt are imaginary.

confined and ductile, whereas salt exposed in the intervening interchannel relief is jointed and interpreted (Talbot 1998) as dilated (Hunsche 1998).

The foliation is axial planar to generally recumbent folds that are gentle in the flanks of the summit dome but increase in amplitude to isoclinal nappes downslope. The grain size of the salt generally decreases downslope along foliation-parallel shear zones that probably root to inflections in the red band or irregularities in the bedrock channel. Minor recumbent folds with mylonites of halite only *c.* 10 cm thick in their lower limbs (Talbot 1998, figs 3–5) emphasize that the flow profiles of allochthonous salt sheets can be very irregular (Talbot 1979; Talbot & Alavi 1996, fig. 7).

We know of no systematic rock-mechanic measurements on Iranian salt. However, laboratory experiments with many different types of rock salts from elsewhere (Spiers *et al.* 1989; Cristescu & Hunsche 1998) indicate that the distributions of impurities (which depends upon strain history) influence creep rates in similar conditions. Rock salt in which impurities are still dispersed within halite grains after low strains creep more slowly than the same salt in which inclusions have collected along grain boundaries or coagulated within the grains as a result of large strains. Zones of high strain in Hormuz salt extrusions are marked by fine-grained mylonites and are expected to creep about an order of magnitude faster than their surrounding protoliths of coarse-grained gneissic salt (Cristescu & Hunsche 1998, pp. 49–55 and 68–69).

Method of measurement

Three theodolite stations overlooking the west, south and east flanks of Jahani were established in June 1994. Two are sited on bedrock overlooking the salt (Fig. 4a). In the absence of suitable bedrock, the northwest theodolite station is sited on salt and located by triangulation between distant bedrock peaks.

Cairns tend to collapse, so we document (relative to baselines tied to bedrock markers) sight-lines to markers such as isolated bushes, insoluble blocks or salt pinnacles on the salt. We report here the apparent displacements of 43 such markers dispersed uniformly over much of Jahani (Fig. 4a). Apparent displacements of markers are calculated by multiplying the sine of the angular difference (in degrees) between successive theodolite readings by the distance between the theodolite and the marker measured on a 1:50 000 map. Calculated displacements are usually close to the average of estimates by two or three independent observers (through the theodolite). Taking our errors in lengths of sight-lines to reach 5%, and in theodolite readings to be ± 0.005 radians, suggests that the apparent displacements we report may be in error by as much as 10%.

Results

Our results are presented in two ways. The three increments of vertical (Fig. 4b–e) and horizontal (Fig. 5a–c) apparent displacements (in centimetres) are written alongside the location of each marker on outline maps of Jahani. Scaled vectors for three apparent displacement increments for each visible marker are also indicated (in centimetres) beside arrows on a panorama of the southern flanks of Kuh-e-Jahani (Fig. 6).

None of the markers could be identified from more than one theodolite station. Thus we only know those components of the horizontal displacement vectors that are to the left or right of each sight-line; the actual displacement direction could differ by 90° from the arrows showing horizontal components in Fig. 5a–c. However,

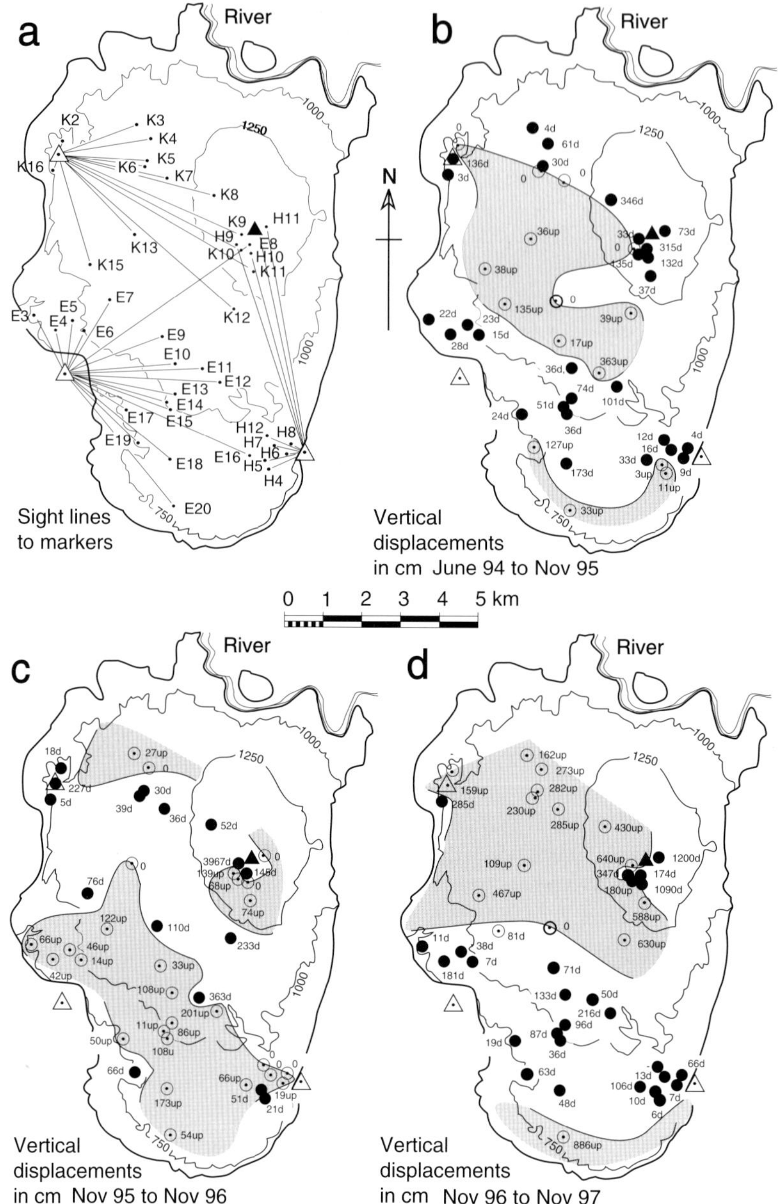

Fig. 4. Outline maps of Kuh-e-Jahani (with topographic contours at 250 m intervals) showing: (**a**) three theodolite stations (△; ▲, summit) and lines of sight to 42 labelled markers (·) on the salt of Kuh-e-Jahani. Markers missing from numerical sequences refer to distant mountain peaks. (**b**–**e**) Vertical displacements (cm) beside circled markers for the time increments indicated. ○, moved up in shaded areas; ●, moved down in areas left unshaded.

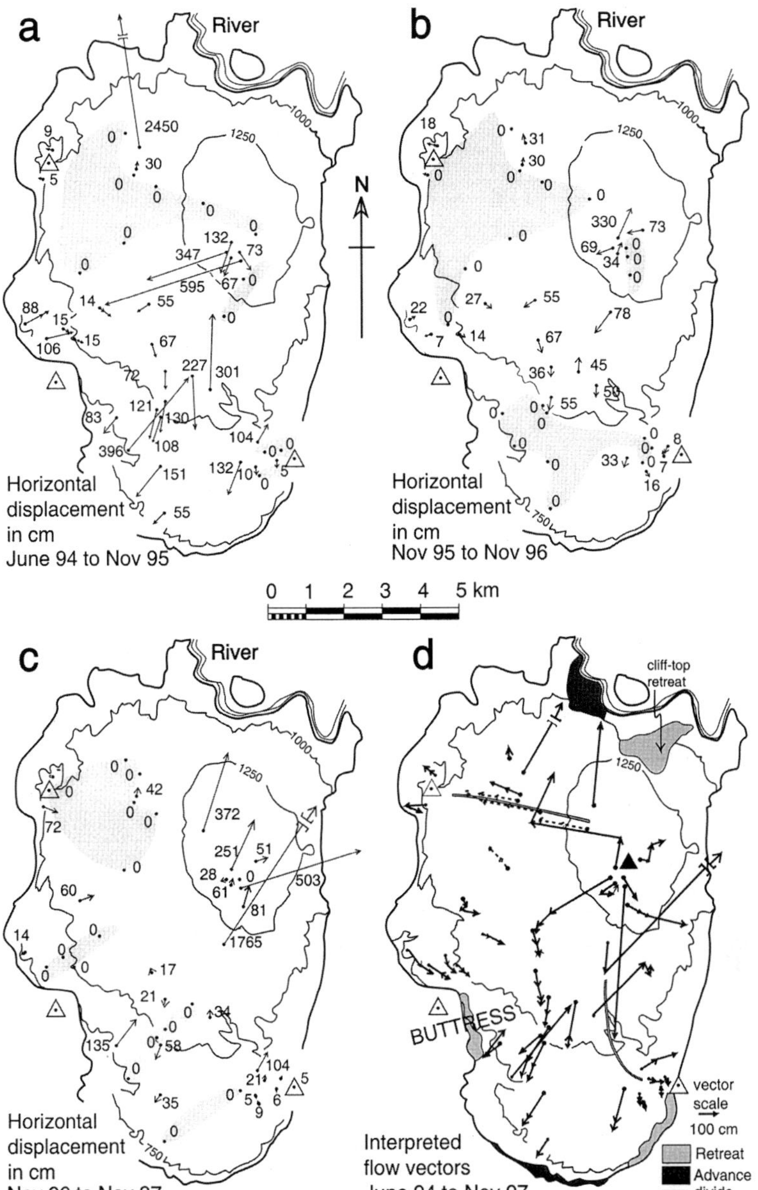

Fig. 5. Outline maps of Kuh-e-Jahani (with topographic contours at 250 m intervals) showing (**a**–**c**) scaled arrows with values (in cm) showing apparent horizontal displacements of markers for the time increments indicated. Markers that did not cross sight-lines have zero beside them and indicate flow divides (shaded); fallen or lost markers have dashes near them. (**d**) Dogleg arrows showing likely vectors (scaled as indicated in key) of up to three increments of horizontal displacements interpreted using apparent horizontal components and salt contours. The first increments are shown closest to the point representing the marker. Also shown are drainage divides in the flowing salt and areas of significant differences between contour maps dated 1952 (Iranian Topographic Survey) and 1977 (Oil Service Company of Iran).

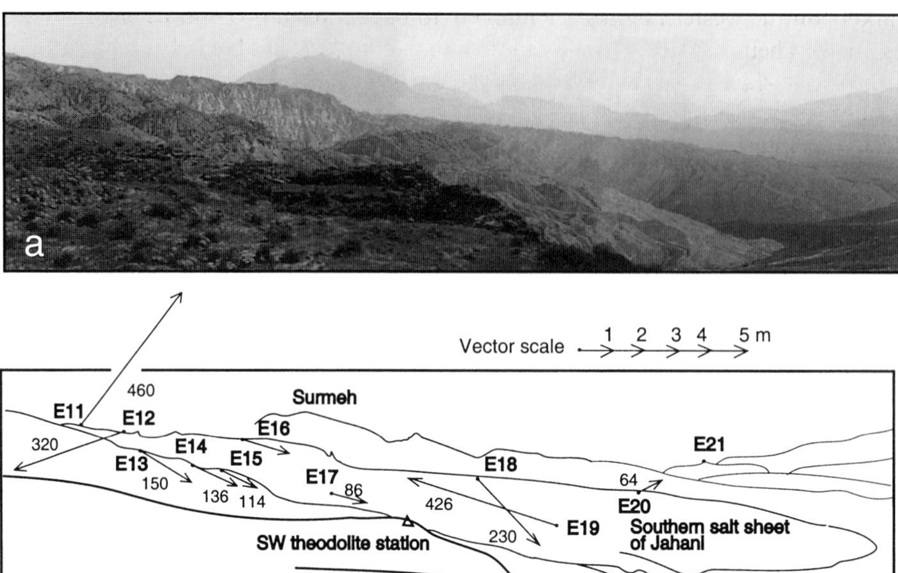

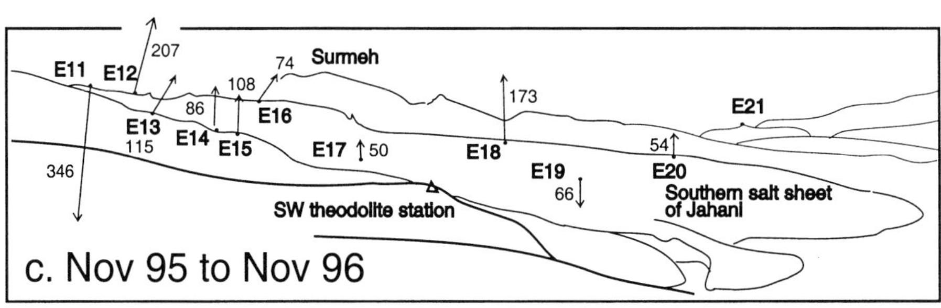

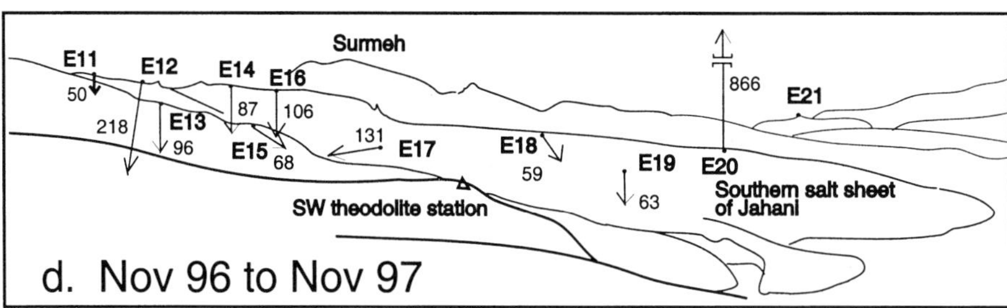

Fig. 6. (**a**) Photographic panorama of the southern slopes of Kuh-e-Jahani looking south-east, with the southwestern theodolite station visible on the limestone platform in the foreground (△ in **b**–**d**). (**b**–**e**) Values of apparent displacement (cm) beside scaled vectors for each visible marker after each survey. These vectors integrate the horizontal and vertical components shown in Figs 4 and 5 and thus differ from those values.

markers on the western flanks are unlikely to have moved upslope, as is apparent in Fig. 5a–c. Their displacements are more likely to have been obliquely down the slope of the salt surface, around the buttress of country rocks on which the southwestern theodolite is located and over its eastern extension beneath the salt. Figure 5d shows our interpretation of up to three increments of likely horizontal displacement of markers downslope across contours on the salt.

Thirteen of the markers on the salt displayed no horizontal movement across sight-lines in the first measurement, 19 displayed none to the second measurement (relative to the first), and five showed none to the third (relative to the second). Most of these markers probably moved toward the theodolite stations viewing them, along divides in salt flow (indicated on Fig. 5d). Despite the limitations of having only components of lateral flow perpendicular to sight-lines, the displacements apparent to each measurement fit consistent but different patterns.

Of our eight markers on the summit dome (>1250 m), the first measurement found all eight to have moved downward between 346 and 0 cm (Fig. 4b). Meanwhile, five appeared to have moved outward from the summit between 595 and 67 cm, whereas three did not leave the line of sight (Fig. 5a). Of the same eight markers, the second reading found three to have moved between 396 and 52 cm downward, two to show no vertical change, and three to have risen between 139 and 68 cm. Meanwhile, only four moved outward between 206 and 93 cm from a centre that appeared to have moved several hundred metres since the first reading (compare Fig. 5a and b). The third reading found three markers on the western half of the summit dome to have risen between 6.4 and 1.8 m, whereas five on the eastern half sank between 1.74 and 12 m. The third reading also found that six markers moved between 372 and 28 cm outward, whereas one (E8) did not move from the sight-line, and another rotated 81 cm backward.

Of the 21 markers on the upper salt sheet (between 1250 and 1000 m), the first measurement found that six on the upper southwest slopes had risen between 363 and 36 cm, four showed no vertical change, and the remaining 11 (on the lower slopes) sank between 136 and 3 cm (Fig. 5a and 6b). Of these 21 markers, seven did not cross sight-lines, and 12 of the remainder moved downslope between 227 and 5 cm, whereas one on a pinnacle rotated backward 301 cm (E12 in Figs 5a and 6b). The second measurement found nine markers further down the southwest slopes to have risen between 201 and 14 cm, two to show no vertical change, and 10 to have moved downward between 363 and 5 cm. Meanwhile, 11 of them appeared to have moved horizontally between 788 and 14 cm downslope (Figs 5b and 6c). The third reading found the northwestern slopes of the upper salt sheet swollen between 630 and 109 cm, and the southern slopes to have subsided between 106 and 7 cm (Figs 5c and 6d). Whereas five markers did not cross sight-lines, eight moved backward between 72 and 17 cm, and only two moved outward by 58 and 42 cm.

Of the initial 14 markers on the lower salt sheet (<1000 m), our first measurement showed 10 to have sunk between 173 and 4 cm. Meanwhile, four in the distal southern lobe rose between 127 and 9 cm (Figs 5a and 6b). Apparent horizontal displacements on the lower slopes ranged from 151 and 15 cm, and marker E19 in Figs 5a and 6b (a salt pinnacle) rotated 396 cm backward. In the second interval, three of these 14 markers neither rose nor fell along a divide in the southeast that had moved since the first measurement, three markers sank between 66 and 21 cm, and the remainder rose between 173 and 9 cm, and only six moved distally in a subdued range of 33 and

7 cm (Fig. 6c). The third reading found all of the lower slopes to have subsided between 181 and 6 cm, apart from marker E20 near the distal salt terminus, which appears to have risen nearly 9 m (Fig. 6d). Horizontal movements in the third increment were also small, ranging between 104 and 5 cm downslope, with one moving 135 cm toward the summit and another not moving from the sight-line.

The most spectacular displacement on Jahani was an apparent horizontal displacement of the second most northern marker by an extraordinary first increment of 24.5 m, when it also sank 61 cm. The second horizontal increment of this marker was only 31 cm without any vertical change, and the third was a rise of 273 cm without any horizontal displacement from the sight-line. However, the first lateral increment, at an average rate of 1.36 m per month over 18 months, was downslope, toward the top of cliffs nearly 300 m high eroded deep into Jahani by a river (Figs 4 and 5; for a photograph of these cliffs, see Talbot & Alavi 1996, fig. 8f). This drastic displacement is reinforced by comparison between contours on 1:250 000 maps dating from 1952 (Iranian Topographic Survey) and 1977 (Oil Service Company of Iran), which indicate that these cliffs advanced *c.* 400 m below the marker (at an average of 16 m a^{-1} over the nominal 25 years between publication) whereas their tops retreated about 500 m and rose to about 400 m above river level further east (Fig. 5d).

Discussion of results

In some of the other mountains we are monitoring (Fig. 1), most markers consistently sink a few centimetres a year. Kuh-e-Jahani is characterized by vertical and horizontal movements that commonly exceed a metre. Although cairns (e.g. markers K2 and H12) and pinnacles collapse, changes in the scenery as a result of salt dissolution are hardly noticeable (although house-sized blocks fallen from the northern river cliffs were present on both sides of the river in November 1996). The rainfall in the region is 200–400 mm a^{-1} and, if all this drained off the salt fully saturated, it could potentially dissolve a vertical thickness of 3–7 cm a^{-1} (Talbot & Jarvis 1984). In fact, much of the precipitation is evaporated on the salt or the cloaking soils, and the runoff is not fully saturated with salt. The range of 3–7 cm a^{-1} for the potential dissolution of salt at Jahani is therefore judged to be on the high side of reality and, even then, small compared with the displacements of our markers. It is therefore clear that the flow of salt on Jahani dominates salt dissolution.

Gravity acting on the 4500 m column of overburden (density 2800 kg m^{-3}) can drive a column of pure salt (density 2160 kg m^{-3}) at Jahani 5833 m high above its source and 1333 m above its bedrock orifice (taken as level with the plain to the south, 585 m above sea level; see Fig. 3b). Because the dissolution rate is relatively small, buoyancy alone can therefore account for the height reached by the salt on Jahani.

Salt in the summit dome of Jahani gravity spread at an average rate of about 1 m a^{-1} outward and 0.5 m a^{-1} downward in the first 18 months, and moved downward about 0.75 m as it accelerated to 1.3 m outward from a different centre in the following year (compare Fig. 5a and b). Salt high on the southwest flank backed up 1 m behind the southwest buttress from June 1994 to November 1995, and was deflected east and southward around this obstruction. By November 1997, the western half of the summit dome had risen by about 4 m whereas the eastern half had sunk as much as 12 m, and the bulge on the southwestern flank had migrated over 2 km downslope and inflated most of the southern salt lobe. This bulge appears to have migrated at least a further 2 km downslope and largely dissipated (to ogives, surface folds in a glacier, Fig. 5c) by the third measurement, which found the northwestern flanks of Jahani apparently inflating (Fig. 4d).

Flow was generally fastest down the main southern stream. Interpreted vectors of total displacements during the first 18 months (Fig. 5d) are about 6 m (i.e. 4 m a^{-1}) on the southern lip of the summit dome, halve to 3 m (2 m a^{-1}) in the upper middle reaches, halve again on the lower slopes, and remain significant (0.5 m, i.e. 0.3 m a^{-1}) behind a steep and tall terminus that inflated *c.* 0.6 m. The steep snout of southern Jahani, which inflated in all increments, appears to have advanced about 200 m at a rate near 8 m a^{-1} averaged over '25 years' according to the two maps dated 1952 and 1977 mentioned above (Figs 5d and 6a). The first increments of horizontal displacements of salt on Jahani were within two orders of magnitude of ice velocities for valley glaciers of the same size. Displacements along the main salt stream (Fig. 6) in the first 18 months simulate classical gravity spreading in that the upper slopes extended laterally as they spread downward over a convex-upward slope whereas parts of the toe rotated backward as they rose as a result of compression from the rear (Fig. 6b). Local advance of the distal termini of salt sheets is indicated by the worn planforms of former goat paths rotating to smooth traces diagonal across slopes too steep for even goats to climb.

Our results are insufficiently detailed to show any discontinuities in flow along the mylonites. None the less, they are the first to constrain the first-order rates of flow of an active extrusion of salt. Annual increments of salt dissolution appear to be negligible compared with annual increments of salt flow at Jahani, both upward out of the bedrock orifice and downslope from the summit dome. Variations in our measurements much larger than our estimated errors emphasize that flow was far from steady. The migration of a bulge down the southwestern flanks of Jahani in 1995 and 1996, and the swelling of the northwestern flanks in 1997 suggest episodic pulses of gravity spreading.

Water weakens salt (e.g. Cristescu & Hunsche 1998, pp. 76–80), so rainfall can be expected to accelerate the gravity spreading of surficial salt (within 20 min according to Talbot & Rogers (1980)) as well as its dissolution. Our larger increments of salt flow have occurred after relatively wet periods as reported by local shepherds and indicated by rivers running in November. The response times of the diapir and its deep source to changes in gravity loading are likely to be very slow compared with changes in lateral tectonic forces in this seismically active region. However, it remains to be seen what proportions of the unsteadiness in flow we have measured at Jahani over 4.5 years reflect changes in the extrusion rate of deep salt out of the bedrock orifice in response to changes in tectonic forces and variations in surficial salt flow as a result of changes in rainfall. It is thus not yet clear whether such salt extrusions are more sensitive gauges of local rainfall or tectonic pressure.

Modelling

The shape and dimensions of a viscous fountain being dissolved over its top surface are a measure of its vigour of extrusion and age. Figure 7 illustrates a profile through an axisymmetric numerical model of a viscous fluid extruding out of a circular orifice,

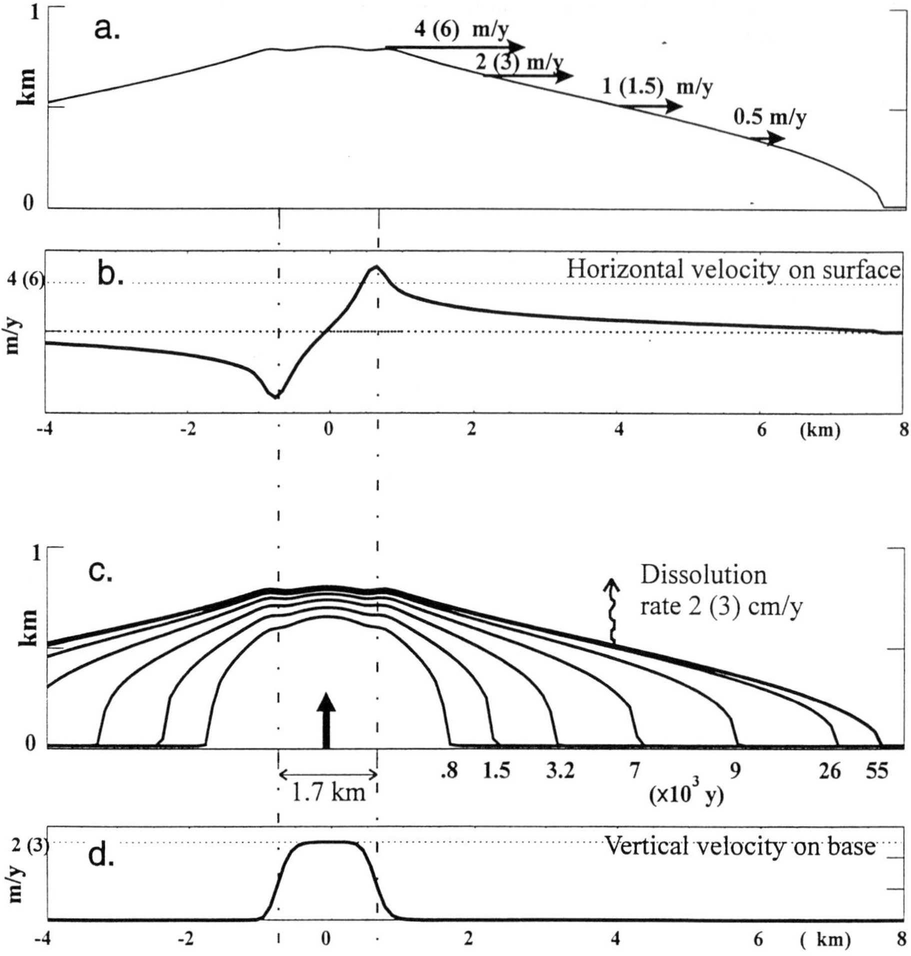

Fig. 7. Numerical model dimensionalized to the profile of the main southern stream of Kuh-e-Jahani assuming an orifice 1.7 km in diameter and a vertical salt dissolution rate of 1–3 cm a^{-1}. (**a**) Present profile, showing characteristic horizontal velocities measured from June 1994 to November 1995 (in cm a^{-1}), compares well with (**b**), which is a plot of velocities to the right from 0 km, the centre-line of the model extrusion. (**c**) Evolution of salt profiles with time (in 10^3 years). (**d**) Plot of vertical velocity above base-line, showing velocity profile up through the bed-rock orifice.

which specifies the ratio between the diameters of the extrusion and the orifice with time (Medvedev 1998). The fluid spreads under gravity as it dissolves over all its top surface at a constant vertical rate. This model greatly simplifies nature by assuming steady dissolution and flow of the salt in a consistent pattern over a smooth horizontal surface. (A very thin layer of salt has to be assumed along the top of the country rocks for the emerging salt to spread along their top surface.) Profiles in time (Fig. 7c) simulate well the young hemispherical domes of salt in the Zagros but not the upward concavity around the summit dome of salt fountains such as

Jahani. There must be some direct relationship between the diameter of the summit dome, the vigour of salt rise and the dimension of the orifice; but as we do not know it, the 1.7 km width of the orifice, one of the most sensitive variables in the current model, is merely 'slightly smaller than the summit dome' in Fig. 7.

The model is non-dimensional but the profiles in Fig. 7 have been fitted to the maximum width and height, and the maximum horizontal velocity, on a north–south centred profile along the main stream of Jahani in November 1995 (Fig. 7a). It is not only the maximum velocity (between 4 and $6\,\mathrm{m\,a^{-1}}$ at the lip of the summit dome) that fits the model in Fig. 7b; the other velocities compare well, both qualitatively and quantitatively.

Dimensionalization of the model allows us to constrain unknown quantities. Our current simple model implies that: the (constant) vertical dissolution rate of the salt surface is between 2 and $3\,\mathrm{cm\,a^{-1}}$; the salt at Jahani has a viscosity in the range 10^{16}–$10^{17}\,\mathrm{Pa\,s^{-1}}$; the vertical extrusion rate of salt out of the orifice is between 2 and $3\,\mathrm{m\,a^{-1}}$; Jahani is close to an equilibrium shape and has taken something like only 55 ka to reach its present size; as growth slows toward equilibrium. The 55 ka age of the current extrusion event at Jahani is imprecise ($\pm$20 ka).

All the inferred values appear to be acceptable, those for viscosities lying within the range of laboratory measurements at surface temperatures and pressures (Spiers *et al.* 1989; Cristescu & Hunsche 1998, p. 55).

The rate of extrusion of salt out of the orifice and the rate of surface dissolution are linked so that the former would be between 1 and $2\,\mathrm{m\,a^{-1}}$ if the latter were between 1 and $2\,\mathrm{cm\,a^{-1}}$. Talbot & Jarvis (1984) estimated that it would take 21 ka of current rainfall to dissolve a mountain of static salt less voluminous than Jahani. The rate of extrusion constrained by field readings here is an order of magnitude faster than the estimate of $17\,\mathrm{cm\,a^{-1}}$ budgeted for that other active salt fountain further south (Talbot & Jarvis 1984). Velocities like those we report here account for the slowest parts of salt sheets (their fronts) having advanced beneath the Gulf of Mexico at average rates up to $275\,\mathrm{mm\,a^{-1}}$ from Mid-Miocene to Plio-Pleistocene time (Wu *et al.* 1990).

Implications

Current understanding of the rates at which rock salt can flow in nature fall into two categories. Measurements of the closure rates of boreholes and mined cavities are consistent with measurements of specimens deformed in the laboratory and indicate that salt flow into mined openings slows to rates of $\mathrm{cm\,a^{-1}}$ to $\mathrm{dm\,a^{-1}}$ over years from having flowed 100 times faster over days and weeks (Cristescu & Hunsche 1998). In great contrast, backstripping of the surrounding country rocks indicates that salt in diapirs rises much more slowly (fractions of $\mathrm{mm\,a^{-1}}$). This is not surprising if the backstripping is of salt pillows and diapirs that are blind beneath roof rocks. Conformable pillows of Hormuz salt under the Persian Gulf rose very slowly ($0.01\,\mathrm{mm\,a^{-1}}$ averaged over 102 Ma) whereas early subsurface salt diapirs in the Zagros rose as rapidly as 0.3–$2\,\mathrm{mm\,a^{-1}}$ (Edgell 1996).

However, our observations in Iran emphasize that after a salt diapir erupts through the restraint of its cover rocks, the liberated salt can extrude into the sky or sea three or four orders of magnitude faster, at rates of metres a year. Salt extrusion as vigorous as this is likely only from a pressurized source through an open conduit. At an extrusion rate of 2–3 m a^{-1}, Jahani extrudes salt at about 5×10^6 m^3 a^{-1} through its orifice 1.7 km in diameter. In its lifetime of *c*. 55 ka, the current extrusion at Jahani has therefore already expelled a batch of something like 200 km^3 of salt from the local geological record. This is likely to have closed its probable source (taken as 1 km thick over an area of radius of 8 km) to a sub-horizontal primary weld (if lateral forces have not already closed the conduit to a steep secondary weld). We therefore infer that episodes of extrusion at the rates we have measured and modelled are comparatively short lived, perhaps less than 10^5 years. Brief episodes of first- and second-generation extrusion, which end when they starve a local compartment in their source layer (reticulated by Zagros folds and transfer faults), account for how sufficient Hormuz salt remains in undepleted compartments of the deep source to feed third-generation salt extrusions behind the Zagros front.

Zirngast (1996) used excess volumes of sediments in areas of salt withdrawal to constrain the vertical extrusion rate of salt from the Gorleben salt diapir in Germany. Zirngast's fastest rate of rise, of 0.145 mm a^{-1} from Cenomanian to Paleocene times, may give a false impression of the rate at which salt can rise because Zirngast could only average his rate over 39 Ma. For perhaps a small fraction of one of those 39 Ma, the salt at Gorleben may have resembled that now at Kuh-e-Jahani, vented to the free surface and extruded vertically at rates nearer a metre per year.

The most startling apparent displacement we measured, of about 1.36 m per month above steep river cliffs carved almost into the core of Jahani, may have important lessons for Quaternary erosion of salt at Gorleben and elswehere. An exploratory mine is being excavated 840 m below the surface in Gorleben salt to decide in 2005 whether radioactive waste will be isolated there by the year 2030. Potash salts are replaced by gravels down to 380 m below sea level in a narrow inclined slot. This slot is currently interpreted as a narrow opening in the floor of a tunnel valley incised into the Gorleben salt by overpressured melt water flowing beneath Elserian ice between 300–500 ka ago (Zirngast 1996).

If the river currently carving the northern edge of Jahani were to bypass the current salt cliffs, our 1995 measurement suggests that the 400 m high river cliffs could fall and flow into the river gorge and ride up the remnant salt body 1 km beyond in <100 years. (Most of the remains of a major cliff fall had been carried downstream by a still-active river by November 1996.) Asymmetric flow would close the current valley and entrap river gravels in a narrow inclined body beneath a hanging wall in which stratigraphic markers can be interpreted as indicating that salt flowed asymmetrically to fill a previously open channel, as we interpret the situation at Gorleben. This logic suggests that far more salt may have been lost from a subglacial channel at Gorleben much faster over a brief interval than indicated by the rise rate of 0.03 cm a^{-1} that Zirngast (1996) calculated from Miocene to Quaternary time but had to average over the last 23 Ma.

Our measurements of flow rates on Jahani also contrast strongly with a recent study of how salt flow might affect wells drilled to hydrocarbons beneath the 270–330 m thick Enchilada sheet of allochthonous salt buried beneath the Gulf of Mexico (Diggs *et al.* 1997). Unlike many more-recent salt bodies that shape the floor of the Gulf of Mexico, the Enchilada salt sheet is now buried beneath clastic sediments in which only two of the many faults are still active. Using the average diameters of halite subgrains etched in thin sections of rotary sidewall samples from two wells, Diggs *et al.* calculated the differential stresses through the salt sheet during the last few thousand years. Interpreting their preliminary results into velocity profiles led to expectations that these wells would be carried sideways only tens of centimetres in smooth arcs over the next 15 years, a comforting result that needs to be substantiated.

Profiles of grain or sub-grain size and translations interpolated between isolated samples through the Enchilada salt sheet are suspiciously smooth. They may miss the few weak mylonites only 10 cm thick between nappes of gneissic salt hundreds of metres thick in Kuh-e-Jahani and several other salt extrusions in southern Iran (Talbot 1979, 1998). Development wells through sheets of allochthonous salt that might still be moving deserve 100% coring through all the salt sections.

We appreciate the help given by many of our colleagues in Uppsala and Tehran, particularly in the field. U. Hunsche, M. Jackson and G. Eisenstadt are thanked for helpful reviews. The Zagros Halokinetics research programme is supported by the Geological Survey of Iran and the Swedish Natural Science Foundation.

References

ALA, M. A. 1974. Salt diapirism in southern Iran. *AAPG*, **58**, 1758–1770.

CRISTESCU, N. & HUNSCHE, U. 1998. *Time Effects in Rock Mechanics*, Wiley, Chichester.

DIGGS, T. N., URAI, J. L. & CARTER, N. L. 1997. Rates of salt flow in salt sheets, Gulf of Mexico: quantifying the risk of casing damage in subsalt plays. *Terra Nova*, **9**, abstract 11/4B29.

EDGELL, H. S. 1996. Salt tectonism in the Persian Gulf basin. *In*: ALSOP, G. I., BLUNDELL, D. J. & DAVISON, I. (eds) *Salt Tectonics*. Geological Society, London, Special Publications, **100**, 129–151.

HUNSCHE, U. 1998. Determination of the dilatancy boundary and damage up to failure for four types of rock salt at different stress geometries. *In*: AUBERIN, M. & HARDY, H. R., JR (eds) *The Mechanical Behaviour of Salt IV; Proceedings of the 4th Conference (MECASALT IV), Montreal, June 1996*. TTP Trans Tech., Clausthal, 163–174.

KENT, P. E. 1958. Recent studies of south Persian salt plugs. *AAPG*, **422**, 2951–2972.

—— 1970. The salt plugs of the Persian Gulf region. *Transactions of the Leicester Literary and Philosophical Society*, **64**, 56–88.

MEDVEDEV, S. 1998. Thin sheet approximations for geodynamic applications. *Acta Universitatis Upsaliensis*, **368**.

PLAYER, R. A. 1969. *Salt Plug Study*. Iranian Oil Operating Companies, Geological and Exploration Division, Report **1146**.

SPIERS, C. J., PEACH, C. J. BRZESOWSKY, R. H., SCHUTJENS, P. M., LIEZENBERG, J. L. & ZWART, H. J. 1989. *Long-term rheological and transport properties of dry and wet salt rocks*. Final Report, Nuclear Science and Technology, Commission of the European Communities, **EUR 11848 EN**.

TALBOT, C. J. 1979. Fold trains in a glacier of salt in southern Iran. *Journal of Structural Geology*, **1**, 5–18.

—— 1993. Spreading of salt structures in the Gulf of Mexico. *Tectonophysics*, **228**, 151–166.

—— 1998. Extrusions of Hormuz salt in Iran. *In*: BLUNDELL, D. J. & SCOTT, A. C. (eds) *Lyell: the Past is the Key to the Present*. Geological Society, London, Special Publications, **143**, 315–334.

—— & ALAVI, M. 1996. The past of a future syntaxis across the Zagros. *In*: ALSOP, G. I., BLUNDELL, D. J. & DAVISON, I. (eds) *Salt Tectonics*. Geological Society, London, Special Publications, **100**, 89–109

—— & JARVIS, R. J. 1984. Age, budget and dynamics of an active salt extrusion in Iran. *Journal of Structural Geology*, **6**, 521–533.

—— & ROGERS, E. A. 1980. Seasonal movements in a salt glacier in Iran. *Science, Washington*, **208**, 395–397.

WU, S., BALLY, A. W. & CRAMEZ, C. 1990. Allochthonous salt, structure and stratigraphy of the north-eastern Gulf of Mexico. Part II: structure. *Marine and Petroleum Geology*, 7, 334–370.

ZIRNGAST, M. 1996. The development of the Gorleben salt dome (northwest Germany) based on qualitative analysis of peripheral sinks. *In*: ALSOP, G. I., BLUNDELL, D. J. & DAVISON, I. (eds) *Salt Tectonics*. Geological Society, London, Special Publications, **100**, 203-226.

Salt tectonics in and around the Nile deep-sea fan: insights from the PRISMED II cruise

VIRGINIE GAULLIER[1], YOSSI MART[2], GILBERT BELLAICHE[3], JEAN MASCLE[3], BRUNO C. VENDEVILLE[4], TIPHAINE ZITTER[5] & SECOND LEG PRISMED II SCIENTIFIC PARTY[6]

[1]*CEFREM, Université de Perpignan, 66860 Perpignan, France (e-mail: gaullier@univ-perp.fr)*

[2]*Recanati Center for Marine Studies, Haifa University, 31905 Haifa, Israel*

[3] *Observatoire Océanologique, Géosciences-Azur, 06235 Villefranche-sur-Mer, France*

[4] *Bureau of Economic Geology, University of Texas at Austin, Austin, TX 78713-8924, USA*

[5] *Free University, Faculty of Earth Sciences, 1081 HV Amsterdam, The Netherlands*

[6]*J. Benkhelil, G. Buffet, L. Droz, M. Ergun, C. Huguen, A. Kopf, R. Levy, A. Limonov, Y. Shaked, A. Volkonskaia and J. Woodside*

Abstract: The recent PRISMED II geophysical survey has documented various styles of salt tectonics in and around the Nile deep-sea fan (Eastern Mediterranean Sea). The first main type of salt-related structures comprises listric normal growth faults and grabens, trending roughly perpendicular to the slope line of the Nile Cone. These faults and associated salt structures result from thin-skinned extension, driven by gravity gliding and spreading as a result of sediment loading of the Plio-Quaternary overburden above the Messinian evaporites, which acted as a décollement layer. The second major type of salt structures consists of lineaments that obliquely intersect the continental slope of the Nile deep-sea fan. These structures may have had some strike-slip movement, and salt diapirs grew reactively or were deformed by fault-block movement. In the western distal part of the Nile deep-sea fan, compressional tectonics of the adjacent Mediterranean Ridge caused the formation of a series of salt-cored folds and reverse faults above the Messinian evaporites. In the eastern distal part of the Nile Cone, sediment progradation progressively expelled salt northward, first forming small folds and tight diapirs, then a scarp of 400 m height around the Eratosthenes Seamount, corresponding to the basinward limit of salt deformation.

Desiccation of the Mediterranean Sea in Messinian (late Miocene) times had two major geological consequences. First, seawater evaporation led to deposition of a thick sequence of halite, gypsum and other evaporites found in three distinct units, the lower evaporites (carbonates and evaporitic sediments), the 'salt' (halite and

From: VENDEVILLE, B., MART, Y. & VIGNERESSE, J.-L. (eds) *Salt, Shale and Igneous Diapirs in and around Europe*. Geological Society, London, Special Publications, **174,** 111–129. 1-86239-066-5/00/$15.00

anhydrite) and the upper evaporites (marls and gypsum) (Ryan *et al.* 1973). Second, a drop in sea level of >1000 m caused intense river erosion onshore. Deep canyons carved into the bedrock, and large volumes of detrital debris were transported offshore.

Salt diapirs that rose through Plio-Quaternary sedimentary overburden in the Mediterranean Sea are one of the most spectacular consequences of deposition of such a thick evaporitic sequence and its overburden. Many of these diapirs currently emerge at the sea floor throughout the Mediterranean deep basin and form bathymetric domes in some places and collapse–dissolution depressions in others (e.g. Ryan *et al.* 1970; Ross & Uchupi 1977; Ryan 1978; Pautot *et al.* 1984; Gaullier 1993; Bellaiche & Mart 1995; Gaullier & Bellaiche 1996). Diapirs in the Mediterranean are commonly elongate salt walls (Ben-Avraham & Mart 1981; Gaullier & Bellaiche 1996), or they form a series of subcircular salt stocks preferentially aligned along lineaments, which suggests that the rise of these diapirs was structurally constrained by faulting. Faults in the Plio-Quaternary section can be grouped into two families. The first family comprises graben faults and listric normal growth faults whose traces are parallel or normal to the slope direction of large Mediterranean deep-sea fans. These faults are commonly caused by thin-skinned, gravity-driven gliding–spreading above salt. Faults of the second family commonly trend obliquely with respect to the direction of the continental margins and may have formed in response to regional thick-skinned tectonics.

Salt diapirs in the southeastern Mediterranean Basin were discovered during early reconnaissance surveys, and their origin was attributed to the presence of a thick Messinian evaporitic sequence (Neev *et al.* 1976; Woodside 1977; Mart *et al.* 1978; Ryan 1978). GLORIA long-range (13 km) side-scan sonar showed that many of these diapirs are elongate structures trending NW to NE (Kenyon *et al.* 1975). Diapirs were known to exist mainly in the eastern part of the Nile deep-sea fan, but the reasons for this preferential distribution were poorly understood (Ross & Uchupi

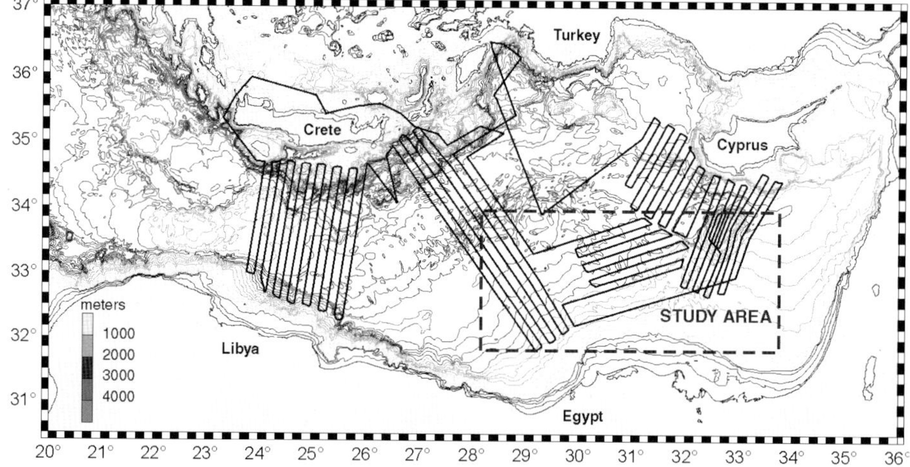

Fig. 1. Bathymetric map of the Eastern Mediterranean Basin showing the location of the study area and of the geophysical profiles recorded during the PRISMED II cruise (R.V. *l'Atalante*, February 1998).

1977; Maldonado & Stanley 1978; Kempler *et al.* 1996). The recent PRISMED II geophysical survey conducted south of Cyprus, revealed that such structures are widespread throughout the Nile deep-sea fan area. The main purpose of this paper is to highlight the principal styles of salt deformation in this part of the Mediterranean on the basis of this new data set.

Dataset

The present study was conducted aboard the R.V. *l'Atalante* in February 1998 using the EM12-Dual/Simrad multibeam sounding system, which surveyed along strip tracks 5–6 times wider than the water depth, which ranges between 1400 and 2700 m there. The general objective was to comprehensively analyse the morphological, structural and sedimentary characteristics of the overall Eastern Mediterranean Basin, whose evolution has been affected by the incipient continental collision between Africa and Europe, with special emphasis on the Nile Cone, which represents the largest submarine fan in the Mediterranean Sea. The survey provided nearly complete bathymetric and back-scattering imagery coverage of the area south and west of Cyprus (Fig. 1).

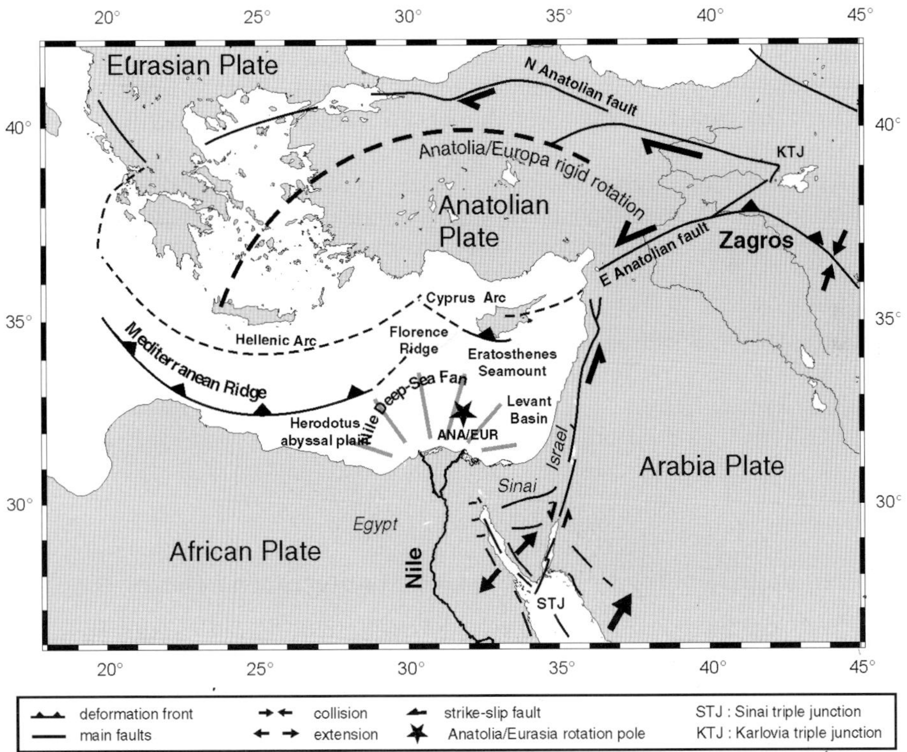

Fig. 2. Geodynamic map of the Eastern Mediterranean Basin (modified after Le Pichon & Gaulier (1988) and Reilinger *et al.* (1997)). The star indicates the Anatolia–Eurasia rotation pole, responsible for extrusion of the Anatolian Plate. The Nile Cone is bounded westward by the Herodotus Abyssal Plain, northward by the eastern part of the Mediterranean Ridge and by the Eratosthenes Seamount, and eastward by the margin of the Levant Basin.

Data were briefly processed onboard using an IFREMER-developed program (Caraibes). Simultaneously, near-surface (upper 20–50 m) sedimentary structures were recorded using a 3.5 kHz profiler. Deeper structures, up to 3 s two-way travel time (TWTT), were also recorded using a six-channel streamer and Sodera GI gun. About 3500 km of continuous geological profiles were acquired over the Nile deep-sea fan area.

Geodynamic and tectonic setting of the Eastern Mediterranean Basin

The tectonic regime of the SE Mediterranean was controlled mostly by the processes that took place along the Cyprus Arc (Fig. 2), where the north edge of the Levant Basin and the Eratosthenes Seamount are being thrust under the Cyprus block (Kempler & Garfunkel 1994; Mart *et al.* 1997; Mart & Robertson 1998). Thrusting under the Cyprus Arc is a part of the overall Africa–Eurasia plate collision, and, in general, does not differ much from the overall subduction of the Ionian crust under southern Europe, which is the process that formed the Mediterranean Ridge. However, the Cyprus Arc is also affected by westward motion of Anatolia (McKenzie 1978; Kempler & Garfunkel 1994; Mart & Robertson 1998), which adds a component of left-lateral slip to the subduction–collision process. An additional tectonic process in the region south of Cyprus is the subsidence of the Nile deep-sea fan as a result of sedimentary loading (Tibor *et al.* 1992). The Nile River has been transporting large sediment loads (the 'Nilotic' sediments) into the Eastern Mediterranean Sea since Miocene time (Said 1981), and its output, like that of all the other Mediterranean rivers, increased substantially during the Messinian desiccation, when sea level dropped by >1000 m and fluvial erosion greatly increased. Accumulation of thick Plio-Quaternary sediments (up to 3.5 km) probably led to additional crustal subsidence. The Nilotic sediments are present over a wide area extending as far as the Levant Basin in the NE and the east sector of the Mediterranean Ridge in the NW (Venkatarathnam *et al.* 1972). Another structural element affecting the geodynamic setting of the SE Mediterranean is a NW-trending fault system that dissects the Nile fan. Although parts of this system of faults and associated diapirs have been previously recognized by local studies (e.g. Kenyon *et al.* 1975; Neev *et al.* 1976; Ross & Uchupi 1977; Ryan 1978; Ben-Avraham & Mart 1981; Kempler *et al.* 1996), their widespread occurrence and distribution were discovered only during the present survey (Mascle *et al.* 1998, 2000; Gaullier *et al.* 1999; Bellaiche *et al.* 1999).

Sedimentological setting of the Eastern Mediterranean Basin

The Nile River, the most prominent river of the Eastern Mediterranean Basin, has controlled the sediment supply of the Levant Basin. During Messinian times, it cut a deep canyon into the bedrock, which subsequently allowed Early Pliocene marine transgression to penetrate inland as far as Aswan in Upper Egypt (Chumakov 1967). The sedimentary accumulation of the Messinian Nile delta, whose elevation was *c.* 1 km or more below the present-day sea level, turned into a large deep-sea fan once the Mediterranean Sea was reconnected to the global ocean system through the Strait of Gibraltar. The sedimentary system of the river was also affected by Pleistocene climate changes because regression of the coastlines placed the river distributary system at the edge of the continental slope. Thus, a large part of the sedimentary load was directly transported onto the continental slope, rather than being deposited on the shelf. Combined sea-level lows and increase in flow volumes of the Nile River, probably as a result of intensification of the monsoon climate system during the ice ages, enhanced sediment transport from the plateaux of East Africa to the Mediterranean basins (Said 1981). The sediment supply of the Nile before the construction of the High Dam in Aswan was estimated to be $150 \times 10^6\,m^3\,a^{-1}$ (Said 1981). The Nile River deposited very large alluvial sequences of deltas, coastal deposits, shelf and slope aprons, and deep-sea fans during Plio-Quaternary time. Onshore the present-day delta, established when the sea level became stable in Holocene time (about 6000 years ago), covers an area of $22\,000\,km^2$. It was the Greek geographer and historian, Herodotus (484–420 BC), who first coined the term 'delta' to describe this particular triangular flood plain. The Nile River affected sediment deposition on the continental shelf, slope and rise of the southeastern Mediterranean, as well as in most of the Levant Basin (Venkatarathnam *et al.* 1972). Nile-derived sediments form thick sedimentary sequences along the continental margin and the deep marine basins offshore from Syria, Lebanon, Israel, Egypt and Libya, and cover the floor of the Levant Basin and the Herodotus Abyssal Plain. The southeast segment of the Mediterranean Ridge also contains significant volumes of Nilotic sediments (Ross *et al.* 1978). These sediments encountered during scientific drilling in the eastern Mediterranean, comprise two fractions, the coarser one of turbiditic sands, and the finer one of hemipelagic clays (Ryan *et al.* 1973; Hsü *et al.* 1978; Emeis *et al.* 1996).

Results of the PRISMED II survey

Recent data obtained during the PRISMED II survey show that the style of salt structures varies between different morphological, tectonic and sedimentological settings within the Nile deep-sea fan and its surroundings. The deep Nile Cone can be divided into three bathymetric provinces, the western, central and eastern provinces (Mascle *et al.* 1998, 2000; Gaullier *et al.* 1999; Bellaiche *et al.* 1999), each having its own characteristics, as follows (Fig. 3).

The western province

The western province (Fig. 4a) has an average slope direction of about N140 and is characterized by a distributive network of NW-trending channels that are clearly visible on swath bathymetry and on the acoustic imagery of the sea floor. These channels can reach 200 km in length and are flanked by sedimentary levees that rise from several tens of metres to 100 m above the thalwegs. Most channels display tight meanders (Fig. 4b) similar to those of the Rhône, Amazon or Zaire deep-sea fans. Some channels are recent and still active, especially in the westernmost part of the area, whereas others are older and now inactive (Bellaiche *et al.* 1999).

The upper continental slope of the western province of the Nile fan is also characterized by numerous listric normal growth faults visible both on the bathymetric map (Fig. 4b) and on seismic-reflection data (Fig. 5). Figure 4b shows that fault traces are either straight or spoon shaped and trend roughly parallel to the slope (i.e. NE–SW).

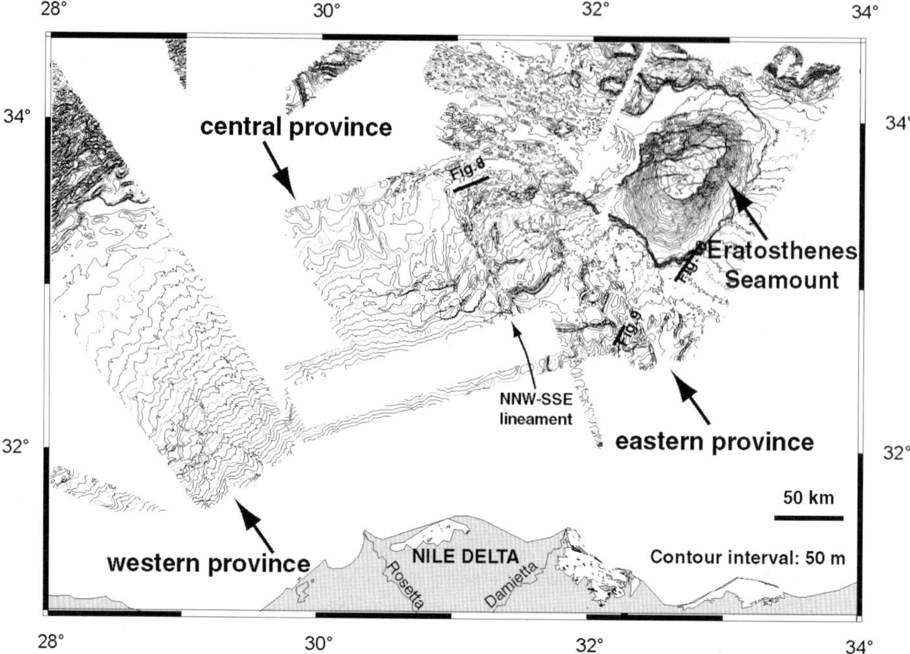

Fig. 3. Swath bathymetry of the western, central and eastern provinces of the Nile deep-sea fan, showing their contrasting morphological and structural patterns (contour interval 50 m). The NE–SW-oriented growth faults are located at mid-slope. The NNW–SSE-trending lineament is the boundary between the central and the eastern provinces (see text for details). Black heavy lines indicate the location of seismic profiles in Figs 8–10.

These listric growth faults dip downslope and bound wedges of tilted Plio-Quaternary syntectonic strata (Fig. 5). The height of the bathymetric fault scarps can reach 150 m, and fault throw can exceed 500 m (0.5 s TWTT). The thickness of the sediments affected by faulting often exceeds 2000 m (2 s TWTT). Growth faults sole out into the Messinian salt layer, whose top corresponds to the strong M reflector (as defined by Ryan *et al.* (1973)) on seismic profiles. In the western province, the M reflector is typically located between 4 and 4.5 s TWTT. The base salt is located at *c.* 5 s TWTT, which corresponds to *c.* 2 km (assuming a seismic velocity for salt of 4000 m s^{-1}). Faults are preferentially located where sediments are thickest. Although growth faulting has continued until the present, the rate of faulting and block tilting seems to have varied through time: Fig. 5 shows recent, horizontal sediments onlapping onto tilted older strata. Listric normal growth faults were triggered by downslope gravity gliding

Fig. 4. (**a**) Computer-generated bathymetric image of the western province of the Nile deep-sea fan. Data are from the EM12-Dual/Simrad multibeam echo-sounder (PRISMED II survey), artificial light comes from the east (N90°). The white heavy line is the seismic profile shown in Fig. 6. (See text for details.) (**b**) Detail of the swath bathymetric map of the western province of the Nile deep-sea fan, illustrating the network of distributive, meandering channels and associated raised levees (contour interval 20 m). The heavy line is a seismic profile crossing the listric normal growth faults discussed in the text (see Fig. 5).

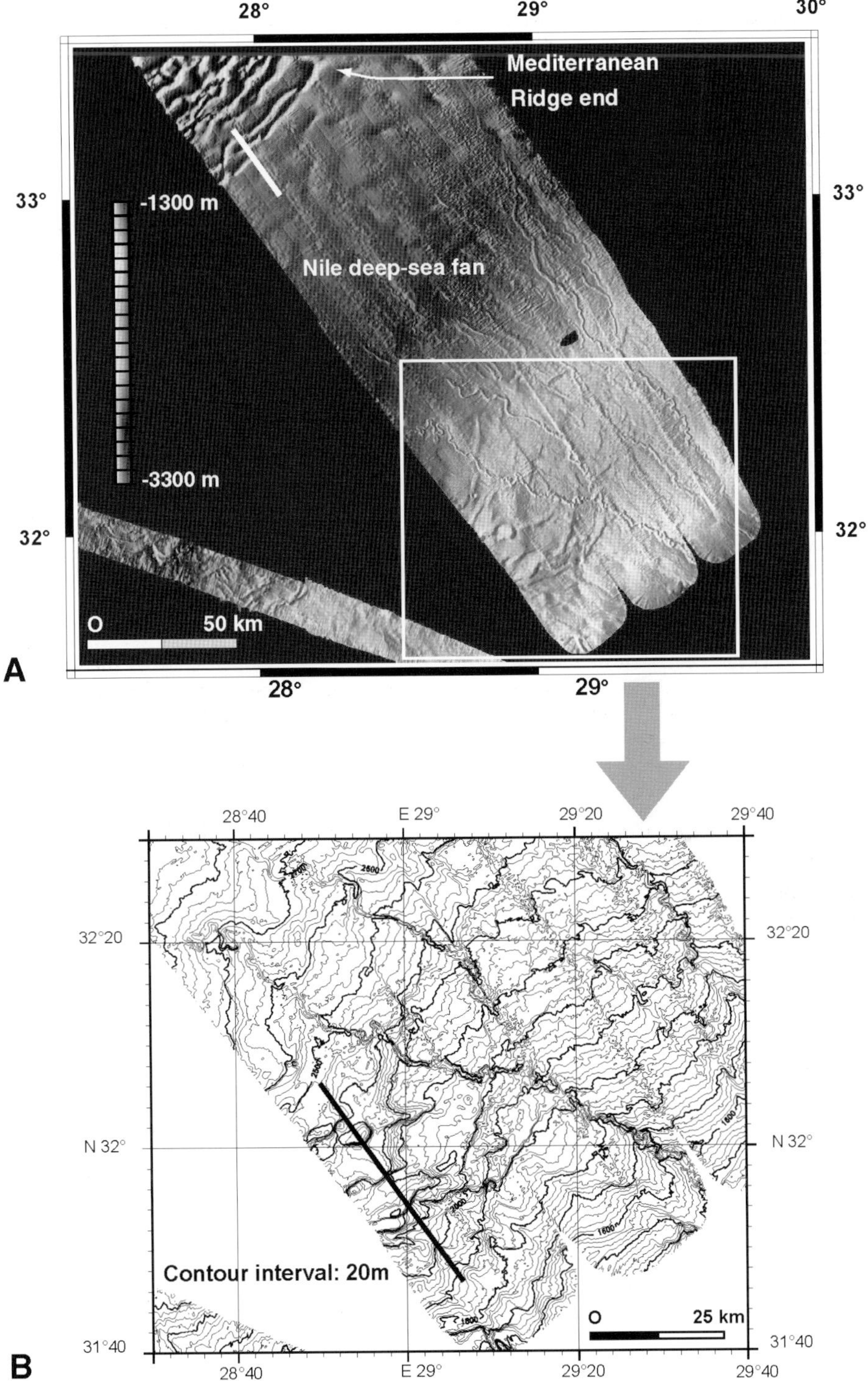

28°
29°
30°
Mediterranean
Ridge end
33°
-1300 m
Nile deep-sea fan
-3300 m
32°
O
50 km
A
28°
29°
28°40
E 29°
29°20
29°40
32°20
N 32°
31°40
Contour interval: 20m
O
25 km
B

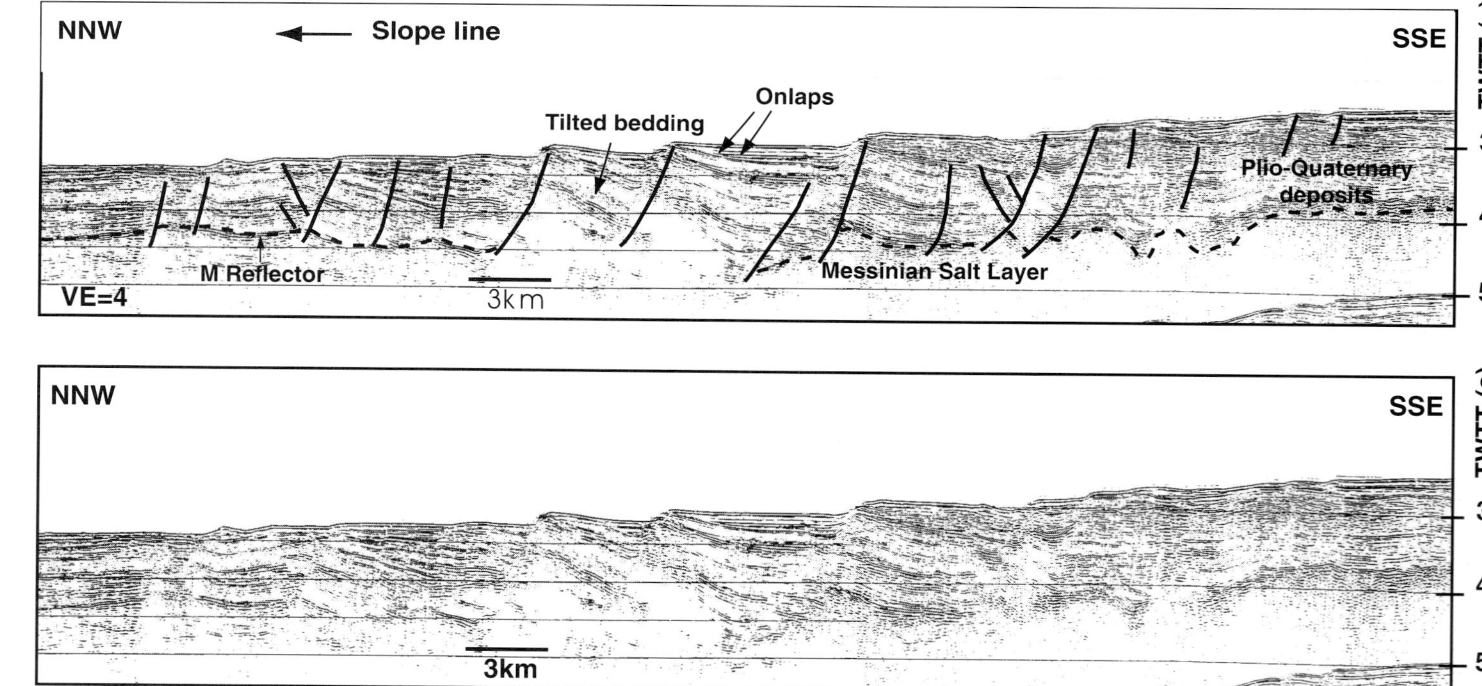

Fig. 5. Uninterpreted and interpreted seismic reflection profiles intersecting listric growth faults along the upper continental slope of the western province. The associated residual salt rollers, stratigraphic wedges and sedimentary rollovers should be noted (see Fig. 4b for location, and text for details). VE, vertical exaggeration.

and spreading of the Plio-Quaternary overburden of the Nile Cone above the Messinian salt layer, which acted as a décollement. Sediments overlying a weak basal salt layer can easily glide and spread seaward under the effect of their own weight where values of the basal and surface slopes are 3° or less (Vendeville 1987). Gravity gliding and spreading induce thin-skinned extension of the overburden and are greatly enhanced by high sedimentation rates (>0.5 mm a^{-1}) and by progradation of depositional sequences. Thin-skinned extension typically forms normal growth faults rooting at depth into the top of the décollement and triangular salt rollers beneath fault footwalls. In the hanging walls, combined faulting and sedimentation result in tilted strata, stratigraphic wedges and rollover anticlines (Fig. 5). The geometry and the dynamics of extensional structures are clearly controlled by the orientation of the continental slope. Listric growth faults and associated salt structures generally trend parallel to the slope.

The Plio-Quaternary sediments consist of three components: Nilotic turbidites, interbedded hemipelagic muds (Maldonado & Stanley 1978) and layers of seismically transparent or chaotic acoustic facies (Fig. 5), which correspond to large mass-movement deposits such as debris flows (Damuth & Embley 1981; Gaullier & Bellaiche 1998). Such layers can be as thick as 700 m (0.7 s TWTT) and can cover areas as vast as 30 000 km^2. These large mass-movement deposits may have been triggered by critical failure of the sediments along the upper continental slope, possibly enhanced by earthquakes (Nur & Ben-Avraham 1978). Growth faults locally offset these debris flows.

In the most distal section of the western province, northeast of the Herodotus Abyssal Plain, the transition from the Nile deep-sea fan to the Mediterranean Ridge is abrupt (Fig. 4a). There, the distal turbidites of the Nile deep-sea fan are folded and locally reverse faulted. Shortening results from a combination of tectonic compression in the Mediterranean Ridge and gravity-driven, thin-skinned contraction at the distal toe of the Nile deep-sea fan balancing thin-skinned extension and seaward translation updip (Fig. 6). This obvious morphological transition clearly corresponds to a change in the tectonic regime from extension and subsidence to horizontal compression. The progressive increase in shortening toward the Mediterranean Ridge internal domain is clearly illustrated by the transition from large salt-cored anticlines and synclines to asymmetric faulted folds and diapirs.

The central province

The central province (Figs 3 and 7) has a mean slope direction of about N160 and is also characterized by a belt, of 200 km length, of numerous normal growth faults located mid-slope (Fig. 7a). Fault traces, sinuous and typically concave seaward, trend roughly E–W to NE–SW. Bathymetric scarps along these faults are *c*. 100–200 m high (Fig. 7b). Unfortunately, we could not properly image them, as our seismic profiles were oriented parallel to the fault trends. However, features in the western Mediterranean Sea having similar bathymetric signatures (Gaullier 1993; Gaullier & Bellaiche 1996), as well as experimental models (Cloos 1955, 1968; Vendeville & Cobbold 1987; Gaullier *et al.* 1993), suggest that these faults, like growth faults in the western province, formed by thin-skinned extension above the Messinian salt layer. In the central province, the Plio-Quaternary sediments have a time thickness of *c*. 2 s TWTT, and the M reflector lies at 3.5 s TWTT.

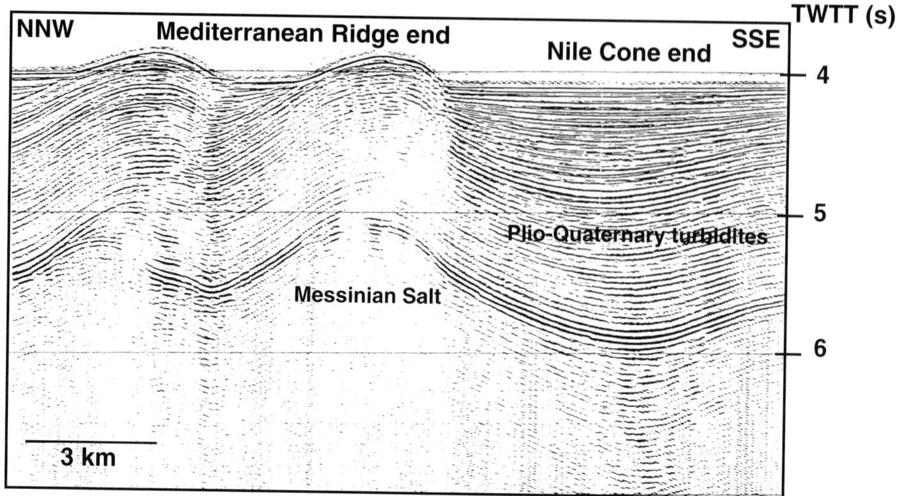

Fig. 6. Seismic reflection profile across the transition zone between the western province of the Nile deep-sea fan and the Mediterranean Ridge. The abrupt morphological transition corresponds to a change in the tectonic regime, from extension and subsidence to lateral compression. A noteworthy feature is the folds affecting the Plio-Quaternary, distal turbidites of the Nile Cone, locally reverse faulted (see Fig. 4a for location).

The eastern province

Swath bathymetry (Fig. 3) and acoustic imagery indicate that there is a significant contrast between the central and western provinces and the eastern province of the Nile deep-sea fan.

Evidence of intense tectonic activity and diapirism in this latter area was reported by Ross & Uchupi (1977), Woodside (1977), Ross *et al.* (1978), Ryan (1978) and Kempler *et al.* (1996). Diapirs in the Levant Basin had previously been studied by Ryan *et al.* (1970), Kenyon *et al.* (1975) and Ross & Uchupi (1977). The sea floor of the eastern province was known to be intensively deformed by fault scarps, grabens and elongate salt ridges trending approximately NW–SE and NE–SW (Kenyon *et al.* 1975; Ben-Avraham & Mart 1981; Kempler *et al.* 1996). The NE–SW lineaments were interpreted as normal faults, the NW–SE ones as strike-slip faults (Kempler *et al.* 1996).

The newer PRISMED II dataset shows that the transition between the central and eastern provinces is abruptly marked by a NNW–SSE-trending lineament of >100 km length (Fig. 3). In the south, this lineament comprises long and narrow depressions that presumably formed by extensional or transtensional faulting. In the north, the lineament merges in a zone of small-amplitude, small-wavelength buckle folds marking the distal edge of the Nile deep-sea fan. This lineament marks the western edge of a fault system that is bounded to the east by the Eratosthenes Seamount.

In the eastern province, the average slope varies from N160 in the west to N180 in the east, between the Eratosthenes Seamount and the African margin. The Messinian evaporite sequence is shallower (M reflector at 2.5 s TWTT) and thinner (<0.5 s

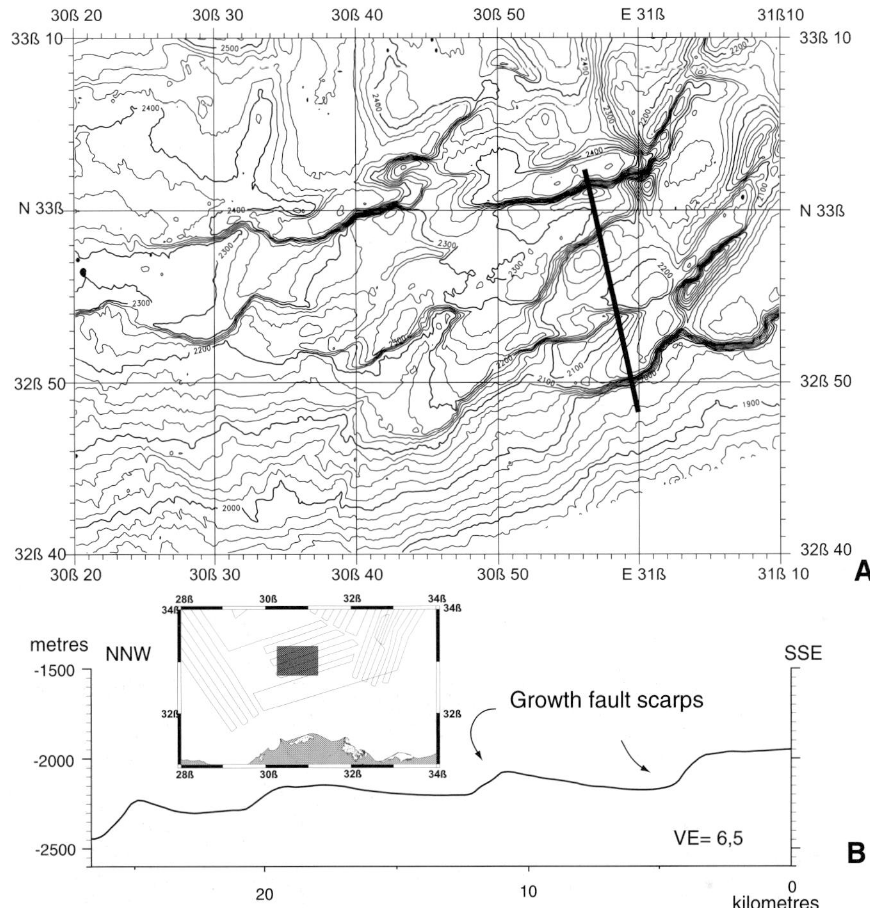

Fig. 7. (**a**) Detail of the bathymetric map of the central province illustrating growth faults (contour interval 20 m). The heavy line indicates the location of Fig. 7b. (**b**) Bathymetric transverse profile across the growth faults showing their offset of *c.* 200 m.

TWTT), suggesting that the Eratosthenes Seamount was already a bathymetric high during Messinian times (Mart *et al.* 1997). Upslope, the Plio-Quaternary overburden thickness decreases eastward, from 2 s TWTT in the west to 0.8 s TWTT in the east, whereas the overburden thickness also decreases northward to *c.* 0.4 s TWTT downslope.

Our PRISMED II data reveal that the area bounded to the west by the NW–SE lineament and to the east by the Eratosthenes Seamount comprises four groups of salt-related structures, each having specific preferential orientation and morphostructural characteristics.

In the first group, structures trend NE–SW to NNE–SSW or ENE–WSW, and mainly correspond to large sinuous grabens cored by Messinian salt walls whose crests are at or near the sea floor. As in the western and central provinces, these grabens were primarily caused by northwestward extension driven by gravity gliding and spreading of the Plio-Quaternary sediments above the Messinian evaporitic décollement. Some diapirs have crestal-collapse grabens that form elongated bathymetric depressions 150–200 m deep and that are the result of thin-skinned extension (reactive or falling diapirs, as described by Vendeville & Jackson (1992*a*,*b*)), possibly augmented by some additional salt dissolution. Brine pools are common in the bathymetric depressions in the eastern area (Kempler *et al.* 1996).

Along strike, these grabens and salt walls are either intersected by, or terminate against, the second group of salt-related structures, which consist of narrow faulted grabens parallel to the NW–SE lineament. Seismic reflection data (Fig. 8) suggest that these faults may have had some amount of strike-slip movement and that salt diapirs along such transtensional structure grew reactively or were deformed by fault-block movement (Fig. 9). The response of salt diapirs to extension and contraction is well documented (e.g. Jackson & Vendeville 1992; Vendeville & Nilsen 1995; Nilsen *et al.* 1996), but the response to transtension is much less known. It is yet to

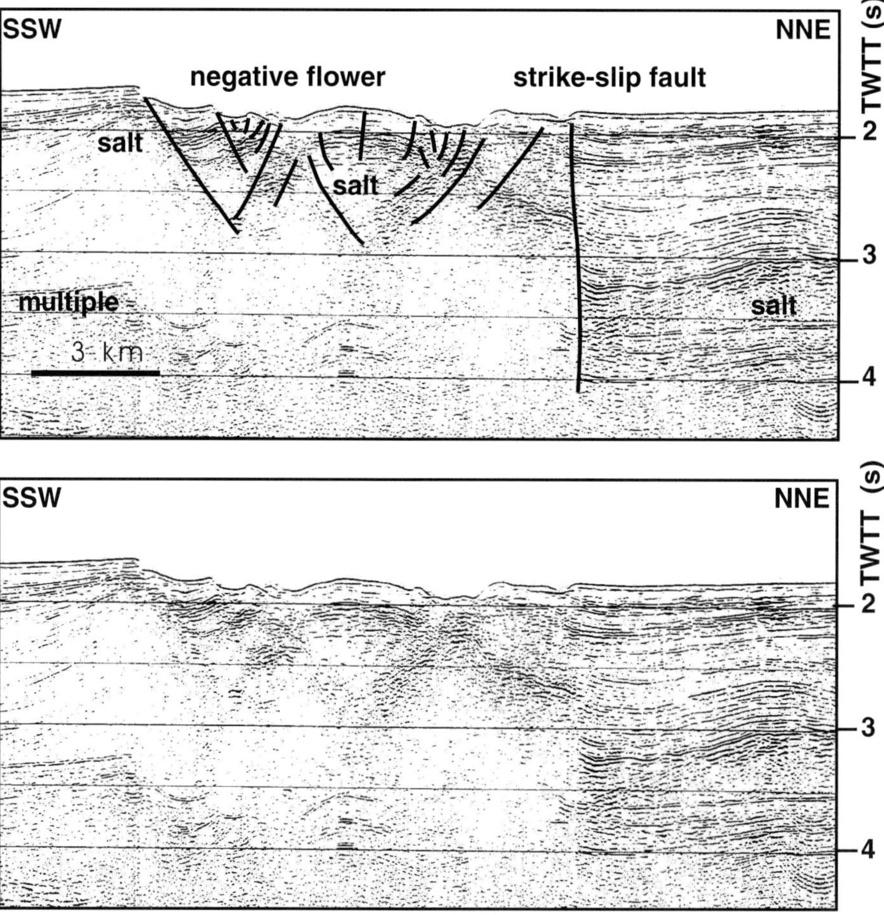

Fig. 8. Uninterpreted and interpreted seismic reflection profiles across the transition between the central and eastern provinces of the Nile Cone.

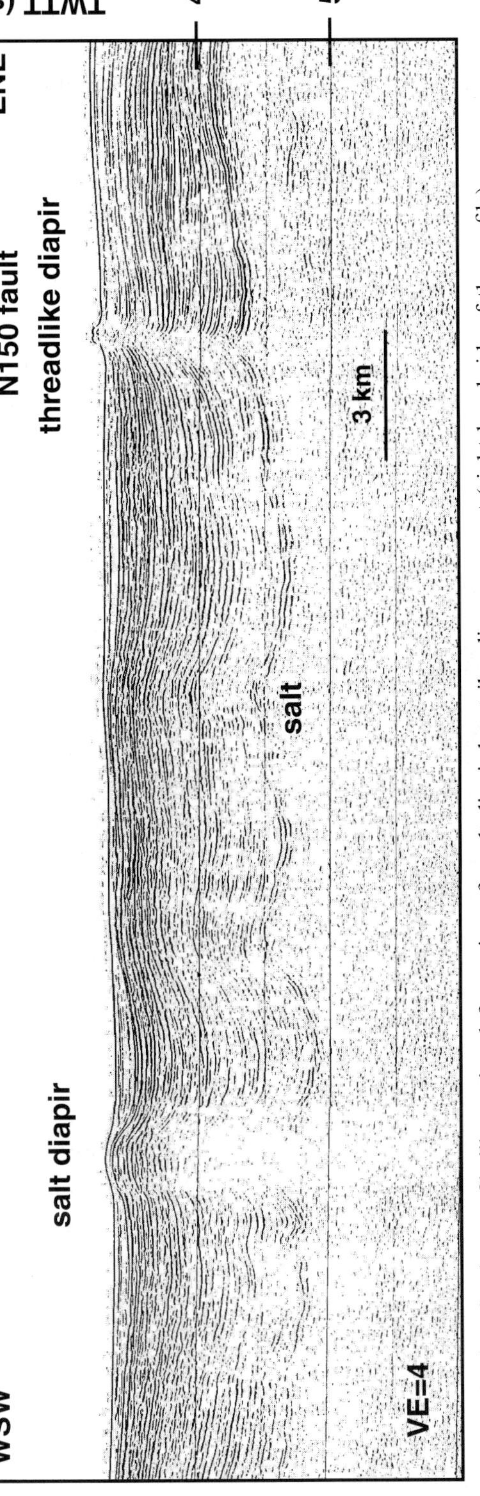

Fig. 9. Seismic reflection profile illustrating deformation of a salt diapir by strike-slip movement (right-hand side of the profile).

be determined whether these NW–SE fault zones initially formed as a radial set of thin-skinned extensional faults that compartmentalized the spreading overburden or whether they formed during a later stage. The narrow width of these structures suggests that they acted, at least during the latest stages, as strike-slip faults bounding corridors within which the overburden was extending and gliding northwestward above the salt layer. Why extension and gravity spreading there are oriented northwestward, and hence oblique to the direction of the slope, remains unknown. Possible explanations include the influence of subsalt structures in the basement or the lack of accommodation space available on the NE side, where the Eratosthenes Seamount acts as a buttress and prevents salt and sediments from spreading northeastward.

The third family of structures is found in the distal part of the eastern province, near the Eratosthenes Seamount. Acoustic imagery and seismic-reflection data (Fig. 10) indicate that small folds and narrow diapirs deform the Plio-Quaternary sediments from the Nile deep-sea fan. The sediment thickness there is *c.* 0.5 s TWTT (i.e. 500 m). This area represents the distalmost part of the prograding wedge, where gravity spreading and gliding shortened and thickened the salt and its overburden.

Finally, the fourth type of structures is found at the easternmost edge of the eastern province, near the Eratosthenes Seamount. The base of the slope is marked by a steep and narrow bathymetric scarp, >400 m high (Fig. 3), that wraps around the base of the Eratosthenes Seamount. Seismic data suggest that this scarp has morphological and structural characteristics similar to those of the Sigsbee Scarp in the northern Gulf of Mexico (Amery 1969; Worral & Snelson 1989; Wu *et al.* 1990; Diegel *et al.* 1996) and corresponds to the present-day limit of the salt basin. There, the salt layer was expelled and inflated by the combined loading and seaward translation of the prograding sediments of the Nile deep-sea fan. The inflating salt layer probably encroached onto the base of the seamount, and hence, the present-day limit of this partly allochthonous salt is higher and farther than the initial, depositional limit of the autochthonous salt basin.

Discussion and conclusions

The recent PRISMED II survey reveals clear evidence of vigorous salt tectonics in and around the Nile deep-sea fan. The salt layer, regardless of its age (Messinian or locally Tortonian), played a critical role in the deformation of the whole Eastern Mediterranean Basin. Moreover, the survey allows the first comprehensive analysis of the distribution of various types of salt structures throughout the area and shows how such structures formed in response to gravity-driven thin-skinned tectonics, geodynamics or a combination of both.

Structures in the Nile deep-sea fan and its surroundings are mainly controlled by three major geological factors. First, the Nile River fed a large supply of sediments that accumulated along the NE African margin and the Levant Basin. Second, the thick layer of weak Messinian evaporites allowed the overlying sediments to move and deform. Third, the collision–subduction regime affecting the Eastern Mediterranean Basin led to shortening in the Mediterranean Ridge and, possibly, subsidence of the Herodotus Abyssal Plain. The interaction between these three factors gives the Nile deep-sea fan and its surroundings their unique geomorphological and structural configuration.

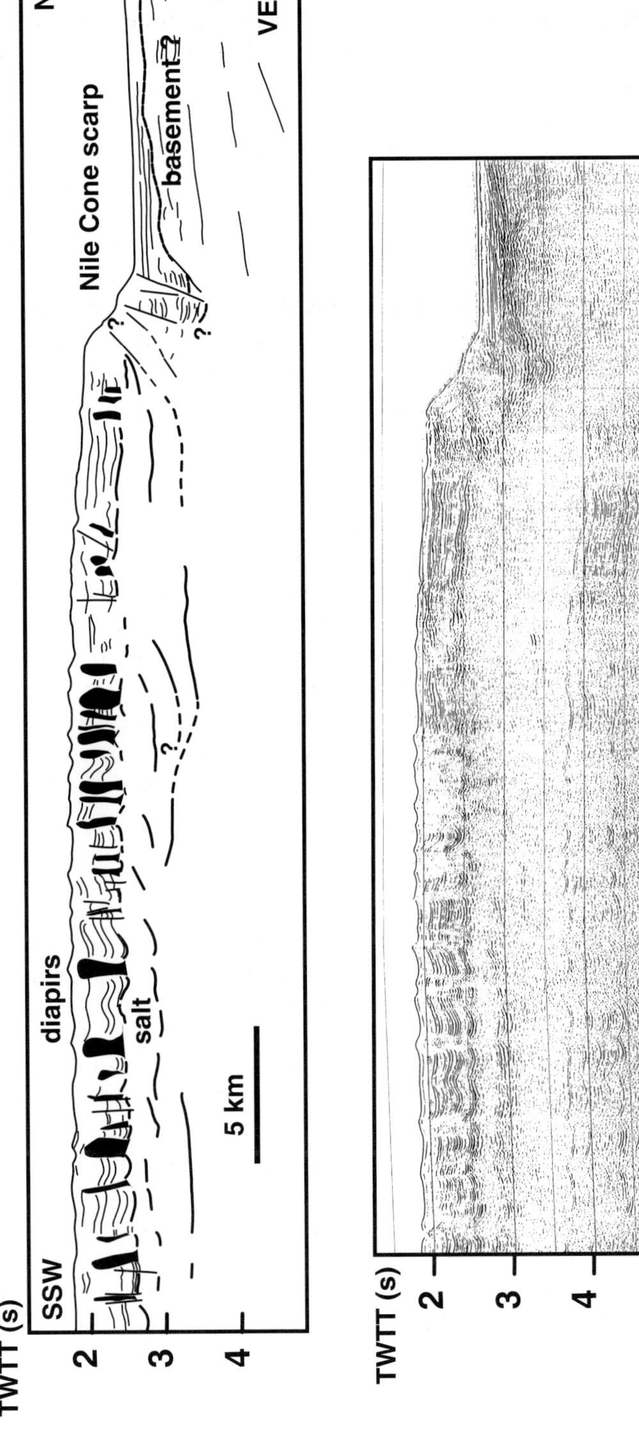

Fig. 10. Seismic reflection profile and line drawing across the prograding front of the Nile deep-sea fan, south of the Eratosthenes Seamount. (See text for details.)

The proximal section of the continental slope comprises listric normal growth faults, grabens, salt rollers and salt walls caused by seaward gravity gliding and spreading of the thick Plio-Quaternary sediment package above the Messinian salt. These features are comparable with faults and diapirs found in many other large deep-sea fans, such as the Niger, Rhône, Amazon and Mississippi deltas, where the rates of sediment deposition and subsidence are high. In the western and central provinces, growth faults are associated with huge debris flows downslope, whose movement was probably triggered or enhanced by flow of the underlying salt.

The distal part of the western Nile province was affected by horizontal contraction in the Mediterranean Ridge area, which acted as an active buttress against which Plio-Quaternary sediments were folded and faulted. The incorporation of Nile sediments into the Mediterranean Ridge has shortened the northwestern part of the Nile deep-sea fan, and the Messinian salt, acting as a lubricant, was intimately involved in this deformation.

Basinward, salt tectonics creates narrow diapirs and tight folds that deform even the sea floor. The easternmost limit of the Nile deep-sea fan corresponds to a scarp that represents the basinward limit of gravity gliding and spreading of the Nile Cone, where the salt has been inflated and the overburden has been shortened, thickened and raised above the regional datum.

A significant sedimentary difference between the western and central provinces and the eastern province of the Nile deep-sea fan is that Plio-Quaternary sediments in the distal part of the eastern province are significantly thinner than in other two provinces. A first hypothesis would be that all three provinces have had a similar supply of sediments provided by the Nile River throughout the delta's history and that all were affected by the same eastward-flowing Mediterranean geostrophic current. If so, the considerably thicker sediment sequence in the western province could be the result of the interference of the faults and the diapirs of the central section with the downslope sediment distribution. By contrast, salt diapirs are more numerous in the eastern than in the western province, even if the salt layer there is thinner in places. On the other hand, although the present-day axial symmetry of the delta suggests that the sedimentary discharge does not currently show any preference between the three provinces, the geographical distribution of sediment accumulation may have varied significantly during the delta's history. Migration of the Nile delta mouth, burial of channel–levees systems and fluctuations of the Mediterranean geostrophic currents during Plio-Quaternary times could have caused large variations in overburden thickness between the three provinces. More detailed studies are necessary for a definite conclusion to be reached.

The three depositional provinces and their distinctly different morphostructural patterns also probably result from two interacting structural processes: subsidence of the Herodotus Abyssal Plain and gravity-driven thin-skinned extension above the Messinian salt layer. The present-day sedimentary distributary system of the Nile deep-sea fan has been determined in detail during the PRISMED II cruise. Feeder channels are widely distributed and exhibit complex meandering patterns. Seismic data have provided evidence of westward channel migration in the western province. This pattern may suggest overall subsidence of the western part of the deep-sea fan and uplift of the eastern part (Bellaiche *et al.* 1999). The central and the western provinces of the Nile deep-sea fan seem similarly constrained because they both receive ample sediment supplies. Nevertheless, these provinces are very different. The western province is transected by several channels, where a major part of the sedimentary transport seems to occur. Such channels are scarce in the central province, where slumps and debris flow are predominant. Furthermore, sinuous faults at mid-slope and salt diapirs associated with them are common in the central province, but are rare in the western province. The distinct difference between the western and central provinces is the way they merge with the Mediterranean Ridge. The transition in the western province is direct, and the distal part of the continental rise, at the edge of the slope, is folded, as a transition to the compressional regime of the ridge. To the west, the Herodotus Abyssal Plain separates the Nile deep-sea fan from the Mediterranean Ridge. There are grounds to presume that a small abyssal plain between a major deep-sea fan and a major accretionary ridge could result from local subsidence. Subsidence could have enhanced the sediment flow rate down the continental slope toward the abyssal plain, thereby favouring formation of channels. The channels, in turn, would have reduced the occurrence of unstable sediment loads along the slope, and hence the occurrence of slumps in the western province.

We are grateful to the crew of the R.V. *l'Atalante* and the technical GENAVIR team. M. G. Rowan and R. Nelson provided constructive comments and careful reviews that helped us improve an earlier version of this manuscript. This paper is Contribution no. 283 of Géosciences-Azur (UMR 6526).

References

AMERY, G. B. 1969. Structure of the Sigsbee scarp, Gulf of Mexico. *AAPG Bulletin*, **53**, 2480–2482.

BELLAICHE, G. & MART, Y. 1995. Morphostructure, growth patterns and tectonic control of the Rhône and Nile deep-sea fans: a comparison. *AAPG Bulletin*, **79**, 259–284.

——, ZITTER, T., DROZ, L., GAULLIER, V., MART, Y., MASCLE, J. & SHIPBOARD SCIENTIFIC PARTY 1999. The Nile Cone: main results of the 'Prismed II' cruise of the R/V 'l'Atalante'. *Comptes Rendus de l'Academie des Sciences, Paris, Serie IIa*, **329**, 727–733.

——, ——, GAULLIER, V., DROZ, L., MART, Y. & Shipboard Scientific Party 1999. Active sedimentary processes on the Nile deep-sea fan. *Terra Abstracts*, **11**, 756.

BEN-AVRAHAM, Z. & MART, Y. 1981. Late Tertiary structure and stratigraphy of North Sinai continental margin. *AAPG Bulletin*, **65**, 1135–1145.

CHUMAKOV, I. S. 1967. Pliocene and Pleistocene deposits of the Nile Valley in Nubia and Upper Egypt. *Trudy Instituta Geologichekikh Nauk, Akademiya SSSR*, **170**, 1–110 [in Russian].

CLOOS, E. 1955. Experimental analysis of fracture patterns. *Geological Society of America Bulletin*, **66**, 241–256.

—— 1968. Experimental analysis of Gulf Coast fracture patterns. *AAPG Bulletin*, **52**, 420–444.

DAMUTH J. E. & EMBLEY, R. W. 1981. Mass-transport processes on Amazon Cone: western equatorial Atlantic. *AAPG Bulletin*, **65**, 629–643.

DIEGEL, F. A., KARLO, J. F., SCHUSTER, D. C., SHOUP, R. C. & TAUVERS, P. R. 1996. Cenozoic structural evolution and tectono-stratigraphic framework of the northern Gulf Coast continental margin. *In*: JACKSON, M. P. A., ROBERTS, D. G. & SNELSON, S. (eds) *Salt Tectonics: a Global Perspective*. *AAPG* Memoirs, **65**, 109–151.

EMEIS, K. C., ROBERTSON, A. H. F., RICHTER, C. *et al.* 1996. *Proceedings of the Ocean Drilling Program, Initial Reports*, 160. College Station, TX, Ocean Drilling Program.

GAULLIER, V. 1993. *Diapirisme salifère et dynamique sédimentaire dans le Bassin Liguro-Provençal: données sismiques et modèles analogiques*. Thèse de Doctorat de l'Université Paris VI.

—— & BELLAICHE, G. 1996. Diapirisme liguro-provençal: les effets d'une topographie résiduelle sous le sel messinien. Apports de la modélisation analogique. *Comptes Rendus de l'Académie des Sciences, Paris, série IIa*, **322**, 213–220.

——, —— 1998. Near-bottom sedimentation processes revealed by echo-character mapping studies, Northwestern Mediterranean Basin: *AAPG Bulletin*, **82**, 1140–1155.

——, BRUN, J. P., GUÉRIN, G. & LECANU, H. 1993. Raft tectonics: the effects of residual topography below a salt décollement. *Tectonophysics*, **228**(3–4), 363–381.

——, MART, Y., BELLAICHE, G., MASCLE, J. & Shipboard Scientific Party 1999. Salt response to the Nile deep-sea fan thin-skinned tectonics and to the convergent geodynamics in the Eastern Mediterranean Sea. *Terra Abstracts*, **11**, 760.

HSÜ, K. J., MONTADERT, L. *et al.* 1978. *Initial Reports of the Deep Sea Drilling Project, 42(1)*. Washington, DC: US Government Printing Office.

JACKSON, M. P. A. & VENDEVILLE, B. C. 1992. Regional extension as a geologic trigger for diapirism. *Geological Society of America Bulletin*, **106**, 57–73.

KEMPLER, D. & GARFUNKEL, Z. 1994. Structures and kinematics in the northeastern Mediterranean: a study of an irregular plate boundary. *Tectonophysics*, **234**, 19–32.

——, MART, Y., HERUT, B. & MCCOY, F. 1996. Diapiric structures in the southeastern Mediterranean Sea: possible indication of extension in a zone of incipient collision. *Marine Geology*, **134**, 237–248.

KENYON, N.H., STRIDE, A. H. & BELDERSON, R. H. 1975. Plan view of active faults and other features on the lower Nile Cone. *Geological Society of America Bulletin*, **86**, 1733–1739.

LE PICHON, X. & GAULIER, J. M. 1988. The rotation of Arabia and the Levant fault system. *Tectonophysics*, **153**, 271–294.

MALDONADO, A. & STANLEY, D. J. 1978. Nile cone depositional processes and patterns in the late Quaternary. *In*: STANLEY, D. J. & KELLING, G. (eds) *Canyons, Fans and Trenches*. Stroudsburg, PA: Dowden, Hutchinson and Ross, 239–257.

MART, Y. & ROBERTSON, A. H. F. 1998. Eratosthenes Seamount: an oceanographic yardstick recording the late Mesozoic–Tertiary geological history of the eastern Mediterranean. *Proceedings of the Ocean Drilling Program, Scientific Results*, 160. College Station, TX: Ocean Drilling Program, 701–708.

——, EISIN, B. & FOLKMAN, Y. 1978. The Palmahim structure—a model of continuous tectonic activity since the upper Miocene in the southeastern Mediterranean off Israel. *Earth and Planetary Science Letters*, **39**, 328–334.

——, ROBERTSON, A. H. F., WOODSIDE, J. M. & ODP Leg 160 Shipboard Science Party 1997. Cretaceous tectonic setting of Eratosthenes Seamount in the eastern Mediterranean Neotethys: initial results of ODP Leg 160. *Comptes Rendus de l'Académie des Sciences, Paris, série IIa*, **423**, 127–134.

MASCLE, J. & Shipboard Scientific Party 1998. New data in Eastern Mediterranean from the PRISMED II survey (R/V *Atalante*). *Annales Geophysicae*, **16**(Supplement I), C305.

——, ZITTER, T., BELLAICHE, G., DROZ, L., GAULLIER, V., MART, Y. & PRISMED II Scientific Party 2000. The Nile deep sea fan: preliminary results from a swath bathymetry survey. *Marine and Petroleum Geology*, in press.

MCKENZIE, D. P. 1978. Active tectonics of the Alpine–Himalayan belt, the Aegean Sea & surrounding regions. *Geophysical Journal of the Royal Astronomical Society*, **55**, 217–254.

NEEV, D., ALMAGOR, G., ARAD, A., GINZBURG, A. & HALL, J. K. 1976. The geology of the southeastern Mediterranean. *Geological Survey of Israel Bulletin*, **68**, 1–51.

NILSEN, K. T., VENDEVILLE, B. C. & JOHANSEN, J. T. 1996. Influence of regional tectonics on halokinesis in the Nordkapp Basin, Barents Sea. *In*: JACKSON, M. P. A., ROBERTS, D. G. & SNELSON, S. (eds) *Salt Tectonics: a Global Perspective*. *AAPG* Memoirs, **65**, 413–436.

NUR, A. & BEN-AVRAHAM, Z. 1978. The eastern Mediterranean and the Levant: tectonics of continental collision. *Tectonophysics*, **46**, 297–311.

PAUTOT, G., LE CANN, C., COUTELLE, A. & MART, Y. 1984. Morphology and extension of the evaporitic structures of the Liguro–Provençal Basin: new sea-beam data. *Marine Geology*, **55**, 387–409.

REILINGER, R. E., MACCLUSKY, S. C., ORAL, M. B., KING, R. W., TOKSOZ, M. N., BARKA, A. A. & KINIK, W. L. 1997. GPS deformation in the East Mediterranean. *Journal of Geophysical Research*, **102**, B5.

ROSS, D. A. & UCHUPI, E. 1977. The structure and sedimentary history of the southeastern Mediterranean Sea. *AAPG Bulletin*, **61**, 872–902.

——, ——, Summerhayes, C. P. & Koelsch, D. E. 1978. Sedimentation and structure of the Nile Cone and Levant platform area. *In*: Stanley D. J. & Kelling, G. (eds) *Sedimentation in Submarine Canyons, Fans and Trenches*. Stroudsburg, PA: Dowden, Hutchinson and Ross, 261–274.

Ryan, W. B. F. 1978. Messinian badlands on southeastern margin of the Mediterranean Sea. *Marine Geology*, **27**, 349–363.

——, Hsü, K. J. *et al.* 1973. *Initial Reports of the Deep Sea Drilling Project, 13*. Washington, DC: US Government Printing Office.

——, Stanley, D. J., Hersey, J. B., Fahlquist, D. A. & Allan, T. D. 1970. The tectonics and geology of the Mediterranean Sea, in Maxwell, A. E. (ed.) *The Sea, 4-II*. New York: Wiley, 387–492.

Said, R. 1981. *The Geological Evolution of the River Nile*. New York: Springer.

Tibor, G., Ben-Avraham, Z., Steckler, M. & Fligelman, H. 1992. Late Tertiary subsidence history of the southern Levant margin, Eastern Mediterranean Sea, and its implications to the understanding of the Messinian event. *Journal of Geophysical Research*, **97**, 17593–17614.

Vendeville, B. C. 1987. *Champs de failles et tectonique en extension: modélisation expérimentale*. Mémoires et Documents du Centre Armoricain, d'Etudes Structurales des Socies (CAESS), Rennes, **15**.

—— & Cobbold, P. R. 1987. Syn-sedimentary gravitational sliding and listric normal growth faults: insights from scaled physical models. *Comptes Rendus de l'Académie des Sciences, Paris, série IIa*, **305**, 1313–1319.

—— & Jackson, M. P. A. 1992*a*. The rise of diapirs during thin-skinned extension. *Marine and Petroleum Geology*, **9**(4), 331–353.

—— & —— 1992*b*. The fall of diapirs during thin-skinned extension. *Marine and Petroleum Geology*, **9**(4), 354–371.

—— & Nilsen, K. T. 1995. Episodic growth of salt diapirs driven by horizontal shortening. *In*: Travis, C. J., Vendeville, B. C., Harrison, H., Peel, F. J., Hudec, M. R. & Perkins, B. F. (eds) *Salt, Sediment, and Hydrocarbons*. Society of Economic Paleontologists and Mineralogists, Gulf Coast Section, 16th Annual Research Conference Program and Extended Abstracts, 285–295.

Venkatarathnam, K., Biscaye, P. E. & Ryan, W. B. F. 1972. Origin and dispersal of Holocene sediments in the Eastern Mediterranean Sea. *In*: Stanley, D. J. (ed.) *The Mediterranean Sea*. Stroudsburg, PA: Dowden, Hutchinson and Ross, 455–469.

Woodside, J. M. 1977. Tectonic elements and crust of the Eastern Mediterranean Sea. *Marine Geophysical Research*, **3**, 317–354.

Worral, D. M. & Snelson, S. 1989. Evolution of the northern Gulf of Mexico, with emphasis on Cenozoic growth faulting and the role of salt. *In*: Bally, A. W. & Palmer, A. R. (eds) *The Geology of North America: an Overview*. Geological Society of America Decade of North American Geology, **A**, 97–138.

Wu, S. A., Bally, A. W. & Cramez, C. 1990. Allochthonous salt, structure and stratigraphy of the northeastern Gulf of Mexico, part II: structure. *Marine and Petroleum Geology*, **7**, 334–370.

Influence of extension and compression on salt diapirism in its type area, East Carpathians Bend area, Romania

MIHAI STEFANESCU[1], OPREA DICEA[2], & GABOR TARI[3]

[1]*Forest Romania Corporation, Sos. Colentina 27 A, 72244 Bucharest, Romania*

[2]*Prospectiuni S.A., 1 Coralilor Str., Bucharest 1, Romania*

[3]*Vanco Energy, One Greenway Plaza, Houston, TX 77046, USA*

Abstract: The East Carpathians Bend area has a very complex structure characterized by the presence of nappes, their post-tectonic cover and salt diapirs. The salt forming the studied diapirs is Early Miocene (Burdigalian) in age. After its accumulation the salt was more or less continuously involved in alternating extensional and compressional stages that deformed it from its original tabular position to the present-day diapir. Five stages of salt deformation have been established: initial, pre-nappe emplacement, nappe emplacement, post-nappe emplacement and Wallachian. During all of these stages the salt was configured into different shapes: it formed a truncated cone during the initial stage, a mushroom head during the pre-nappe emplacement stage, and an increasingly more tapered shape with nappe emplacement and during the post-nappe emplacement stages. Finally, it was squeezed out and refashioned by strike-slip faulting during the Wallachian compressional stage of Pleistocene age.

In the East Carpathians Bend area the numerous salt exposures as well as salt-water springs reveal the presence of subsurface salt bodies. Historically, the salt has been of major economic importance. However, scientific interest regarding the geometry of the salt became important at the beginning of this century when the connection was discovered between the salt bodies and hydrocarbon fields. Thus, during the first decade of this century important papers on salt geometry and its kinematics were published. In the 'pioneering era' *sensu* Jackson (1996), Mrazec (1907) introduced the term 'diapir' into the geological literature for all types of piercement structures related to salt, clay and magmatic rocks.

Scientific attention given to salt slowly shifted from the study of surface exposures to the subsurface part of the salt 'icebergs', especially because of oil industry interest. Initially well data and, later, gravity and seismic reflection information contributed to the imaging of the salt geometry. At first, the gravimetric approach was used successfully to determine the presence of subsurface salt, its general outline and the area of its maximum thickness. Later, seismic-reflection research substantially improved the image of salt configuration at depth, especially in defining the top of the salt and its diapir limb geometry. Unfortunately, because of rugged

From: VENDEVILLE, B., MART, Y. & VIGNERESSE, J.-L. (eds) *Salt, Shale and Igneous Diapirs in and around Europe*. Geological Society, London, Special Publications, **174**, 131–147. 1-86239-066-5/00/$15.00

topography and potentially very high cost, 3D seismic-reflection data have not been acquired in the East Carpathians Bend area to date. Therefore the present paper is based on the interpretation of selected 2D seismic-reflection profiles. The lines presented were recorded and processed in the 1980s by Prospectiuni S.A., Romania.

Generally, as a result of its plastic character and unusual deformational characteristics, salt makes the geologist's life more challenging and hydrocarbon exploration more expensive. This statement also applies to the East Carpathians Bend Zone, but here the situation is even more complicated. In this area the surface geology clearly demonstrated (Olteanu 1951; Popescu 1951) that the salt formations (massive salt bodies usually associated with sedimentary breccias) accumulated at two different stratigraphic levels of the Lower (Burdigalian) and the Middle Miocene (Badenian) succession. This observation complicated the subsurface interpretation until Albu & Baltes (1981) demonstrated that the subsurface salt in the diapiric fold zone is only Early Miocene (Burdigalian) in age. Diapiric salt developed along four lineaments from north to south: overturned, exaggerated, attenuated and cryptodiapirs (Atanasiu 1948). This ranking of diapirs is still used in spite of the fact that seismic data showed it does not match the real subsurface salt geometries. The seismic lines presented in Fig. 1 cross all the above-mentioned lineaments as follows: Line A crosses the overturned diapir, Line B crosses the exaggerated and attenuated diapirs, Line C crosses an attenuated diapir, and Line E crosses the attenuated and the cryptodiapirs. Strike Line E was selected to offer the opportunity to picture an 'attenuated' diapir in 3D.

The present paper focuses primarily on the diapirs in the East Carpathians Bend. The salt-related geology of the surrounding regions is also briefly discussed.

Geological framework

The western half of the East Carpathians Bend Zone (Fig. 1) shows a complex structure in which the external wedge of the Carpathian thrust–fold belt, its foreland the Moesian Platform, and their mutual post-tectonic cover are involved (Fig. 2). A short stratigraphic description of the units imaged on the seismic-reflection sections is presented below (Fig. 3).

Moesian Platform

The stratigraphic data from the Moesian Platform came from a large number of industry wells. On the basis of these data, Paraschiv (1979) published the best documented synthesis on the Moesian Platform. From these data only those for the area located just south of the study area were considered, because they may be extrapolated downdip below the Carpathian overthrust.

Late Jurassic time is represented by a calcareous sequence composed of poorly bedded neritic limestones, locally in reef facies, and dolomites. The thickness of this unit varies between 200 and 700 m.

Stratigraphic relations in Upper Jurassic–Cretaceous strata are not uniform over the foreland. The lack of Tithonian strata in the eastern Moesian Platform indicates a break of sedimentation at the Jurassic–Cretaceous boundary. In the study area, the seismic sections (Figs 4 and 5) show a low-angle unconformity at this boundary. The

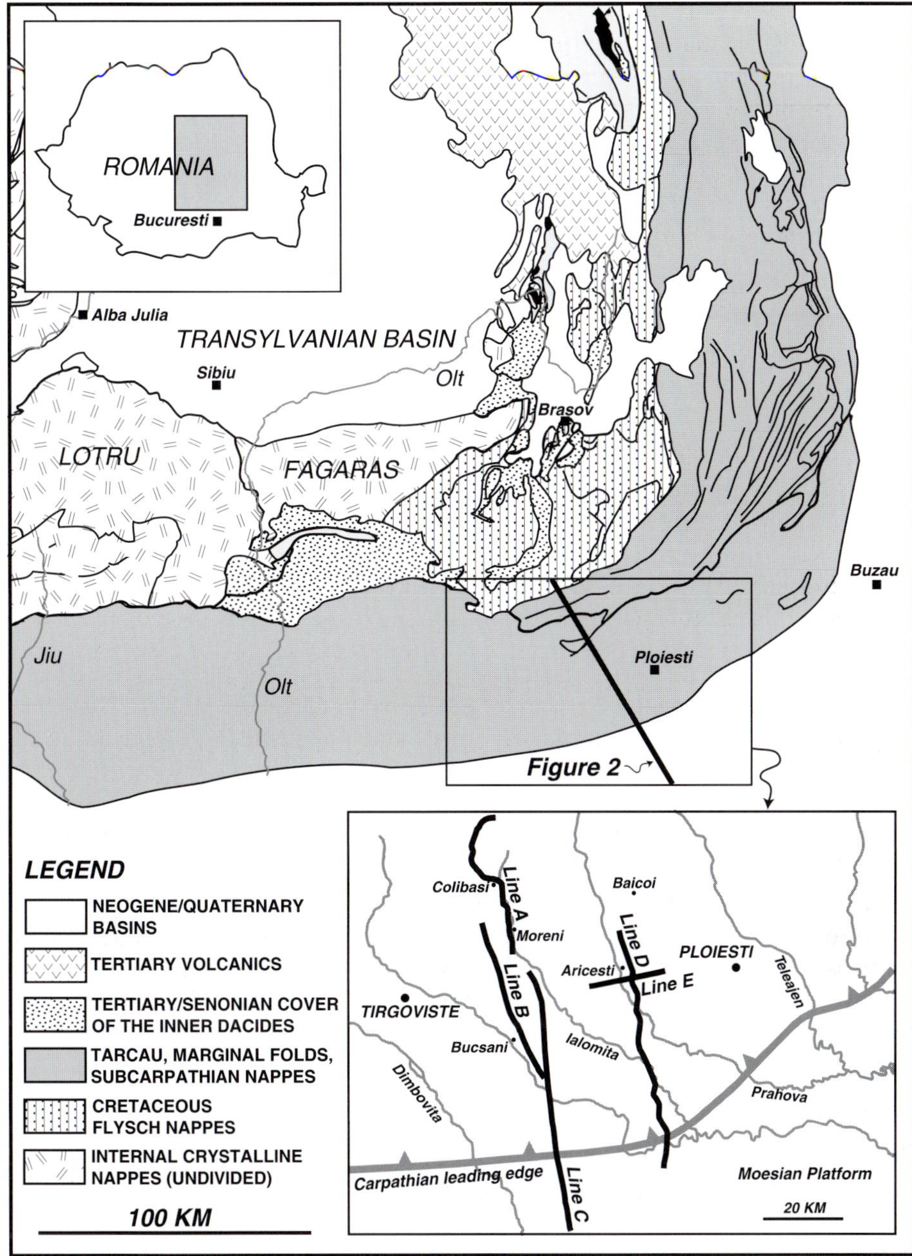

Fig. 1. Location of the study area within the Carpathian Bend area and index map of the seismic lines shown in Figs 4–8.

Cretaceous sequence begins with limestones, more or less massive, and reef-like in places (Lower Cretaceous units). After a break during Aptian time, sedimentation resumed with glauconitic calcareous sandstones, marls and locally with bedded limestones (Middle–Upper Cretaceous units). The thickness of the entire Cretaceous sequence reaches a maximum of 1200 m.

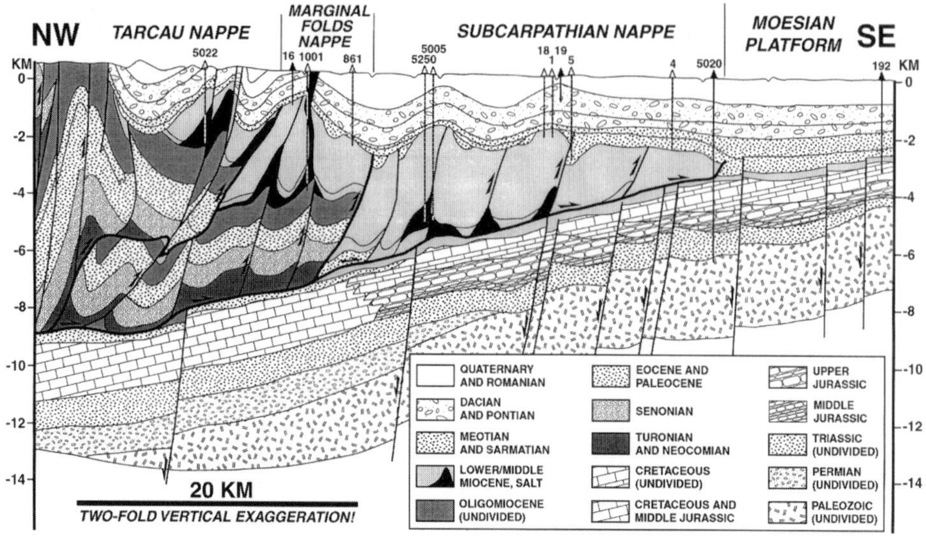

Fig. 2. Regional structure transect through the Carpathian Bend area adapted from Stefanescu *et al.* (1985). (For location see Fig. 1.)

After the latest Cretaceous (Senonian) time, the Moesian Platform experienced regional uplift and subaerial exposure, which lasted until Middle Miocene (Badenian) time, with the exception of some small areas containing Middle Eocene sediments. During Badenian time, several metre-thick thin-bedded neritic limestones, calcareous sandstones, siltstones and marls were deposited.

The Sarmatian (Upper Miocene) deposits within the Carpathian sedimentary wedge can be divided into three distinct subunits, labelled from bottom to top sequences A, B and C (Figs 5 and 6). In this section of the paper only the Sarmatian A unit is described, because this is the only part of the Sarmatian sequence that unconformably overlies the foreland and does not seal the leading edge of the Carpathians. The Sarmatian A succession is made up of siltstones, marls, sands and sandstones, and its thickness varies from a few metres to 250 m towards the Carpathian front.

Outer edge of the Carpathian thrust–fold belt

The outer edge of the Carpathian thrust–fold belt has a very complex internal structure involving Cretaceous to Middle Miocene (Badenian) strata. The stratigraphic description presented below is limited to only those interpreted on the seismic reflection sections as well as those that typically underlie the salt (Fig. 3).

Two different interfingering facies developed during the Late Oligocene–Early Miocene (Burdigalian) time (Grigoras 1955; Stefanescu *et al.* 1993). The southern, external facies consists of up to 1000 m thick of black, black–brown, bituminous shales with interlayers of cherts, siliceous sandstones, grey marls that alternate with calcareous sandstones, dacitic tuffs, siltstones and locally coarse sandstones. This type of section is known as ‘Bituminous facies with Kliwa Sandstones’. Further to the north and west, the internal ‘Puciosa facies with Fusaru Sandstones’ replaces it

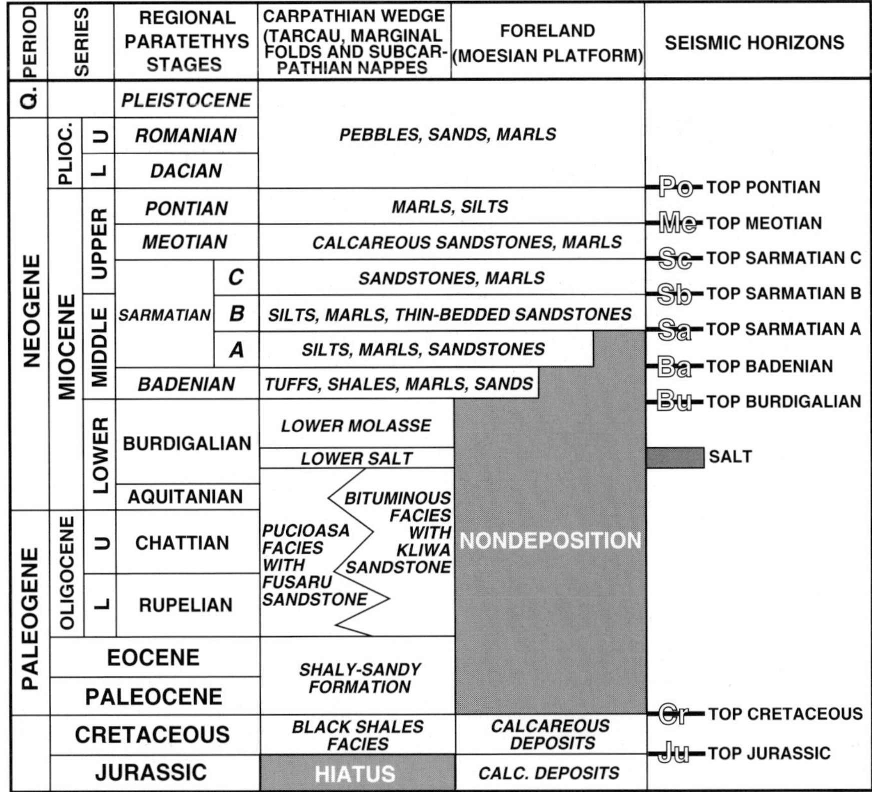

Fig. 3. Simplified stratigraphic column and legend for the seismic illustrations shown in Figs 4–8.

and has a thickness of 1200 m. The dominant lithology of the Puciosa facies with Fusaru Sandstones consists of grey marls and occasional black clays with interlayers of micaceous sandstones, centimetres to metres thick. The lower and upper parts are represented by packages of bituminous shales.

The above-described successions are normally covered by a complex sequence of the Lower Miocene (Burdigalian) Lower Salt Formation. This sequence is composed of more or less massive salt bodies associated with sedimentary breccias, which occasionally replace the salt completely. It should be noted that the salt has clay and/or breccia intercalations that are expressed as internal reflectors on seismic-reflection sections.

In a normal sequence the salt is unconformably overlain by a thick (1800–2300 m) molasse sequence of Early to Mid-Miocene (Burdigalian–Badenian) age that consists of massive polymictic sandstones and/or conglomerates, marls and clays alternating with medium- to thin-bedded micaceous sandstones, gypsum layers (up to 20 m thick) and tuffs, and locally contains thin stromatolitic calcareous shales.

Middle Miocene (Badenian) strata form a characteristic succession that begins with a package of white–greenish dacitic tuffs and white Globigerina marls covered by sedimentary breccia or pebbly mudstones associated with salt (Upper Salt Formation on the outcropping Tarcau nappe), black, bituminous radiolarian shales, grey marls and locally sandstones. The maximum thickness of this sequence reaches 500 m. The

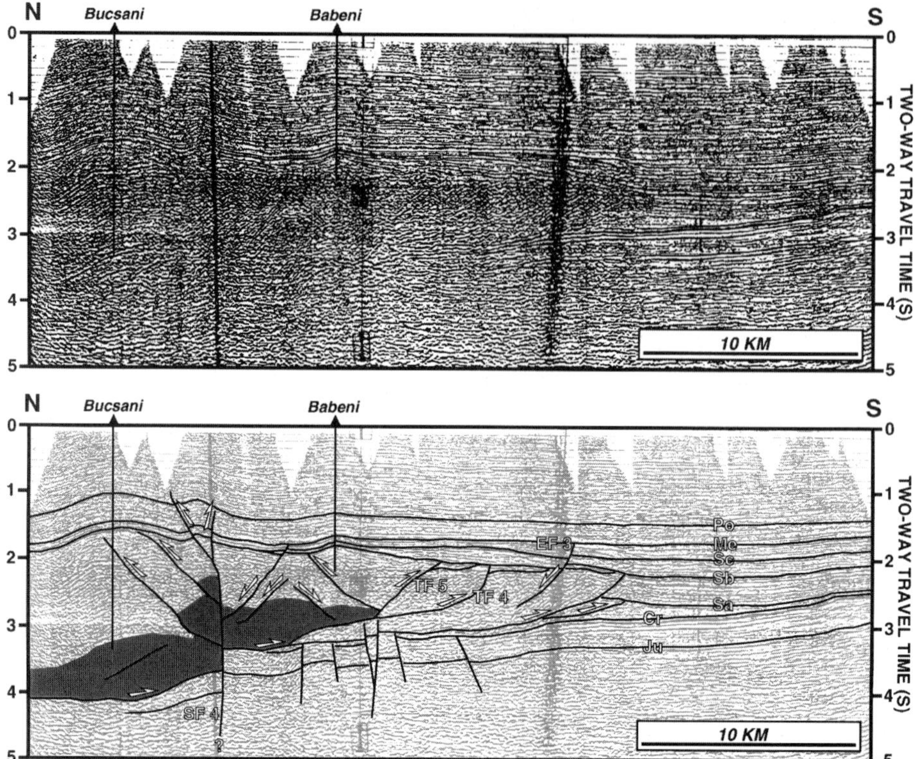

Fig. 4. Seismic-reflection section (line C) across the Bucsani and Babeni structures (for location, see Fig. 1; for legend, see Fig. 3). EF, normal fault; SF, strike-slip fault; TF, thrus fault.

Mid-Miocene (Badenian) age of the Upper Salt Formation was very well documented (Olteanu 1951; Popescu 1951) using surface data. Since then, the age of the subsurface salt bodies existing in the Diapiric Fold Zone was much debated, until it was demonstrated (Albu & Baltes 1981) that all the salt is of Early Miocene (Burdigalian) age. This point will be revisited below.

Sarmatian A strata are involved in the nappe structure of the Carpathian wedge, especially in the larger and deeper synclines. It consists of marls and thin-bedded sandstones. Because of insufficient well data in the study area these strata on the interpreted seismic lines have not been correlated for this paper, except in the outermost part of the Subcarpathian nappe.

Post-tectonic cover

The post-tectonic cover units of Upper Miocene–Pliocene formations include the Sarmatian B and C units covering the frontal thrusts of the Carpathian wedge as well as the Meotian, Pontian and Dacian units that are obviously common (Figs 4 and 5) to the thrust–fold belt and the foreland.

The Sarmatian B unit, with an average thickness of 350 m, uniformly consists mostly of marls and thin- to medium-bedded sandstones. The Sarmatian C unit is about 300 m thick and is sandier (medium- to thick-bedded sands and sandstones),

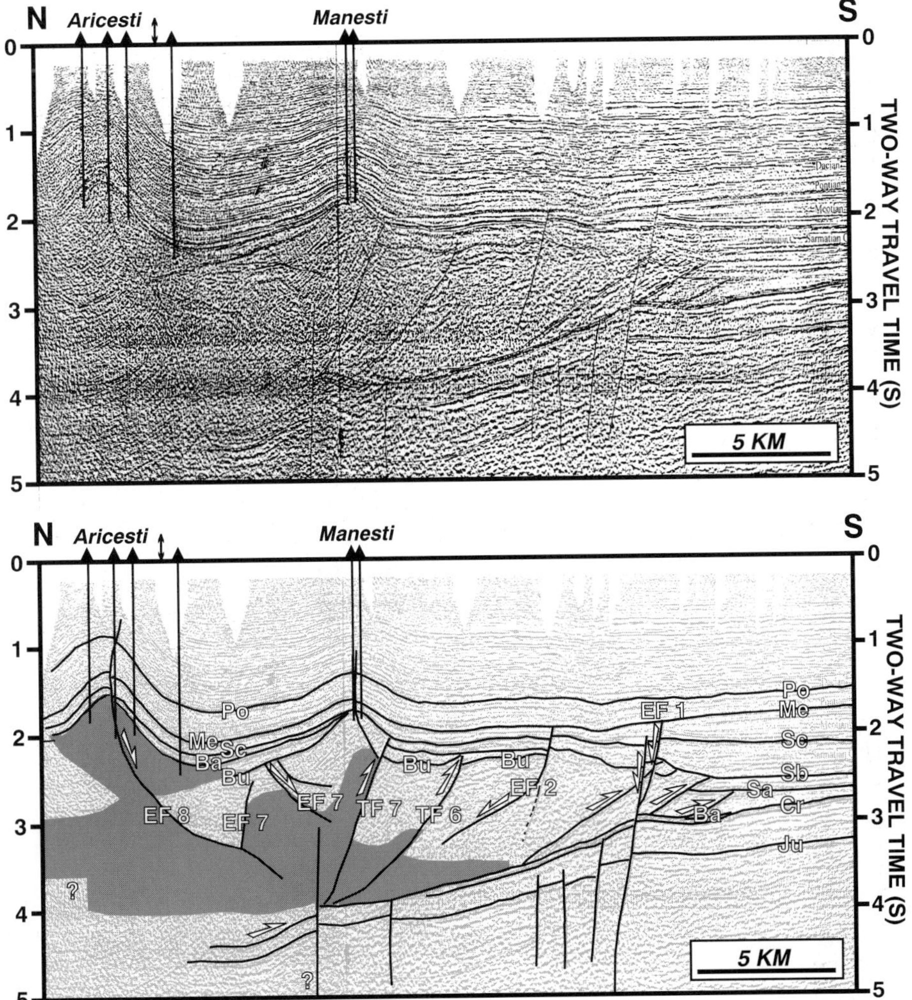

Fig. 5. Seismic-reflection section (line D) across the Aricesti and Manesti diapirs (for location, see Fig. 1; for legend, see Fig. 3).

and towards its upper part contains a sequence of calcareous sandstones and limestones.

Meotian strata are unevenly developed, thicker to the south (300 m) and thinner to the north (50 m). They consist of calcareous sandstones, sand and sandstones, silts and marls. Pontian strata consist of a very monotonous sequence of marls and silts, 400–500 m thick. Occasionally they contain lens-like sand bodies. Overlying Dacian–Quaternary strata are composed of >1000 m of sands, silts, marls, coal beds and pebbles.

Tectonic position of the diapirs

The inner structure of the outer sedimentary wedge in the Carpathian Bend Zone is characterized by the presence of three distinct nappes: from internal to

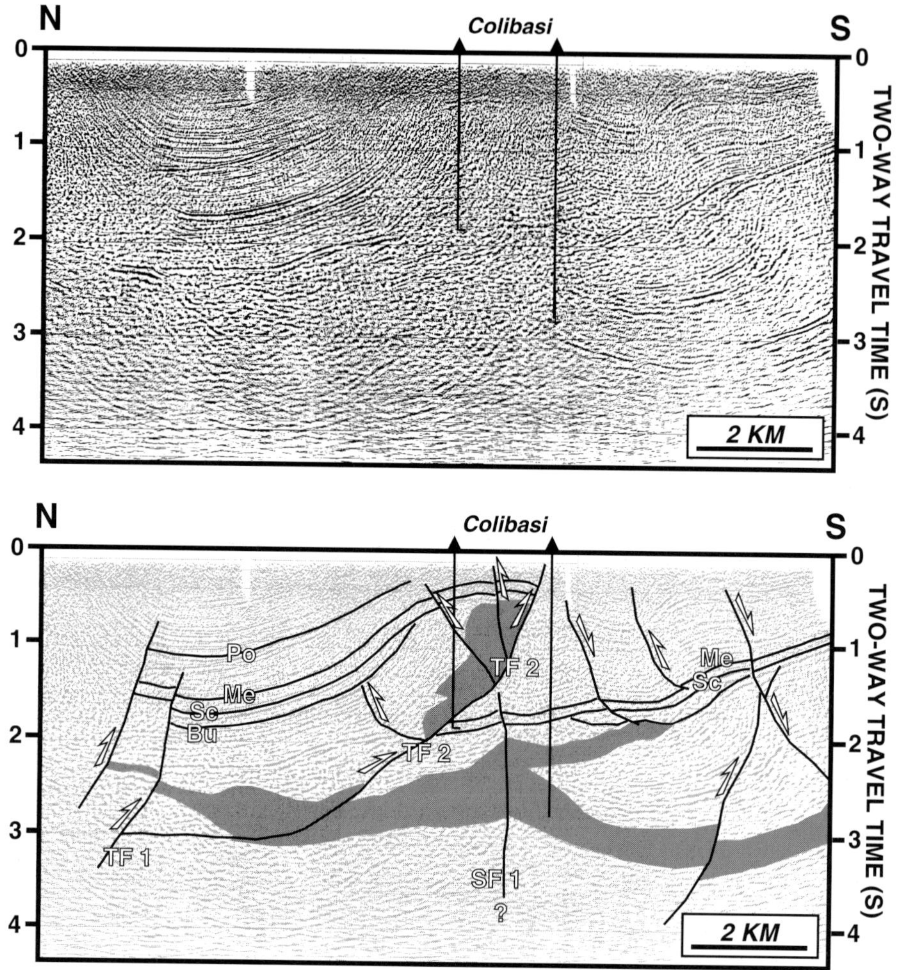

Fig. 6. Seismic-reflection section (line A) across the Colibasi diapir (for location, see Fig. 1; for legend, see Fig. 3).

external, the Tarcau, Marginal Folds and Subcarpathian nappes. In the study area only two, the Tarcau and the Subcarpathian nappes, crop out. Thick post-tectonic deposits cover the tectonic contact between these two nappes, making it very difficult to locate the position of the contact between them and, consequently, to determine to which of these nappes the salt diapirs belong. Only the Oligocene–Lower Miocene facies distribution helps resolve these problems. The Oligocene–Lower Miocene deposits in a facies characteristic for the outermost part of the Tarcau Nappe certainly (proved by wells) developed to the position of the Moreni diapir. By inference, the leading edge of the Tarcau Nappe extends farther to the south and is located either below or just south of the Moreni diapir lineament. Consequently, the Colibasi and Moreni diapirs are developed within the Tarcau Nappe, whereas the more external ones are located within the Subcarpathian Nappe.

Tectonic evolution

Following the Cimmerian orogeny, the Moesian Platform was uplifted and subaerially exposed during Early Jurassic time. During Late Jurassic and Early Cretaceous time the platform formed a north-facing Tethyan passive margin, which continuously subsided and in which marine deposits accumulated. During Aptian time, a short period of uplift occurred, then the platform subsided until the end of Cretaceous time when the platform area was uplifted and remained subaerially exposed until Mid-Miocene time.

The subsidence history of the external wedge of the Carpathians (outer part of the flysch basin) is more difficult to reconstruct because some strata have been removed along the overthrust planes. However, well data west of the study area indicate that the oldest deposits present in the area are Early Cretaceous in age. Combining the well and surface data one can infer that, beginning in Early Cretaceous time, a deep marine basin with flysch deposits continuously subsided until Early Miocene (Burdigalian) time. The subsidence rate gradually decreased and, consequently, the water depth also decreased. Eventually salt deposition occurred in an extremely shallow basin. After a short phase of deformation, subsidence resumed and Lower Molasse (Lower–Middle Miocene) strata weredeposited in an extensional setting. Basin water depths remained shallow as only molasse deposits accumulated until Late Miocene (Sarmatian) time.

During most of the Badenian and Sarmatian A time an important deformational stage took place when the outer Carpathians were overthrust onto the foreland. One well, located westwards of the study area, passed through the allochthonous Carpathian units and penetrated into the autochthonous platform, proving an overthrust of *c.* 20 km. Seismic lines B and D (Figs 5 and 7) show a horizontal displacement of at least 25 km. On the regional geological cross-sections (e.g. Section A-20, Stefanescu *et al.* 1985) the amount of horizontal displacement was interpreted to be a minimum of 100 km. According to the results of a 2D balanced cross-section, drawn to the NE of the study area, the post-Oligocene shortening is 130 km (Roure *et al.* 1993). Consequently, it may be inferred that, after its deposition, the salt was transported, with any other deposits belonging to the nappe bodies, for a distance of >100 km. After this important shortening episode the independent evolution of the platform and outer Carpathians ended.

Following the emplacement (after Sarmatian B time) of the Subcarpathian Nappe, the thrust–fold belt and its post-tectonic cover were deformed together by younger tectonic events. During post-thrusting deformation the leading edge of the nappe was no longer displaced. It remained sealed by the post-tectonic sequences (Figs 4 and 5). Younger tectonic events formed out-of-sequence thrust faulting west of the leading edge of the Subcarpathian Nappe (Stefanescu & Dicea 1995).

On seismic lines C and D normal faults (EF) may be seen (Figs 4 and 5). They affect all the deposits from Early Miocene to Meotian age and are sealed by Pontian sediments (EF 1 and 2 in Fig. 5 and EF 3 in Fig. 4). One of these faults, EF 1 is deeply rooted into the foreland (Fig. 5). In contrast to other coeval normal faults, EF 1 cuts the sole thrust. Other faults that intersect and displace the sole thrust (SF 2 and 3 in Fig. 7) have a character typical of a strike-slip fault (SF). The strike-slip faults are younger than the pre-Pontian EF 1 fault and were active until Late Pliocene time (SF 2 and 3 in Fig. 7).

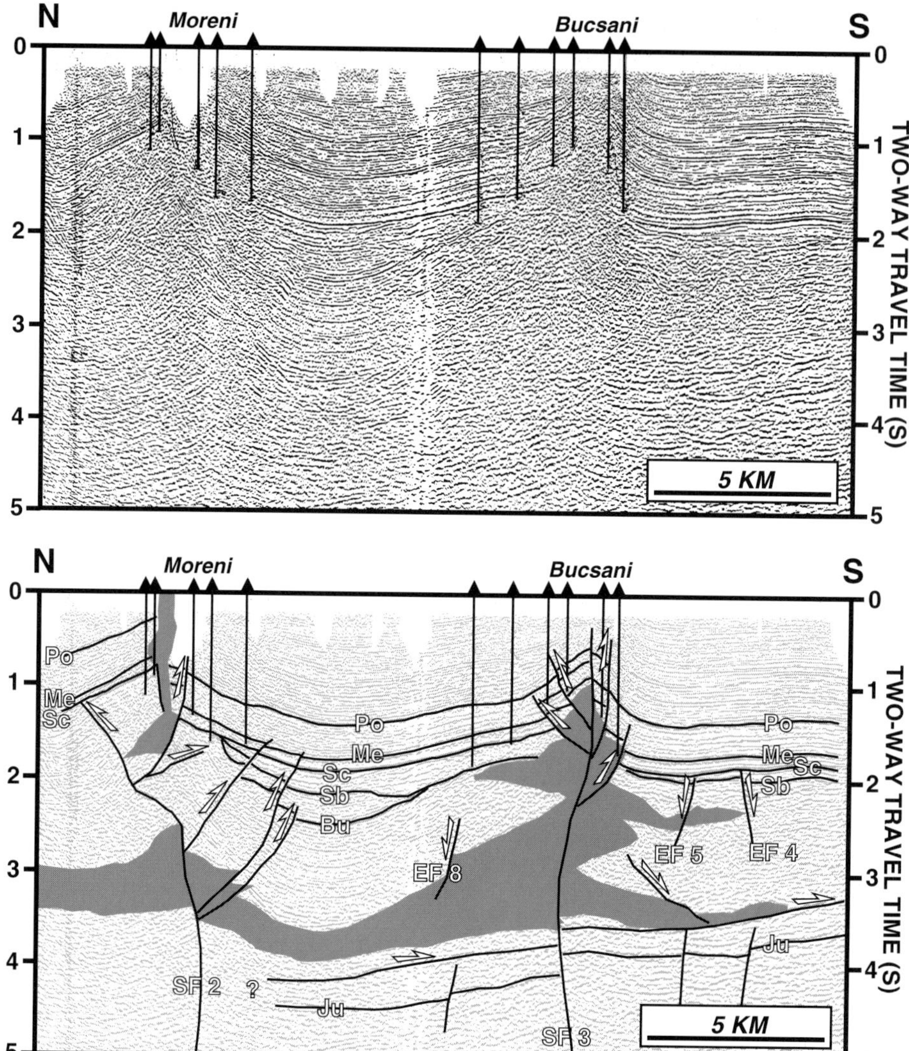

Fig. 7. Seismic-reflection section (line B) across the Moreni and Bucsani (western part) Diapirs (for location, see Fig. 1; for legend, see Fig. 2).

The study area is located in the western segment of the East Carpathians Bend area, west of which the Getic Depression was developed. The seismic lines C and D display a different tectonic aspect of the Carpathian wedge. On Line C (Fig. 4) the structural geometry is similar to that in the Getic Depression with a thin and horizontal Subcarpathian Nappe, whereas on Line D (Fig. 5) a thicker and north-dipping Subcarpathian Nappe is similar to the structural style in the East Carpathian Bend. It should be noted that the diapirs are located in the transitional zone between the Getic Depression and the East Carpathians, where the Subcarpathian Nappe becomes progressively deeper from the west towards the east and northeast.

Evolution of salt diapirism

When Mrazec (1907) described the diapirs from the East Carpathian Bend, he considered that the horizontal compressional forces were the driving mechanism that squeezed up the plastic salt into the anticline axes. The seismic information illuminates a more complex salt history controlled by the evolution of the outer sedimentary wedge of the Carpathians. The salt diapirs evolved in successive stages (from oldest to youngest): accumulation stage, initial stage, pre-nappe emplacement stage, nappe emplacement stage, post-nappe emplacement stage and Wallachian stage.

Accumulation stage

During Late Oligocene–Early Miocene time, the Tarcau Nappe was overthrust by the more internal Cretaceous flysch nappes (Figs 1 and 2). As a result of these movements a shallow basin was formed, in front of the thrust sheet in which the Lower Miocene (Burdigalian) salt accumulated. The topographic relief at the inner basin margin was pronounced enough to cause terrigeneous material and sedimentary breccias to be shed into the salt basin.

Initial stage

Normal faults affecting the salt and the early molasse deposits are visible on the seismic lines C and D, which clearly show the seismic character of the extensional faults (see Diegel *et al.* 1996; Liro & Coen 1996; Rowan 1996; Schuster 1996). Some of these faults (e.g. EF 8 in Fig. 7) are sealed within the Lower–Middle Miocene deposits. Other extensional faults, with a different, i.e. western, dip, such as EF 7 (Fig. 5), apparently affected the same deposits. However, they are younger in age, active until the end of Early–Mid-Miocene time when sealed by the Middle Miocene (Badenian) strata.

From the available data on the interpreted seismic lines it is difficult to exactly determine when extension started. One extensional fault (EF 9 in Fig. 7) was active in a very early stage of basin evolution, just after the deposition of *c.* 100 m of strata (visible in its northward neighbouring syncline). Because rocks overlying the salt were not thick enough to explain the salt movement by gravitational instability one should consider that the fault was generated by extension. Evidence of an early extensional period suggests that tensional forces were active from the beginning of Early–Mid-Miocene time when the lower molasse accumulated. This fact agrees with the Burdigalian extension, of similar age, in the eastern side of the Getic Depression documented by Matenco (1997). Consequently, it may be inferred that extension was the driving mechanism for the diapirism even during the early stage of salt displacement. During the initial stage the salt moved up in a reactive fashion along or between extensional faults (*sensu* Vendeville & Jackson 1992). As a result of this movement the salt obtained a truncated cone shape between the opposite vergent extensional faults (Fig. 5).

Pre-nappe emplacement stage

It is often very difficult to draw a sharp boundary between two deformational periods, such as between the initial and the pre-nappe emplacement (main) stage of salt evolution in the studied area. Conventional thought holds that the stage boundary is related to the transformation of diapirs, in a series of events, into either mushroom-shaped structures or lateral salt flow sheets. The transition from one stage to the next occurred because, during the initial stage, enough salt was inflated into the diapir to start its active growth to the surface.

The upward growth of the salt kept up with the sediment accumulation rate. Consequently, the salt was continuously exposed at the basin floor on the crest of the anticlines. As the salt continued to rise it formed a salt dome above the basin floor. As a result of the positive bathymetric relief the salt started to flow laterally. During this stage the salt sheet and the mushroom head-like structure formed as seen in the Aricesti structure (Fig. 5) or Bucsani structure (Fig. 7). The salt continued to prograde laterally until it reached its maximum extent. Then, gradually, it retrograded as the head-shape was closed. One possible explanation for the salt withdrawal could be the depletion of the salt source. An alternative explanation may be suggested if the timing of the withdrawal period is taken into consideration. This period coincides with a change in the stress field from extension to compression. The compression was responsible for a major tectonic event all along the Outer Carpathians, namely the emplacement of the Subcarpathian Nappe. During this interval the compression forces had the tendency to gradually close up the neck between the feeder and the mushroom heads. An example of this type of evolution is displayed in the Aricesti diapir, whose mushroom head-like structure is well illustrated by both dip section (Fig. 5) and strike section (Fig. 8).

The Aricesti structure is also important to show the age of the lower salt horizon. The point of maximum lateral extent at the salt head corresponds to the top of the Lower Molasse (of Early–Mid-Miocene age). On its southern slope the salt is overlapped by Badenian deposits. This geometric relationship could lead to a wrong conclusion, namely that the salt is Mid-Miocene (Badenian) in age, although the salt is clearly from a Lower Miocene (Burdigalian) stratigraphic level. In such a situation only good quality seismic sections can help to correctly determine the age of the intruding salt.

Nappe emplacement stage

During this important compressional event the salt, as part of the Subcarpathian Nappe, was transported east or southeast away from its original depositional setting by at least 100 km. The question remains as to how the compressional forces affected the salt.

First, it should be noted that the salt was cut by thrust faults (TF 1, 2 and 3 in Fig. 6 and TF 4 in Fig. 5). Some of these thrust faults were reactivated in later stages of deformation. They moved the salt up, as in TF 7 (Fig. 5).

The compression modified the shape of the salt. During this compressional stage the mushroom heads became gradually more tapered. This modification is proved by the fact that some salt structure tops are covered either by Sarmatian C (Aricesti diapir, Fig. 5) or even by Meotian strata. These features including the closing up of the salt heads, are all the result of the Middle Sarmatian compression as evidenced by the interpreted seismic-reflection lines. Generally, they are less spectacular than previous works suggested.

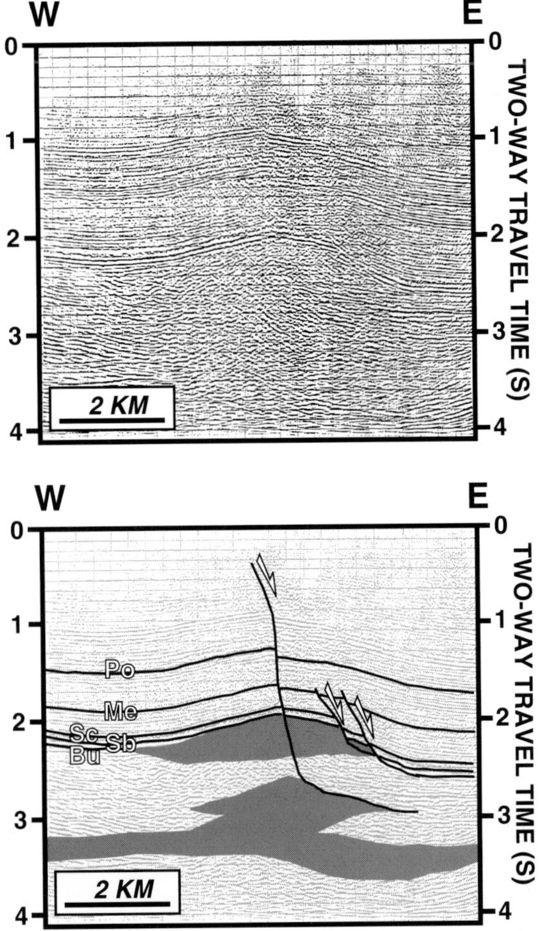

Fig. 8. Strike-oriented seismic-reflection section (line E) along the Aricesti diapirs (for location, see Fig. 1; for legend, see Fig. 3).

Post-nappe emplacement

After the Subcarpathian nappe was emplaced at the Sarmatian A–Sarmatian B boundary, the compression remained active and produced reverse faults within the nappe. During these movements the salt was affected by newly created thrusts (TF 5 in Fig. 4 and TF 7 in Fig. 5). Also, in pre-existing diapirs the salt moved up soon after it was covered by Sarmatian sediments (Figs 5 and 7). The data on the available seismic lines do not reveal what happened to the salt bodies during the Meotian–Pontian extension. That is why no salt activity is shown in Fig. 9. However, it is proposed here that the salt slowly continued to move upwards.

Wallachian stage

This stage of deformation practically corresponds to the Wallachian phase as defined by Dumitrescu & Sandulescu (1964). It can be subdivided into two distinct events that influenced salt evolution: folding and strike-slip faulting.

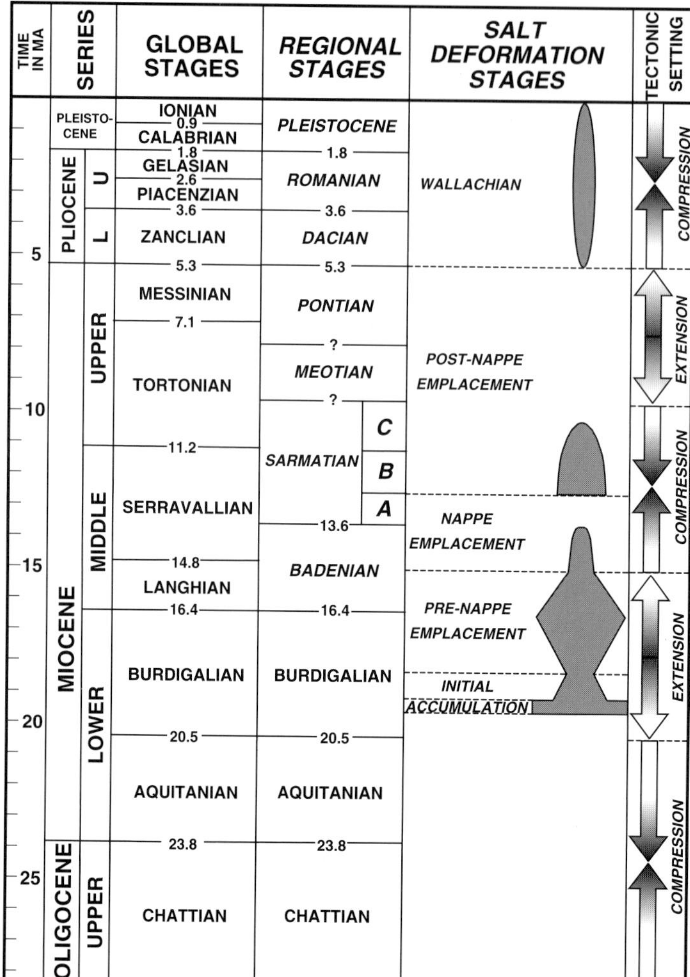

Fig. 9. Simplified diagram showing salt shape evolution through time.

Microstructural studies have documented a N–S (Hyppolite & Sandulescu 1996) or NNW–SSE (Morley 1996) directed compressional stress field responsible for the youngest folding in this area. During folding, the salt bodies from the mushroom heads were squeezed up to their present positions. There is no evidence on the interpreted seismic data for deformation associated with folding at the deeper salt levels.

A system of strike-slip faults developed coeval with the folds. The strike-slip faults affected all rocks up to those of Pleistocene age, regardless of whether they belong to the foreland, the nappe system or their mutual post-tectonic cover (Figs 6 and 7).

One important fault visible on sections is located in the Colibasi area (Fig. 6). This fault, the Campina fault, can be followed for a distance of *c.* 30 km. On geological maps its trace is curved, striking from N40E in its northeastern segment to N40E in the western segment. In its northeastern segment a left-lateral component of movement almost parallel to the fault plane was recorded (Hyppolite & Sandulescu

1996; their site 14). The map-view expression of the oblique-slip movement is estimated to be *c.* 6–8 km in a sinistral sense, using laterally offset geological markers.

The salt body on the cross-section (Line A, Fig. 6) looks like a salt wall. This shape is not only the result of thrusting perpendicular to the strike but also is due to the fact that the salt body represents the remaining part of the original diapir that was displaced by the strike-slip fault and uplifted by the component of thrust movement. The separation of the top Pontian horizon (*c.* 1500 m) is the result of a reverse fault movement and the effect of the juxtaposition of two different compartments as a result of an oblique displacement along the strike-slip fault.

The adjacent fault to the south striking N70E is a strike-slip fault (Matenco 1997) following along the Moreni diapir (Fig. 7) and continuing to the Baicoi diapir (Fig. 1). A separation of *c.* 400 m at the top Pontian level can be demonstrated. During the oblique-slip displacement the salt was completely cut off from its 'reservoir' and squeezed up to the surface.

On lines B and C a strike-slip fault related to the Bucsani diapir can be interpreted. The faults SF 3 (Fig. 7) and SF 4 (Fig. 4) appear to correlate with different effects and timing. The line C meridian (Fig. 5) is located on the southern limb of the Bucsani major structure where the flower structure looks like a parasitic fold. The salt was not very strongly involved, it penetrating only into a part of the Early Molasse sequence. On line B the strike-slip fault affects the axial zone of the structure. If the correlation of the faults is correct then there is a *c.* 35° difference between the major fault axis and the strike-slip fault. The latter fault is oriented N85E and it continues eastwards toward SF 5 forming the same lineament. In the western part of Bucsani structure, the salt had a more complete evolution before the strike-slip fault became active, as clearly shown by its mushroom head-like shape. During the strike-slip movement the salt was squeezed up until it penetrated the lower part of the Meotian strata (Paraschiv 1979).

The tectonic effects along the fault trace are variable. For instance, in the eastern segment of the structure the strike-slip fault generated a small anticline that involves Meotian–Dacian deposits. It has a more important effect at the lower structural levels, where separation of *c.* 200 m occurs in both the foreland and the Subcarpathian Nappe. On the western side of the Bucsani structure the strike-slip fault produced a complex positive flower structure that involves the diapiric salt. This situation can be explained by the salt evolution before the strike-slip fault became active. Thus, although the seismic image on line C (Fig. 4) shows that the salt remained in the initial stage of development on line B (Fig. 7) at least the main stage of evolution with its mushroom head-like structure is well expressed.

Data on the age of the fault come from line C (Fig. 5), where only one fault branch (the northern one) affects the Dacian deposits. It is coeval with the activity shown by the data on line B (Fig. 7) in the axial zone of the Bucsani diapir. The southern branch of the fault was no longer active after Pontian time. Two faults of the same age (pre-Dacian) bounding the diapir to the north and to the south are visible on line B. The three branches in the centre of the strike-slip fault cut the Dacian deposits. On the basis of the data shown by both seismic lines (B and C) it is clear that this strike slip-fault had two stages of movement. The first stage is pre-Dacian and the second is Pleistocene. There is no information about the movement sense along this fault (TF 3, 4 and 5). Taking into consideration its possible effect on the Carpathian leading edge, it could be considered as a dextral strike-slip fault. The structural aspect and the intensity of deformations along this fault suggest that it was not a simple strike-slip fault but had a more complex displacement history. It had a predominant horizontal displacement in the pre-Dacian stage of its evolution and then an important vertical component in its central segment during Pleistocene time.

The Wallachian N–S compression was associated with coeval E–W elongation that caused the extensional faults on the eastern pericline of the Aricesti diapir (Fig. 8). The surface data combined with seismic-reflection sections demonstrate that salt was involved in the Wallachian stage deformations until Pliocene time, playing again a reactive part. It is proposed that the sequence of deformational events involving the Burdigalian salt applies to the entire western segment of the East Carpathians Bend area.

Conclusions

Similar to salt deposits in many other areas of the world, the East Carpathians Bend's salt tectonics produced anticlines with piercing salt cores in the middle of their more or less ellipsoidal map-view outlines. These structures are in sharp contrast to the continuity of the nappe structures (Tarcau and Subcarpathian) of the Carpathian orogen in which the diapirs are embedded. This structural contrast suggests the partially independent evolution of the salt structures within the compressional thrust–fold belt.

Because of the complex tectonic alternation of extension and compression the salt went through a multi-stage structural history before achieving its present-day shape. Emphasis is placed on the fact that the diapirism began very early when the salt was still in its undisplaced basin and continued during and after the salt basin was thrust over the foreland.

Figure 9 summarizes the salt evolution through a 20 Ma period from accumulation stage to the present. As a result of the Early–Mid-Miocene regional extensional setting, the top of the salt began to move to higher positions. Later, the salt overflowed and spread over the surrounding deposits. Extension changed to compression, causing the salt to withdraw its area of expansion. Then, before the Subcarpathian Nappe emplacement, the mushroom head shaped diapirs were formed. Later, but still during the same compressional stage, the diapir heads became more accentuated. Available seismic-reflection data do not indicate the response of the salt to the Meotian–Pontian extension, but it is suggested that the salt continued to move up slowly. Finally, during the Wallachian compressional stage the salt was squeezed to locally reach the surface.

This paper concludes that, in the type area of diapirism (Mrazec 1907; western segment of the East Carpathians Bend), the salt evolution was governed by alternating extensional and compressional periods that exerted markedly different structural influences on the growth of complex salt diapirs. Thus, as is customary in geology, the *locus typicus* of diapirism does not appear to offer simple examples for this geological phenomenon.

We are grateful for the numerous discussions on the geology of the Carpathians with A. Bally, whose ideas and suggestions improved the manuscript. Similarly, a very thorough and constructive review by C. Burchfiel is very much appreciated. Thanks are due to B. Vendeville for his editorial patience.

References

Albu, E. & Baltes, N. 1981. Considération sur l'âge du sel dans la zone des plis diapirs attenués et incipients de la Muntenie et ses implications sur la genèse et la repartition des gisements d'hydrocarbures. *Carpathian–Balkan Geological Association, 12th Congress*, 257–264.

Atanasiu, I. 1948. Romanian oil fields. *Technical Publication of the Romanian Association of Engineers*, **3**, 111–125.

Diegel, F. A., Karlo, J. F., Schuster, D. C., Shoup, R. C., & Tauvers, P. R. 1996. Cenozoic structural evolution and tectono-stratigraphic framework of the Northern Gulf coast continental margin. *In*: Jackson, M. P. A., Roberts, D. G. & Snelson, S. (eds) *Salt Tectonics: a Global Perspective*. *AAPG* Memoirs, **65**, 109–152.

Dumitrescu, I. & Sandulescu, M. 1964. Problèmes structuraux des Carpathes roumains et de leur Avant-pays. *Annuaire du Comité Géologique de la Roumanie*, **36**, 195–218.

Grigoras, N. 1955. Comparative study of the Paleogene facies between Putna and Buzau valleys. *Anuarul Comitetului Geologic*, **XXVIII**, 99–219.

Hyppolite, J. C. & Sandulescu, M. 1996. Paleostress characterization of the 'Wallachian' phase in its type area, southeastern Carpathians, Romania. *Tectonophysics*, **263**, 235–269.

Jackson, M. P. A. 1996. Retrospective salt tectonics. *In*: Jackson, M. P. A., Roberts, D. G. & Snelson, S. (eds) *Salt Tectonics: a Global Perspective*. AAPG Memoirs, **65**, 1–28.

Liro, L. M. & Coen, R. 1996. Salt deformation history and postsalt structural style. *In*: Jackson, M. P. A., Roberts, D. G. & Snelson, S. (eds) *Salt Tectonics: a Global Perspective*. *AAPG* Memoir, **65**, 323–332.

Matenco, L. 1997. *Tectonic evolution of the outer Romanian Carpathians. Constraints from kinematic analysis and flexural modelling*. PhD thesis, Vrije Universiteit, Amsterdam.

Morley, C. K. 1996. Models for relative motion of the crustal blocks within the Carpathian region, based on restorations of the outer Carpathian thrust sheets. *Tectonics*, **15**(4), 885–904.

Mrazec, L. 1907. Despre cute au simbure de shapungere [On folds with piercing cores]. *Society of Stüte Bulletin, Romania*, **16**, 6–8.

Olteanu, F. 1951. Observatii asupra 'breciei sarii', cu masive de sare din regiunea mio-Pliocena dintre valea Teleajenului si piriul Balaneasa (cu privire speciala pentru regiunea Pietrari–Buzau). *Dari de seama ale sedintelor Institutului geologic al Romaniei*, **32**, 12–18.

Paraschiv, D. 1979. *Romanian Oil and Gas Fields*. Institute of Geology and Geophysics. Technical and Economical Studies, A Series, **13**.

Popescu, G. 1951. Observatiuni asupra 'breciei sarii' si a unor masive de sare din zona Paleogena–miocena a judetului Prahova. *Dari de seama ale sedintelor Institutului geologic al Romaniei*, **32**, 3–12.

Roure, F., Roca, E. & Sassi, W. 1993. The Neogene evolution of the outer Carpathian flysch units (Poland, Ukraine and Romania): kinematics of a foreland/fold-and-thrust belt system. *Sedimentary Geology*, **86**, 177–201.

Rowan, M. G. 1996. Structural style and evolution of allochthonous salt, Central Louisiana outer shelf and upper slope. *In*: Jackson, M. P. A., Roberts, D. G. & Snelson, S. (eds) *Salt Tectonics: a Global Perspective*. AAPG Memoirs, **65**, 199–228.

Schuster, D. C. 1996. Deformation of allochthonous salt and evolution of related salt-structural systems, Eastern Louisiana Gulf coast. *In*: Jackson, M. P. A., Roberts, D. G. & Snelson, S. (eds) *Salt Tectonics: a Global Perspective*. *AAPG* Memoirs, **65**, 177–198.

Stefanescu, M. & Dicea, O. 1995. Multiple thrust events at the Carpathian orogen/foreland Contact in Romania. *AAPG. Nice Conference, Abstracts Volume*.

——, Polonic, P., Popescu, I., Balintoni, I., Dinica, G. & Gheuca, I. 1985. *Geological sections across Romania*, scale 1:200 000. Geological Institute of Romania, A-20.

——, Popescu, I., Stefanescu Marina, I. V., Melinte, M. & Stanescu, V. 1993. Aspects of the possibilities of the lithological correlation of the Oligocene–Lower Miocene deposits of the Buzau valley. *Romanian Journal of Stratigraphy*, **75**, 83–90.

Vendeville, B. & Jackson, M. P. A. 1992. The rise of diapirs during thin-skinned extension. *Marine and Petroleum Geology*, **9**, 331–353.

Recrystallization salt fabric in a shear zone (Cardona diapir, southern Pyrenees, Spain)

L. MIRALLES[1], M. SANS[2], J. J. PUEYO[3] & P. SANTANACH[2]

[1]*LIFS, Dept. Geoquímica i Petrologia. Universitat de Barcelona, C/Martí Franqués s/n, Barcelona 08071, Spain (e-mail: lourdes@natura.geo.ub.es)*

[2]*Dept. Geodinàmica i Geofísica, Universitat de Barcelona, C/Martí Franqués s/n, Barcelona 08071, Spain*

[3]*Dept. Geoquímica i Petrologia, Universitat de Barcelona, C/Martí Franqués s/n, Barcelona 08071, Spain*

Abstract: In the southern Pyrenees foreland, the Cardona salt diapir has a 250 m high stem and a small bulb, which is partially exposed. The internal structure of the diapir consists of a main sheath fold divided by a shear zone. This shear zone crosses two salt units with different initial properties. In this paper, a qualitative analysis of the fabric both units crossed by the shear zone and a 3D quantitative morphological and textural analysis of the unit with small grain size were carried out. In this unit, the fabric is characterized by a grain size similar to that of the initial fabric, disappearance of primary structures, a poor grain orientation with the long axis statistically at 20° of the shear boundary, and a strong {100} crystallographic preferred orientation. The unit composed of coarser grains shows a disappearance of primary structures. The fabric changes in both units and the strong decrease in water content during deformation suggest that fluid-assisted synkinematic recrystallization was dominant. The different behaviour of the two studied units on shear deformation also suggests that the initial grain size and water content were the main factors controlling fabric changes.

Comparison between models and natural patterns of crystallographic orientations have suggested that rock salt mainly flows by intragranular slip in domal salt (Schwerdtner 1968). By contrast, equations obtained from creep experiments on polycrystalline aggregates and domal rock salt describe fluid-assisted diffusional creep and dislocation-controlled creep (Carter & Hansen 1983; Spiers *et al.* 1988). Moreover, the presence of water (inherent or added) in natural rock salt (Urai *et al.* 1986) and fine-grained synthetic aggregates (Spiers *et al.* 1988) has a weakening effect on their mechanical behaviour. The stress–strain relationship reveals that both mechanisms would compete over most physical conditions in which shallow rock salt structures form (Carter *et al.* 1990). To sum up, distinct features such as thin brine films and overgrowth indicate a major role of fluid-assisted mechanisms whereas strong crystallographic preferred orientations and the presence of subgrains seem to indicate the operation of intragranular deformation mechanisms.

Salt diapirs are typically characterized by anisotropic gneissose or granoblastic fabrics with a large grain size (from 0.5 to 5 cm) (Balk *in* Talbot & Jackson 1987).

From: VENDEVILLE, B., MART, Y. & VIGNERESSE, J.-L. (eds) *Salt, Shale and Igneous Diapirs in and around Europe*. Geological Society, London, Special Publications, **174**, 149–167. 1-86239-066-5/00/$15.00

The orientation and shape of the grains can define either planar foliations in S-tectonites (Larsen & Lagoni 1984; Talbot & Jackson 1987) or lineations in L-tectonites (Jackson & Talbot 1987). In contrast, finer-grained salt is expected in the shear zones separating different spines (*sensu* Kupfer 1968) in the diapir. In these zones, the complex ductile structures in one or both of the adjoining spines are simplified by internal rotation and deformation to form comparatively simple fabrics in which the foliation is subparallel to the bedding with or without a lineation (Talbot & Jackson 1987).

In this paper, we present the fabric characterization of an internal shear zone of the Cardona diapir. This shear zone crosses two salt units with different grain size, water content and initial fabrics. Changes in the unit with large grain size are easily visible at naked eye; by contrast, for the unit with small grain size, a 3D fabric characterization has been carried out in the laboratory. Our work has focused on the unit with small grain size, as its fabric in less deformed areas of the deposit has been widely studied (Busquets *et al.* 1985; Rosell & Pueyo 1997; Miralles 1999). Both grain shape and crystallographic fabrics will be analysed to discuss their relationship with the macro-scale structure.

Geological setting

In the southeastern Pyrenees foreland, the presence of three evaporitic formations (Beuda, Cardona & Barbastro) determines the location and the geometry of the Triangle Zone of the Pyrenean deformation front (Sans *et al.* 1996*a*) (Fig. 1a and b). The structures developed in the Triangle Zone are detached fault-related folds cored by evaporites with different trends above each evaporitic formation (Vergés *et al.* 1992; Sans & Vergés 1995). In the central part of the Triangle Zone, folds and thrusts have a NE–SW trend and are detached on the upper Bartonian–Priabonian Cardona salt formation. These anticlines have locally been imprinted by diapirism and salt outcrops at the surface (Fig. 1c). The Cardona formation has been well studied from the sedimentological and petrographical viewpoint (Pueyo 1975; Ayora *et al.* 1995; Rosell & Pueyo 1997), and more recently from the structural and fabrics points of view (Miralles & Sans 1996; Sans *et al.* 1996*b*; Miralles 1999; Sans 1999). Fundamental outcrops for these studies are the underground mining works, which extract the potash layers, the mining exploration wells and the excellent outcrop in the western termination of the Cardona diapir (Fig. 2a) (Wagner *et al.* 1971; Riba *et al.* 1983). The Cardona formation is 300 m thick in the centre of the basin (Pueyo 1975). Three units (Fig. 2b) can be distinguished: a basal anhydrite, a Lower Salt Unit and an Upper Salt Unit bearing potash salts. The basal anhydrite unit is 4–5 m thick and laminated. The Lower Salt Unit is 130–200 m thick. It consists of white and grey decimetric banded halite (Busquets *et al.* 1985). The grey bands result from the sedimentary accumulation of abundant hopper crystals (up to 1–2 cm) and diffuse clay. The clear bands have either transparent grains or smaller hopper crystals than the grey bands. The water content in the Lower Salt Unit ranges from 0.04 to 1 wt %, with an average of 0.32 wt %. Intergranular brine accounts for 0.18 wt % whereas brine trapped in fluid inclusions, which vary in size from 10 to 100 μm, accounts for 0.14 wt % (de las Cuevas & Pueyo 1995). Brine of fluid inclusions is dependent on the presence of primary structures such as large (up to 2 cm) hopper crystals. The Upper Salt Unit is 50–100 m thick (Pueyo 1975; Ayora *et al.* 1995). This unit consists of two subunits: the Sylvinite Unit and the Carnallitite Unit. The Sylvinite Unit includes two enriched

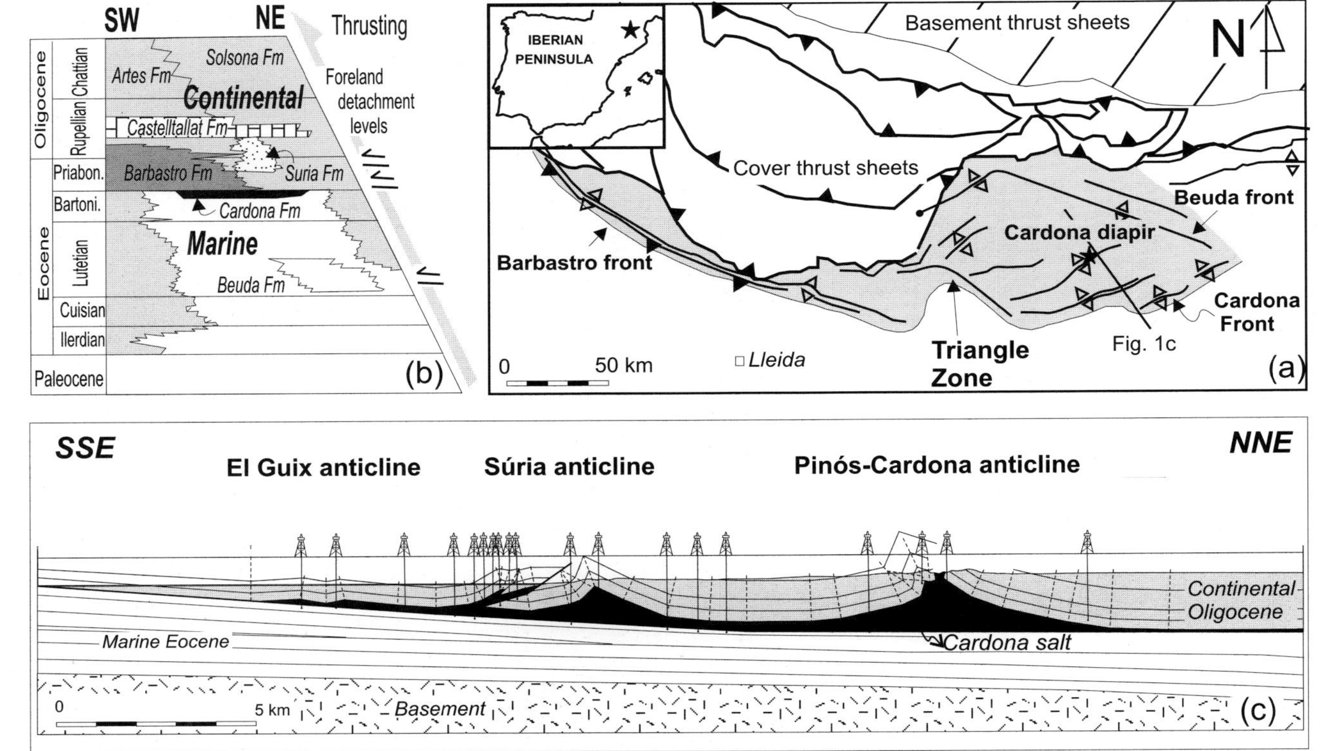

Fig. 1. Geological setting of the Cardona diapir. (**a**) Sketch of the south Pyrenean Triangle Zone (shaded in grey). Three evaporitic levels develop three thrust fronts at their sedimentary pinch-outs (Beuda, Cardona and Barbastro formations). The Cardona diapir is superimposed on the Cardona–Pinós anticline. Location of cross-section in (**c**) is indicated. (**b**) The Cardona salt represents the transition from marine to continental deposition in the Ebro basin infill. The Beuda and Barbastro formations are also good detachments in the Triangle Zone. Cardona salt formation is shown in black. (**c**) Geological cross-section across the Cardona thrust front (modified from Vergés *et al.* 1992).

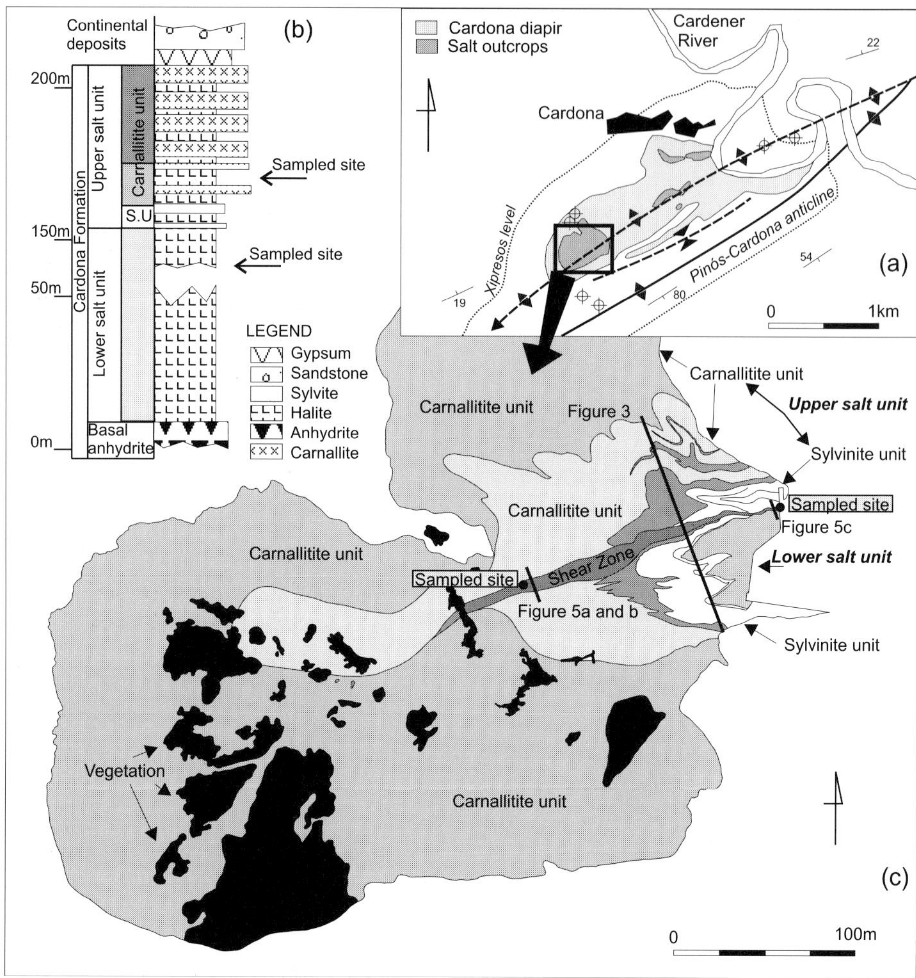

Fig. 2. The Cardona diapir. (**a**) The Cardona diapir pierces the northern limb, close to the hinge, of the Pinós–Cardona anticline and it is elongated parallel to the fold axis. (**b**) Stratigraphy of the Cardona formation. Two sites were sampled, one in the Lower Salt Unit and the other in the Upper Salt Unit. (**c**) Geological map of the southwestern termination of the Cardona diapir. Locations of the photographs in Fig. 5 are shown.

sylvinite layers separated by a metre-scale salt layer (Fig. 2b). The Carnallitite Unit consists of 40–80 m of alternating beds of carnallite (10 cm to 2 m), thinner halite and clay horizons. A lower halite-dominated part and an upper carnallite-dominated part can be distinguished (Fig. 2b and c). The carnallite beds are red and coarse grained, usually brecciated with irregular fragments of anhedral red crystals of carnallite in a pink matrix. These halite cycles consist of small (*c.* 1 mm) clear halite grains. However, alternation of clear halite and small cubic or elongated hopper crystals (200 μm–1 mm), which are related to the enriched clay layers, also occurs (Rosell & Pueyo 1997). The water content of the halite cycles in the Upper Salt Unit ranges from 0.02 to 0.38 wt %, with an average of 0.12 wt %; 0.08 wt % accounts for intergranular brine and 0.04 wt % for fluid inclusions of less than 20 μm (de las Cuevas

& Pueyo 1995). The size of the primary structures such as tabular hopper crystals is small (300 µm) and therefore hinders the development of large fluid inclusions.

Above the Cardona formation (Fig. 1b), the overburden involved in the folds and thrust of the Triangle Zone consists of a sequence of alluvial fans (Artés, Solsona and Súria formations) that prograde into a lacustrine system (Castelltallat and Barbastro formations) (Busquets *et al.* 1985; Sáez 1987; Anadón *et al.* 1989).

The Cardona diapir

The Cardona diapir is superimposed onto the Pinós–Cardona anticline (Figs 1a and c, and 2a). This anticline is a south-vergent detachment anticline with a northern limb dipping 20–30° and a southern limb dipping 70°. It can be mapped in a N55E direction for *c.* 30 km. The Cardona diapir is located close to the eastern termination of this anticline (Fig. 2a) and pierces its northern limb, very close to the anticline hinge, forming a small syncline between its outcrop and the anticline hinge. It is elongated in a NE–SW direction following the anticline trend; it is 2 km long and its width ranges from 0.2 to 0.7 km (Fig. 2a). The diapir has pierced around 300 m of overburden mainly formed by gypsum, marls, some thin limestones and fine to coarse sandstones of the Barbastro, Súria, Castelltallat and Solsona formations. Their preserved thickness is *c.* 400 m in the anticline crest and around 2000 m in the northern syncline and 1200 m in the southern one (Fig. 1c).

At present, the Cardona diapir crops out in a subordinate valley that drains to the Cardener river. It is thus lower than the surrounding topography. The Cardona salt crops out 100 m above the valley floor, only in the western part of the valley, giving a local topographic relief of 100 m (Muntanya de sal). This relief is around 50–100 m lower than the highest surrounding topography. The contact between the overburden and the diapir corresponds to an external shear zone 2–6 m thick formed by a mélange of country rock and sheared salt. Outside this zone several minor normal faults deform the overburden (Wagner *et al.* 1971; Riba *et al.* 1983).

Internal and external structure

The Cardona diapir rises from a salt-cored detached anticline, which is 300 m below the present erosion surface (Fig. 1c). From the crest of the anticline, the diapir has a stem of 250–700 m width and *c.* 250 m height. The present-day erosion level allows us to see outverging folds, indicating the initiation of a bulb (Fig. 3). The internal structure of the salt layers, the complete stratigraphic sequence in a normal position and especially the presence of horizontal fold axial planes in the higher parts of the diapir suggest that a small bulb, 60–80 m high, is preserved close to the diapir walls (Fig. 4). The outcropping internal structure consists of a major sheath anticline cut by a shear zone (Fig. 2c). The second-order anticlines show vertical axial planes and vertical axes near the western terminations of the diapir and vertical axial planes and sub-horizontal axes in the central part. This, together with the crosscutting relationships between the boudins' long axis and the folds' axis, which are parallel in the upper part of the diapir and perpendicular in the western termination of the diapir, suggests that the folds are large sheath folds elongated in a NE–SW direction (Fig. 4). The significance of the shear zone cutting the main salt anticline can be compared with that of the internal shear zones defined by Kupfer (1968, 1976) and

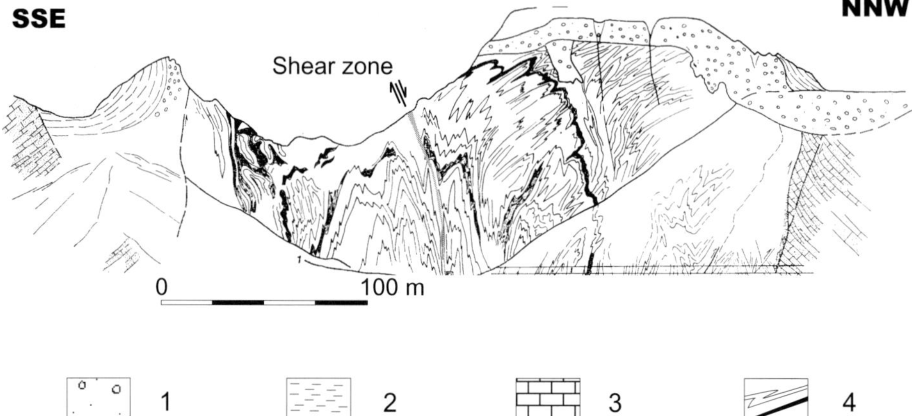

Fig. 3. Sketch of the NW outcrop of the Cardona diapir viewed from the east (modified from Wagner *et al.* (1971) and courtesy of Springer-Verlag and F. Mauthe). 1, Cap rock and salt breccia; 2, Plio-Pleistocene deposits; 3, Oligocene materials; 4, Cardona formation.

indicates different rates of movement on the two sides of the shear zone. In this regard, two main spines in the Cardona diapir could be defined. The southern spine would have exceeded the northern one in height.

Shear zones

Shear zones are located in a band that divides into two the major anticline in the western part of the Cardona diapir outcrop (Figs 2c, and 5a and b). This shear zone can be mapped through the Lower Salt Unit, the Sylvinite Unit, and partially in the Carnallitite Unit in the Upper Salt Unit. In the Lower Salt Unit, this shear zone is vertical, 2.5 m wide and trends 065N–070N; it is therefore subparallel to

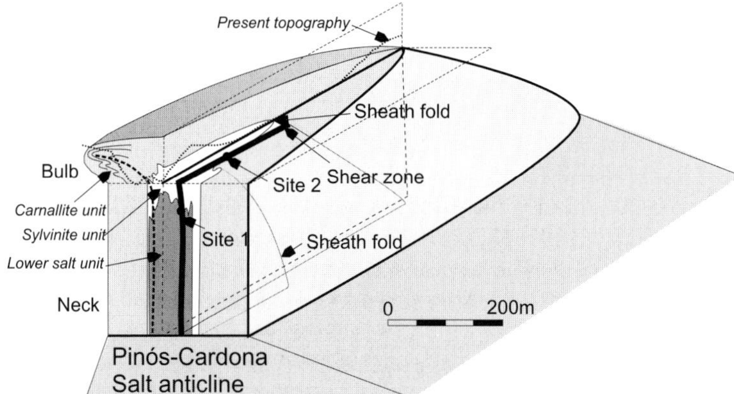

Fig. 4. Sketch of the internal structure and external shape of the Cardona diapir. The present-day topography permits us to see the outverging folds of the bottom part of the bulb. The internal structure consists of a main sheath fold, which is divided by the shear zone into two spines with different vertical movement. The two sampled sites, one in the Lower Salt Unit and the other in the Upper Salt Unit, are located inside the shear zone. These two sites are shown in the sketch.

the folds' axial plane. In the higher parts of the diapir, where the shear zone crosses the Upper Salt Unit, it widens to 12 m. This widening is related to the fan geometry of the diapir towards its higher parts. The single shear band observed in the lower part splits and defines a wider band with more diffuse outer limits. The obliquity between bedding and the shear direction causes the formation of tight folds inside the shear band. These folds have parallel stretched limbs whereas the hinges between them remain less deformed. In this area, the southeastern boundary of the shear band dips *c.* 45° towards the northwest, and the northwestern boundary is steeper and dips *c.* 70° towards the south. The shear direction cannot be inferred directly from any macroscopic kinematic indicator. However, from the field mapping, units cropping out in the southeastern block are older than those in the northwestern one, suggesting that, although the horizontal component of movement is unknown, the southeastern block has had a larger upward vertical motion than the northwestern one.

Fabric analysis

Two sites were compared for the fabric analysis. One site is located in the Lower Salt Unit where the calculated reduction in thickness of the bedding (t_f/t_0) is 15%. In this site, the main characteristic is the large grain size, with single crystals as large as 30 cm. For this reason only a qualitative fabric analysis was carried out. The second site is located in the highest part of the shear zone, in the lower halite levels of the Carnallitite Unit of the Upper Salt Unit (Fig. 2b), where the calculated reduction in thickness is 45% (Fig. 5a). A quantitative analysis of the fabric was performed on these samples, as the small grain size (1 mm) does not allow us to see variations with the naked eye, and because equivalent samples have already been studied in less deformed areas of the deposit (Rosell & Pueyo 1997). The three samples were taken from the same layer in the limb of a fold developed inside the shear band (Fig. 6). The sampled limb attitude is 090/42N, and slightly oblique to the shear-zone boundary (080/42N). The limb is straight and its minimum width is 50 cm. The beds can be followed for over 10 m, and the samples were taken from where the sampled layer is 6.5 cm thick (sample 1) to where the layer is 2.8 cm thick and achieves a maximum thinning of 43% (sample 3).

Methods of fabric analysis

The techniques used to analyse the fabric of the collected samples have been described in detail by Miralles (1999), and only a short description is given here.

The samples were obtained from the diapir outcrop at the surface and were oriented in the field according to an orthogonal coordinate reference system, which will be referred to from here on as the stratigraphic coordinate system (Fig. 6). This reference system was defined as *abc*, (+)*a* being the strike direction of the bedding, (+)*b* being the dip direction of the layer, (+)*c* the perpendicular to the bedding pointing to the top of the layer (Fig. 6). Three mutually perpendicular thin sections were made parallel to the planes defined by the axes of this coordinate system, which were labelled *ac*, *ab* and *bc*. The samples were cut, ground and polished using low-speed techniques to avoid damage and breakage of the grains. The qualitative analysis of the fabrics was performed by optical microscope observations on the three thin sections.

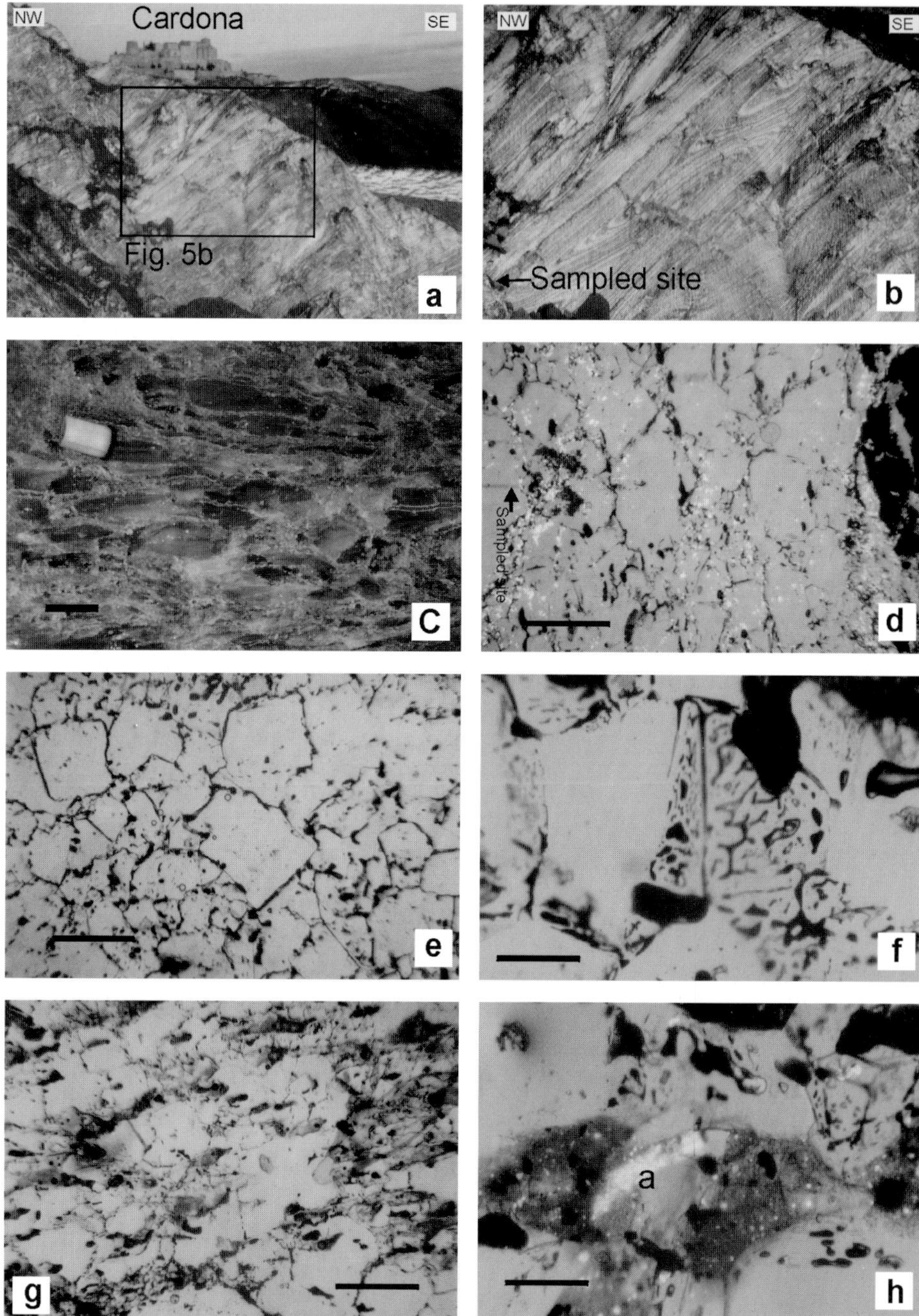

Fig. 5. (**a**) General view, from the west, of the shear zone crossing the Upper Salt Unit in the upper part of the Cardona diapir. (See general location in Fig. 2c.) (**b**) Detailed view of the box in (**a**). (**c**) Outcrop of the Lower Salt Unit affected by the shear zone in the Cardona diapir. The halite cycles separated by thin clay layers show large transparent halite monocrystals (>10 cm). Scale bar represents 5 cm, and top of the layer is on the right. (**d**) Microscopic overview of a halite cycle of the Upper Salt Unit which shows slightly elongated

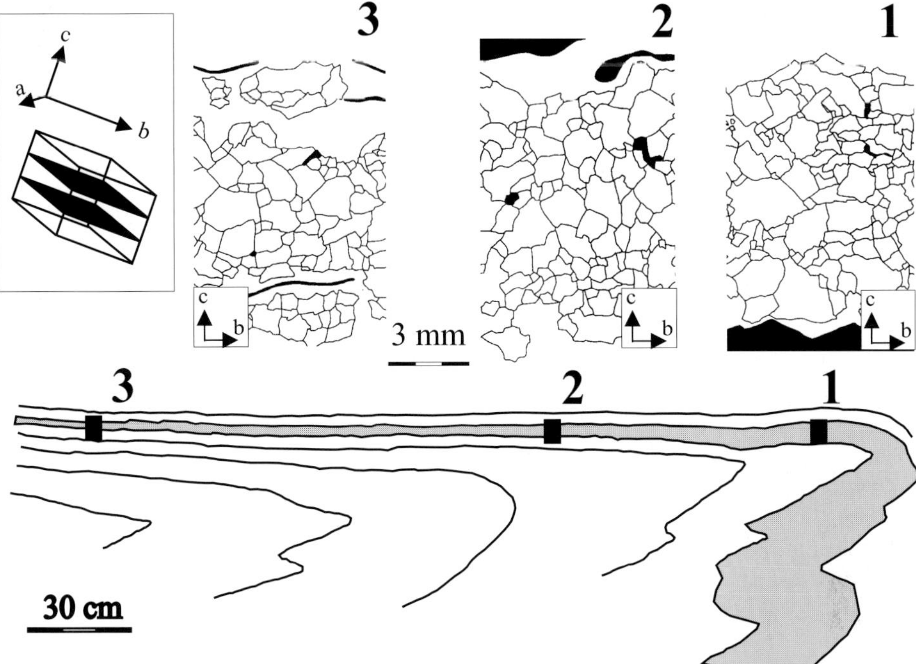

Fig. 6. Sketch of the sampled fold inside the shear zone and location of the samples (sampled layer in grey). The inset shows the orientation of the stratigraphic coordinate system with respect to the stratigraphic elements. This figure is a section on the *bc* plane of the stratigraphic coordinate system. Images of the grain boundary network are shown only for the *bc* section. The clay layers are shown in black. Grain boundaries are often straight, especially in the larger grains.

As the samples were taken from the surface outcrop, they may be affected by meteoric waters. The different fabrics present in the hand samples have to be observed cautiously, and any halite that has been newly precipitated by superficial processes identified. To discriminate the presence of a meteoric fabric in the samples, petrographic analysis is essential. Meteoric and deformational fabrics can coexist in the same hand sample, and their identification was carried out in the qualitative analysis of the fabric. As the meteoric fabric was more abundant in the part of the hand sample that was exposed at the surface, these areas were avoided, and the quantitative fabric analysis was carried out only in those areas where there was no sign of water circulation.

---

halite grains and small anhydrite grains that are disseminated. Scale bar represents 2 mm. (**e**) Synkinematic fabrics, which consist of equant halite grains with straight grain boundaries. (Note the absence of accessory minerals such as anhydrite.) Scale bar represents 2 mm. (**f**) Thin interconnected brine films present in the grain boundary structures of the halite grains. Scale bar represents 200 μm. (**g**) Overview of a halite cycle affected by weathering processes. Greyish areas are turbid halite grains with different concentrations of solid inclusions. Scale bar represents 1 mm. (**h**) Microphotograph taken under half-crossed polarizers, showing a detail of a turbid halite grain. A large elongated anhydrite crystal is marked 'a'. The turbid halite grain has many solid inclusions of small rounded anhydrite grains and clay particles. Scale bar represents 200 μm.

A conventional image analyser was used for measurements of the morphological characteristics (size and orientation of the grain population in the three sections). First, the grain boundary network was drawn using a drawing tube adapted to the petrographic stereomicroscope (from 2.5× to 25×), and a combination of transmitted and reflected illumination conditions was used to obtain a better resolution on those grain boundaries that were badly oriented. The image of the grain boundary network was scanned to obtain an image (580 × 780 pixels), which was processed using the IMAT program (Serveis Cientificotècnics, Universitat de Barcelona 1992). With this program, the area (A), perimeter (P), long axis (l), short axis (s) and the angle Φ were measured for each grain. The long axis is the line between the two most distant points in the grain that crosses the gravity centre, considering the grain as a rigid body of uniform density. The short axis is perpendicular to the long axis and crosses the gravity centre. The angle Φ is the angle between the long axis and a reference line. For sections *ab* and ac the reference is the $(+)a$ axis of the stratigraphic coordinate system, whereas for section *bc* the reference is the $(+)b$ axis. From the measured data, the L/S shape factor of each grain in each section was calculated as the ratio between the long axis and the short axis. All these values were statistically treated to determine the number of populations present in each sample and the mean value for each variable. The Kolgomorov–Smirnov test was applied and the critical value $\alpha = 0.05$ in a two-tailed test was used as a boundary to reject the hypothesis that samples follow a normally distributed population (Davis 1986). The datasets had to be transformed to logarithm (area and perimeter), to square-root (long and short axis), and to reciprocal (L/S), to approximate normal distributions. In the case of angular values, the raw data were used to obtain the mean value and to check the presence of more than one main orientation.

To determine the 3D shape of the grains that characterizes each sample, a shape ellipsoid based on the three perpendicular sections was calculated. This ellipsoid was defined by six independent variables obtained from an overdefined linear system of nine equations with six unknowns which were defined as was done by Milton (1980). These nine equations were designed taking the following variables: long axis, short axis and the angle Φ obtained from the three 2D analysed sections. The excess number of equations with respect to the variables sought allows the least-squares method to be applied to adjust the best matrix. Diagonalization of the resulting symmetric matrix gave the length and orientation of the axes for each sample. Finally, the full crystal orientation of each grain was measured on a U-stage mounted on a petrographic microscope and used in transmitted light conditions. Crystals in thin sections of the *bc* section were cleaved along $\{100\}$ planes by mechanical shock transmitted by high-speed grinding. At least two perpendicular $\{100\}$ planes had to be present to obtain the complete crystal orientation.

Qualitative analysis of the fabrics

The Lower Salt Unit affected by the shear zone consists of large (up to 30 cm) and transparent recrystallized halite monocrystals free of primary fluid inclusions (Fig. 5c) amongst small grains (few millimetres). Occasionally, the thickness of the salt layer is made up of single halite monocrystals, most of which are lensoidal in shape, elongated parallel to the bed (Fig. 5c). A similar fabric to this one has only been described in deep parts of the Cardona diapir (the 920 m level of the Cardona mine) (Pueyo 1975). This type of fabric contrasts with the Lower Salt Unit fabric in less tectonized areas (Busquets *et al.* 1985; Rosell & Pueyo 1997), where it consists of white and dark grey bands of halite grains of 1–3 cm in size. The dark grey bands are predominantly formed by coarse hopper crystals (7–10 mm), which contain primary fluid inclusions (100–200 μm) and scattered clay fragments. The white bands are of clear halite grains and clay fragments are absent. The accessory minerals, anhydrite and polyhalite, are associated with the clay zones and they can also be found scattered throughout the rock salt (either inside the halite grains or at their grain boundaries).

The samples in the Upper Salt Unit were collected at the diapir surface and show two different types of fabric. The first consists of turbid halite grains located at the grain boundaries or at triple grain junctions. The intensity of the brown colour of the grains depends on the density of the trapped solid inclusions, which consist of small clay particles and anhydrite crystals (Fig. 5g). These crystals are small (*c.* 10 μm), show rounded edges and are randomly distributed inside the halite grains; occasionally, they can be as large as 300 μm (Fig. 5h). This type of fabric is related to the external alteration zone of the diapir.

The second type of fabric consists of a mosaic of small clear halite crystals (*c.* 1 mm) with no hopper crystals. The halite grains have no preferred grain shape orientation and only a slight grain shape elongation (Fig. 5d). Some of the grains tend to be of euhedral form with straight grain boundaries (Fig. 5e) showing thin brine film structures (Fig. 5f). The thin layers of clay minerals that separate the different cycles are boudinated and the main accessory mineral in the halite cycle consists of disseminated, small anhydrite grains of euhedral form. Occasionally, anhydrite is included in the halite grains. The water content in the halite cycle is very low, <0.01 wt % in the form of intergranular water, and no water from fluid inclusions is detected. In contrast to the fabric described in less deformed areas (Busquets *et al.* 1985; Ayora *et al.* 1995; Miralles & Sans 1996; Rosell & Pueyo 1997), this type of fabric shows no primary structures such as small hopper crystals (300 μm).

Quantitative analysis of the fabrics

The quantitative analysis of the fabrics was carried out on the three samples from the Upper Salt Unit affected by the shear zone (Fig. 6). The measurements were performed in those areas with clear halite representative of the second type of fabric.

The 2D analysis shows that there is only one grain population which follows a normal distribution (Fig. 7). The mean values of the grain size variables (area, perimeter, long axis and short axis) obtained for each section of the three samples show a similar tendency. Section ac has the lowest values for all the parameters with a slight increase from sample 1 to sample 2 whereas in section *ab* the mean values remain nearly constant with a slight increase from sample 1 to sample 3 (Fig. 8). Finally, the values of section *bc* show an increase from sample 1 to sample 2, and remain constant or decrease slightly to sample 3. In general, differences in the mean values between the three samples are small, with area being the most sensitive factor (Fig. 8). Mean values of the grain shape ratio L/S are also similar in all sections and samples, varying between 1.56 and 1.67, sample 3 showing the largest

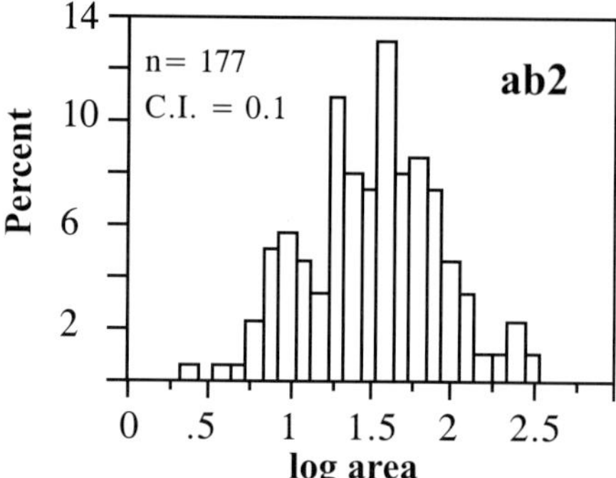

Fig. 7. Frequency distribution for grain area of sample 2 (Upper Salt Unit) measured in section *ab*. The mean grain area has been calculated for a normal distribution of log area. C.I., coefficient interval of the bars, which is 0.1 of log area. *n*, number of grains measured.

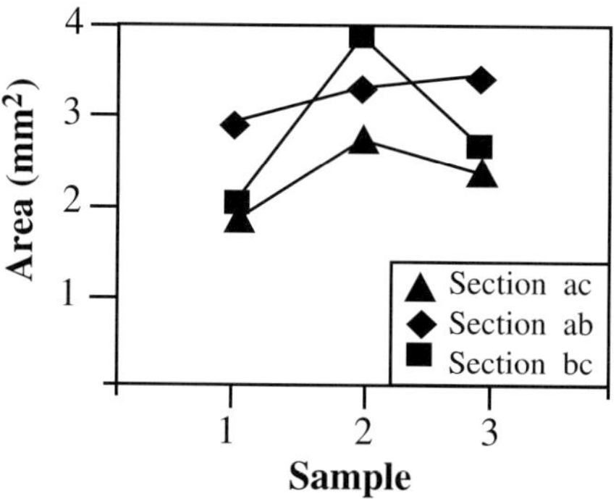

Fig. 8. Plot of the mean values of the area for the three sections measured. All other variables measured (perimeter, long axis and short axis) show a similar pattern.

variations (Table 1). The grains' long axes are not markedly oriented and, in general, they tend to be grouped around the mean value (Fig. 9).

The 3D analysis shows that the representative ellipsoid has only very slight changes in volume in the three samples, sample 1 being smaller than samples 2 and 3. This is

Table 1. *Mean values of the variables measured in the three thin sections*

Sample	n^*	Area (mm^2)	Perimeter (mm)	Long axis (mm)	Short axis (mm)	*Rf*	Φ
1*ab*	155	2.818	2.512	0.847	0.484	1.67	−67.90
2*ab*	177	3.236	2.630	0.847	0.497	1.64	−5.50
3*ab*	173	3.388	2.690	0.864	0.511	1.64	69.90
1*bc*	162	1.995	2.042	0.713	0.420	1.64	0.40
2*bc*	134	3.802	2.690	0.90	0.552	1.56	31.60
3*bc*	163	2.630	2.344	0.790	0.458	1.67	19.30
1*ac*	190	1.862	1.995	0.681	0.40	1.64	0.80
2*ac*	189	2.754	2.399	0.790	0.467	1.64	15.90
3*ac*	188	2.455	2.239	0.734	0.458	1.56	37.20

Mean values calculated from the normal distribution of log area, log perimeter, $\sqrt{}$(long axis), $\sqrt{}$(short axis) and $1/Rf$.

* *n* is the number of grains measured for the 2D quantitative analysis.

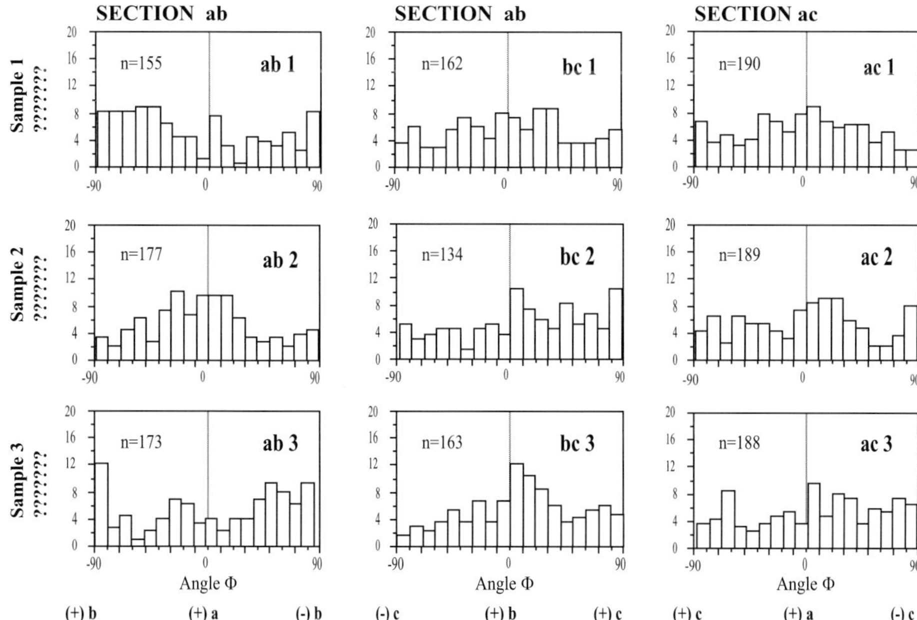

Fig. 9. Frequency distribution for the angular values (Φ) of all samples and the three sections measured. The position of the reference axes of the system with respect to the measured angle is also given at the bottom of each column. C.I. is 10°.

due to an increase in the length of the long axis from sample 2 to sample 3, whereas the medium axis diminishes in a smaller proportion and the short axis remains almost constant (Table 2). The grain shape ellipsoids have a clear prolate shape for samples 1 and 3 whereas sample 2 shows a slight oblate shape (Table 2). This shape is consistent with the predicted $L > S$ fabric for the upper part of the stem or the bulb below the divergent zone in the diapir crest (Talbot & Jackson 1987). The orientation of the 3D grain shape ellipsoid axes of the three samples (Figs 10 and 11) shows that the short axes are grouped around 143/74, and that the intermediate and long axes define a great circle (053/33N), perpendicular to the short axes. The poor definition of the orientation of the long and intermediate axes has to be treated cautiously given the scant data analysed (three samples) in this outcrop.

The pole density figures of the crystallographical orientations (Fig. 10) show a preferred orientation of the {100} poles with a main maximum quasi-parallel to the *a* axis of the stratigraphic coordinate system. This maximum tends to be more scattered along the *ac* plane than in the plane parallel to stratification (*ab* plane). Moreover, along the *bc* plane there is a girdle of less intensity. This means that there is a degree of freedom in the orientation of the other two {100} poles. Nevertheless, in sample 3, where maximum thinning is achieved (Fig. 6), a sub-maximum can be defined around the (+)*b* axis in the intersection between the bedding plane and the shear zone (Fig. 11). In this way, the degree of freedom shown by the other two {100} poles tends to be fixed. The {100} crystallographic faces are disoriented 64° with respect to the plane defined by the long and intermediate axes of the shape ellipsoid. There is, therefore, no clear link between the morphological and crystallographic preferred orientations.

Table 2. *Size, orientation and shape of the 3D mean grain shape ellipsoid*

		Sample		
		1	2	3
Size				
L (mm)		0.924	0.882	1.243
I (mm)		0.518	0.664	0.548
S (mm)		0.438	0.469	0.442
Volume (mm^3)		0.88	1.15	1.26
Orientation				
L	*a*	0.8269	0.004	0.7483
	b	0.5622	−0.918	−0.505
	c	0.0077	0.396	0.4301
I	*a*	0.561	0.8355	−0.655
	b	−0.826	0.2206	−0.668
	c	0.0554	0.5033	0.3546
S	*a*	−0.038	−0.55	−0.108
	b	0.0415	0.3288	0.5469
	c	0.9984	0.7681	0.8302
Shape				
L/I		1.78	1.33	2.27
I/S		1.18	1.42	1.24

L, *I* and *S* stand for long, intermediate and short axes, respectively; *a*, *b* and *c* are the three axes of the stratigraphic coordinate system.

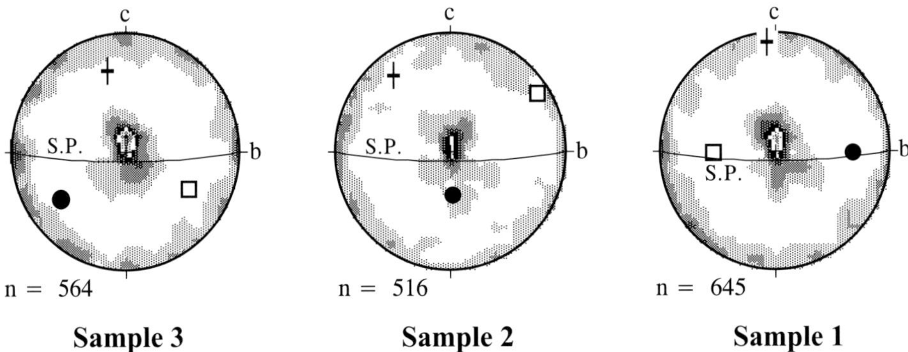

Fig. 10. Equal area stereographic projection in the lower hemisphere of the orientation of axes of the grain shape ellipsoid. ●, long axis; □, intermediate axis; +, short axis of the grain shape ellipsoid. The {100} pole density figures on the *bc* projection plane are shown in area densities of 1% of 1%. S.P., shear plane, which is drawn as a solid great circle.

Discussion

In the Cardona diapir, salt crops out at the surface. For this reason is important to distinguish the meteoric fabric from the deformational fabrics when studying the salt fabrics. The climate conditions of the area consist of alternating dry and humid periods. The humid periods exert a strong influence in the dissolution of the diapir surface. The scarcity of superficial salts precipitated in the form of efflorescences as well as the presence of a karst supports this idea. In the dry periods,

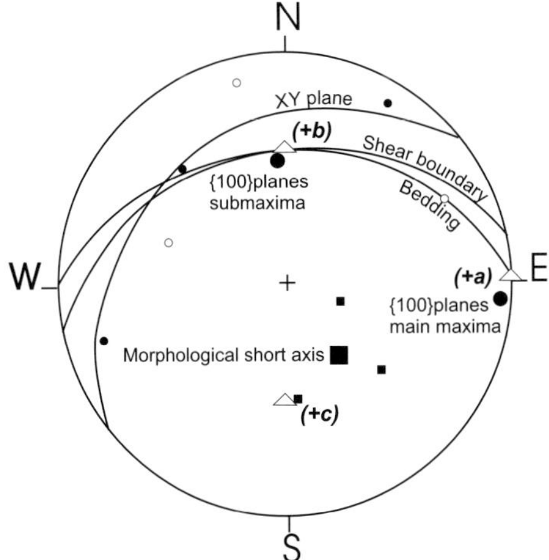

Fig. 11. Equal area stereographic projection in the lower hemisphere of the main tectonic directions, the three grain shape ellipsoid axes and the main maxima of the {100} poles. The long (•) and intermediate axes (○) of the grain shape ellipsoid plot on a great circle that defines the *XY* plane of the strain ellipsoid. This plane is at 20° to the shear boundary. △, axes of the stratigraphic coordinate system; ▪, the short morphological axes; ■, mean short axis orientation. ● maxima of the {100}planes.

halite would precipitate in the small voids created during the humid periods, in the form of turbid halite grains (Fig. 5g). These turbid grains are mostly located at the grain boundary structure of the limpid grains, suggesting a partial dissolution of the latter. Similar fabrics have been described in a petrographic study of nitrate deposits of the Northern Atacama desert, Chile (Searl & Rankin 1993), where the nitratite crystals enclose different densities of smaller solid inclusions (<2 μm diameter) including another type of nitrate salt. Two possible processes could have led to these fabrics: flooding of the ore horizons by brines or the mobilization of significant amounts of soluble salts by fresh rainwater and a rapid evaporation of groundwater (Searl & Rankin 1993). In the Cardona diapir, we suggest that rainwater dissolves the halite and washes the clay layers out of the deposit outcrop creating a turbid saturated brine. Subsequently, a rapid precipitation in the form of turbid halite in the voids created by prior dissolution would trap as solid inclusions the less soluble minerals, which could have undergone partial dissolution.

Once the meteoric fabrics have been identified, the deformational fabrics can be studied. In this paper we have focused on the fabric changes that occurred in the internal shear zone (*sensu* Kupfer 1968) as it crosses two different salt units. The internal shear zone is a zone of higher deformation with respect to the two adjacent spines. The internal structure of the shear zone shows no folding in the lower part of the outcrop, where bedding inside the shear zone is completely parallel to the shear boundaries and there is considerable thinning of the beds. In the upper part of the outcrop, there is some obliquity between the shear boundaries and the bedding, which generates folding (Fig. 5a and b). Moreover, in the lower part of the outcrop the shear zone crosses the Lower Salt Unit whereas in the upper part it crosses the Upper Salt Unit. Bed thinning and grain recrystallization are observed in both salt units, but increase in grain size is not. The Lower Salt Unit undergoes an increase in grain size from 1–2 cm to 7–10 cm, whereas in the Upper Salt Unit grain size remains constant at around 1 mm. In metamorphic rocks, inclusion inhibition by accessory minerals in the form of small particles is often responsible for grain-size variations in adjacent layers with varying contents. However, at relatively high temperatures, the motion of grain boundaries may not be impeded by these accessory minerals (Hobbs *et al.* 1976). In rock salts, trace amounts of brine can act like heat to weaken salt (Talbot & Jackson 1987). From experimental work on small halite grain aggregates (1 mm), water has been suggested to play an important role in the deformation mechanism even when it is present in very small amounts (0.05 wt %) (Spiers *et al.* 1986; Urai *et al.* 1986, 1987). Moreover, water has also been documented to enhance salt deformation in nature as measured in the flow of a namakier (Talbot & Rogers 1980). In the present case, accessory minerals in the form of small particles are present in both salt units in the initial fabrics, whereas the water content and initial grain size are different. The total water content of the Lower Salt Unit of the Cardona formation, in undeformed areas, is high (0.3 wt % average) according to de las Cuevas & Pueyo (1995), and can even be higher if the brine trapped in the fluid inclusions is considered. By contrast, in the Upper Salt Unit, the free intergranular water of primary fabrics is lower (0.12 wt % average according to de las Cuevas & Pueyo (1995), although it is higher than the water content measured in the samples affected by the shear zone in the Cardona diapir (<0.01 wt %). Moreover, the primary grain size of the Upper Salt Unit is one order of magnitude smaller than in the Lower Salt Unit. We suggest that the initial grain size and the available water content are the main controlling factors in this case. However, when water is scarce and grain size small, the accessory minerals may become an important factor controlling grain-size variations. In this sense, the role of accessory minerals as small particles could be more important in the Upper Salt Unit than in the Lower Salt Unit, as the amount of water in this unit is lower.

The movement of the two spines in the Cardona diapir was different and, although the horizontal component is unknown, the southeastern spine had a larger upward vertical movement. This suggests a simple shear component in the shear band. This shear component is also supported by the fabric. The long and intermediate axes of the grain shape ellipsoid plot on a great circle in the stereographic projection (Figs 10 and 11). This plane represents a statistical cleavage that is not observed with the naked eye and that defines the xy plane of the deformation ellipsoid. The angle between the xy plane and the shear boundary (20°) indicates deformation by simple shear although a pure shear component cannot be completely ruled out. Assuming a simple shear mechanism, a shear value of 4.2 for the sampled site is calculated. Other evidence that the clear halite fabric is a synkinematic fabric is the absence of primary features (Fig. 5d and e). In addition, although this fabric has a similar grain size to that of the initial fabric known for these levels in other areas (Miralles 1999) it has a different grain shape. The main difference is the presence of straight grain boundaries and even some grains tending to a euhedral section (Fig. 5e). The amount of available water during deformation and the worm-like grain boundary structures suggest that this fabric results from fluid-assisted diffusional creep (Urai *et al.* 1986). However, a small grain size could also be obtained by the breakage of previously elongated grains after exceeding a critical value of elongation (Talbot & Jackson 1987).

The poles of the {100} planes plot close to the shear boundary plane, suggesting that they could be genetically related. Initially, samples 1 and 2 show a {100} pole figure in which only one {100} face is fixed perpendicular to the shear plane, whereas the other two {100} planes have no orientation restrictions (Figs 10 and 11). However, in sample 3, there is a higher density of {100} poles oriented parallel to the measured shear plane direction. This indicates that the {100} plane perpendicular to the intersection between the layering and the shear zone tends to have a preferred orientation. Similar patterns of {100} crystallographic preferred orientations have been described for domal salt by Muelhberger & Clabaugh (1968) in which a large strain by slip on primary $\{100\}\langle 1\bar{1}0\rangle$ followed by annealing recrystallization was proposed by Schwerdtner (1968). However, in the study case, the {100} crystallographic preferred orientations are coherent with the suggested fluid-assisted synkinematic recrystallization as suggested by the fabric changes and the strong decrease in water.

Conclusions

In the shear zone studied a synkinematic recrystallization of halite has formed an anisotropic fabric.

The intermediate and long axes of the shape ellipsoid are randomly distributed on a plane that statistically defines the cleavage plane (xy). This cleavage is at an angle (20°) to the shear zone boundary, suggesting a simple shear deformation of $\gamma = 4.2$.

For small grain size layers (Upper Salt Unit), the synkinematic recrystallization fabric is characterized by a grain size similar to that of the initial fabrics, straight grain boundaries with thin brine films, poor orientation and no marked elongation.

Initial grain size and available water content are the controlling factors for the final grain size of the synkinematic fabric.

The inferred decrease in water content suggests that recrystallization by fluid-assisted processes was the main operative mechanism.

Maxima of the {100} crystallographic axes indicate that, with increasing deformation, the orientation of the equivalent crystallographic faces that are perpendicular to the shear zone is progressively fixed.

This work has been partially funded by CEE contract FI-1W-0235-E(TT) in the R & D programme on management and storage of radioactive waste—Part B. H.A.W. (High Active Waste) developed for ENRESA. Additional support was provided by CIRIT projects GRQ94-1049 and 1996SGR-0070, and Project 1997SGR00073 of the Geodinàmica i Anàlisi de Conques group. We would like to thank Sal Roja S.L. for permission to have access to the diapiric outcrop and to use their facilities. We also wish to thank ADARO for allowing us to review their internal reports on the Cardona salt mine, and Serveis Cientifico-tècnics from the Universitat de Barcelona for the image analysis. The authors are grateful to S. Galí for his contribution to the mathematical development of the 3D shape ellipsoids and the computer programs to calculate the pole figures, and to M. Gómez and J. M. Casas for their comments on the first versions of the manuscript. The stereographic projections have been performed with the R. Allmendinger STEREONET v.4.9.5. program. We thank J. P. Burg and W. M. Schwerdtner for their careful review of the paper and suggestions, and Springer-Verlag and F. Mauthe for permission to use the published figure (Fig. 3) of the Cardona diapir.

References

ANADÓN, P., CABRERA, L., COLLDEFORNS, B. & SÁEZ, A. 1989. Los sistemas lacustres del Eoceno superior y Oligoceno del sector oriental de la cuenca del Ebro. *Acta Geologica Hispánica*, **24**, 205–231.

AYORA, C., TABERNER, C., PIERRE, C. & PUEYO, J. J. 1995. Modelling the sulphur and oxygen isotopic composition of sulphates through a halite–potash sequence: implications for the hydrological evolution of Upper Eocene South Pyrenean Basin. *Geochimica et Cosmochimica Acta*, **59**, 1799–1808.

BUSQUETS, P., ORTÍ, F., PUEYO, J. J. *et al.* 1985. Evaporite deposition and diagenesis in the saline (potash) Catalan Basin, Upper Eocene. *In*: MILÁ, D. & ROSELL, J. (eds) *6th European Regional Geology Meeting (IAS), Excursion Guidebook*, 13–59.

CARTER, N. L. & HANSEN, F. D. 1983. Creep of rocksalt. *Tectonophysics*, **92**, 275–333.

——, KRONENBERG, A. K., ROSS, J. V. & WILTSCHKO, D.V. 1990. Control of fluids on deformation of rocks. *In*: KNIPE, R. J. & RUTTER, E. H. (eds) *Deformation Mechanisms, Rheology and Tectonics*, Geological Society, London, Special Publications, **54** 1–13.

DAVIS, J. C. 1986. *Statistics and data analysis in geology*. John Wiley & Sons, New York.

DE LAS CUEVAS, C. & PUEYO, J. J. 1995. The influence of mineralogy and texture in the water content of rock salt formations. Its implication in radioactive waste disposal. *Applied Geochemistry*, **10**, 317–327.

JACKSON, M. P. A. & TALBOT, C. J. 1987. External shapes, strain rates and dynamics of salt structures. *Geological Society of America Bulletin*, **97**, 305–323.

HOBBS, B. E., MEANS, W. D. & WILLIAMS, P. F. 1976. *An Outline of Structural Geology*. John Wiley & Sons, Singapore.

KUPFER, D. H. 1968. Relationship of internal to external structure of salt domes. *In*: BRAUNSTEIN, J. & O'BRIEN, G. D. (eds) *Diapirism and Diapirs*. AAPG Memoirs, **8**, 79–89.

—— 1976. Shear zones inside Gulf Coast stocks help to delineate spines of movement. *AAPG Bulletin*, **60**, 1434–1447.

LARSEN, J. & LAGONI, P. 1984. *Zechstein Salt Denmark; Salt Research Project EFP-81, Vol. 3: Fabric Analysis of Domal Rock Salt*. Geological Survey of Denmark Series C, **1**, 1.

MILTON, N. J. 1980. Determination of the strain ellipsoid from measurements on any three sections. *Tectonophysics*, **64**, 19–27.

MIRALLES, L. 1999. *Estudi de la fàbrica de roques d'halita: aplicació a la conca Potàssica Sudpirinenca*. PhD thesis, Universitat de Barcelona.

—— & SANS, M. 1996. Fábrica de rocas salinas en una zona de cizalla (anticlinal de Súria, Barcelona). *Geogaceta*, **20**, 763–766

MUELHBERGER, W. R. & CLABAUGH, P. S. 1968. Internal structure and petrofabrics of Gulf Coast salt domes. *In*: BRAWNSTEIN, J. & O'BRIEN, G. D. #(EDS) *Diapirism and Diapirs*. AAPG Memoirs, **8**, 90–99.

PUEYO, J. J. 1975. *Estudio petrológico y geoquímico de los yacimientos potásicos de Cardona, Suria, Sallent (Barcelona, España)*. PhD thesis, Universitat de Barcelona.

RIBA, O., REGUANT, S. & VILLENA, J. 1983. Ensayo de síntesis estratigráfica y evolutiva de la cuenca terciaria del Ebro. *Geología de España*, **7**, 131–159.

ROSELL, L. & PUEYO, J. J. 1997. Second marine evaporitic phase in the South Pyrenean foredeep: the Priabonian Potash Basin. *In*: BUSSON, G. & SCHREIBER, B. Ch.(eds) *Sedimentary Deposition in Rift and Foreland Basins in France and Spain* (*Paleogene and Lower Neogene*). Columbia University Press, New York, 358–387.

SÁEZ, A. 1987. *Estratigrafía y sedimentología de las formaciones lacustres del tránsito Eoceno Oligoceno del NE de la cuenca del Ebro*. PhD thesis, Universitat de Barcelona.

SANS, M. 1999. *From thrust tectonics to diapirism: the role of evaporites in the kinematic evolution of the eastern south Pyrenean front*. PhD Thesis, Universitat de Barcelona.

—— & VERGÉS, J. 1995. Fold development related to contractional salt tectonics: southeastern Pyrenean thrust front, Spain. *In*: JACKSON, M. P. A., ROBERTS, D. G. & SNELSON, S. (eds) *Salt Tectonics: a Global Perpective*. AAPG Memoir, **65**, 369–378.

——, MUÑOZ, J. A. & VERGÉS, J. 1996*a*. Triangle zone and thrust wedge geometries related to evaporitic horizons (southern Pyrenees). *Bulletin of Canadian Petroleum Geology*, **44**, 375–384.

——, SÁNCHEZ, A. L. & SANTANACH, P. 1996*b*. Internal structure of a detachment horizon in the most external part of the Pyrenean fold-and-thrust belt (N Spain). *In*: ALSOP, G. I., BLUNDELL, D. J. & DAVISON, I. (EDS) *Salt Tectonics*, Geological Society, London, Special Publications, **100**, 65–76.

SCHWERDTNER, W. M. 1968. Intragranular glidding in domal salt. *Tectonophysics*, **5**, 353–380.

SEARL, A. & RANKIN, S. 1993. A preliminary petrographic study of the Chilean nitrates. *Geological Magazine*, **130**, 319–333

SPIERS, C. J., URAI, J. L., LISTER, G. S., BOLAND, J. N. & ZWART, H. J. 1986. The influence of fluid–rock interaction on the rheology of salt rock. *Nuclear Science and Technology*. EUR 10399 EN. Office for Official Publications of the European Communities, Luxembourg.

TALBOT, C. J. & JACKSON, M. P. A. 1987. Internal kinematics of salt diapirs. *AAPG Bulletin*, **71**, 1068–1093.

—— & ROGERS, E. A. 1980. Seasonal movements in a salt glacier in Iran. *Science*, **208**, 395–397

URAI, J. L., SPIERS, C. J., PEACH, C. J., FRANSSEN, R. C. M. W. & LIEZENBERG, J. L. 1987. Deformation mechanisms operating in naturally deformed halite rock as deduced from microstructural investigations. *Geologie en Mijnbouw*, **66**, 165–176.

——, ——, ZWART, H. J. & LISTER, G. S. 1986. Weakening of rock salt by water during long-term creep. *Nature*, **324**, 554–557

VERGÉS, J., MUÑOZ, J. A. & MARTÍNEZ, A. 1992. South Pyrenean fold-and-thrust belt: role of foreland evaporitic levels in thrust geometry. *In*: MCCLAY, K. R. (ed.) *Thrust Tectonics*. Chapman & Hall, London, 255–264.

WAGNER, G., MAUTHE, F. & MENSIK, H. 1971. Der Salzstock von Cardona in Nordostpanien. *Geologische Rundschau*, **60**, 970–996.

Extrusion dynamics of mud volcanoes on the Mediterranean Ridge accretionary complex

ACHIM KOPF[1] & JAN H. BEHRMANN[2]

[1]*Géosciences Azur, B.P. 48, 06235 Villefranche-sur-Mer Cédex, France (e-mail: kopf@.obs-vlfr.fr)*

[2]*Geologisches Institut, Albert-Ludwigs-Universität Freiburg, Albertstrasse 23B, 79104 Freiburg, Germany*

Abstract: Drilling of two submarine mud domes situated in the Olimpi field on the northern flank of the Mediterranean Ridge accretionary complex has documented episodic eruptive activity over the last 1 to >1.5 Ma. Mud extrusion is related to plate convergence between Africa and Eurasia, having caused backthrust faulting of accreted strata containing overpressured mud at depth. The domes mainly consist of mud breccia with up to 65% of polymictic clasts embedded in a clayey matrix. On the basis of modifications of Poiseuille's and Stokes' laws, mud extrusion rates were calculated for Milano and Napoli mud domes. Mud ascent velocities are estimated to be up to 60–300 km a^{-1}, and are comparable with those of silicate magmas. Using physical property, structural and flux data of the mud breccias, and compiled data from mud domes on land, the diameter of the feeder channel and the depth of origin for the overpressured muds could be reliably estimated for the first time. Feeder channels are likely to be only a few metres wide. Gas efflux estimates constrain a source depth to *c.* 1700 ± 50 m below sea floor in the Olimpi field, which is considerably shallower than estimations made from the thermal maturity of solid organic carbon in the mud breccias. The efflux data suggest that the overpressured muds were not mobilized at décollement depth, but at a shallower level within the accretionary prism. A comparison of mud ascent rates (as determined from Poiseuille flow) and the total volumes of mud extruded indicate that only a fraction of the time span constrained from biostratigraphic data (*c.* 1 Ma) is needed to build up the Milano and Napoli mud domes. Durations of 12–58 ka of extrusive activity suggest mud volcanism here to have been a highly episodic phenomenon.

Mud volcanism and diapirism are well-known phenomena both in areas undergoing convergence (e.g. Brown & Westbrook 1988; Shipley *et al.* 1990) and extension (e.g. Limonov *et al.* 1995). The presence of mud domes and ridges along parts of the northern flank of the Mediterranean Ridge accretionary complex is related to its overall collisional tectonics between Africa and Eurasia (Camerlenghi *et al.* 1995). Geophysical data suggest that individual mud domes vary greatly in size (Limonov *et al.* 1994; Camerlenghi *et al.* 1995). The material recovered by traditional coring devices was described as 'mud breccia' (Cita *et al.* 1981), showing variable lithologies and ages of clasts in a clayey matrix (Staffini *et al.* 1993). During Ocean Drilling Program (ODE) Leg 160, the drilling of two transects of deep holes into two mud domes (Milano and Napoli, Fig. 1) permitted insights into their formation and evolution (Robertson & Shipboard Scientific Party of ODP Leg 160 1996). Different types of mud volcano deposits (for convenience summarized as 'mud breccia' in this paper,

From: VENDEVILLE, B., MART, Y. & VIGNERESSE, J.-L. (eds) *Salt, Shale and Igneous Diapirs in and around Europe*. Geological Society, London, Special Publications, **174,** 169–204. 1-86239-066-5/00/$15.00

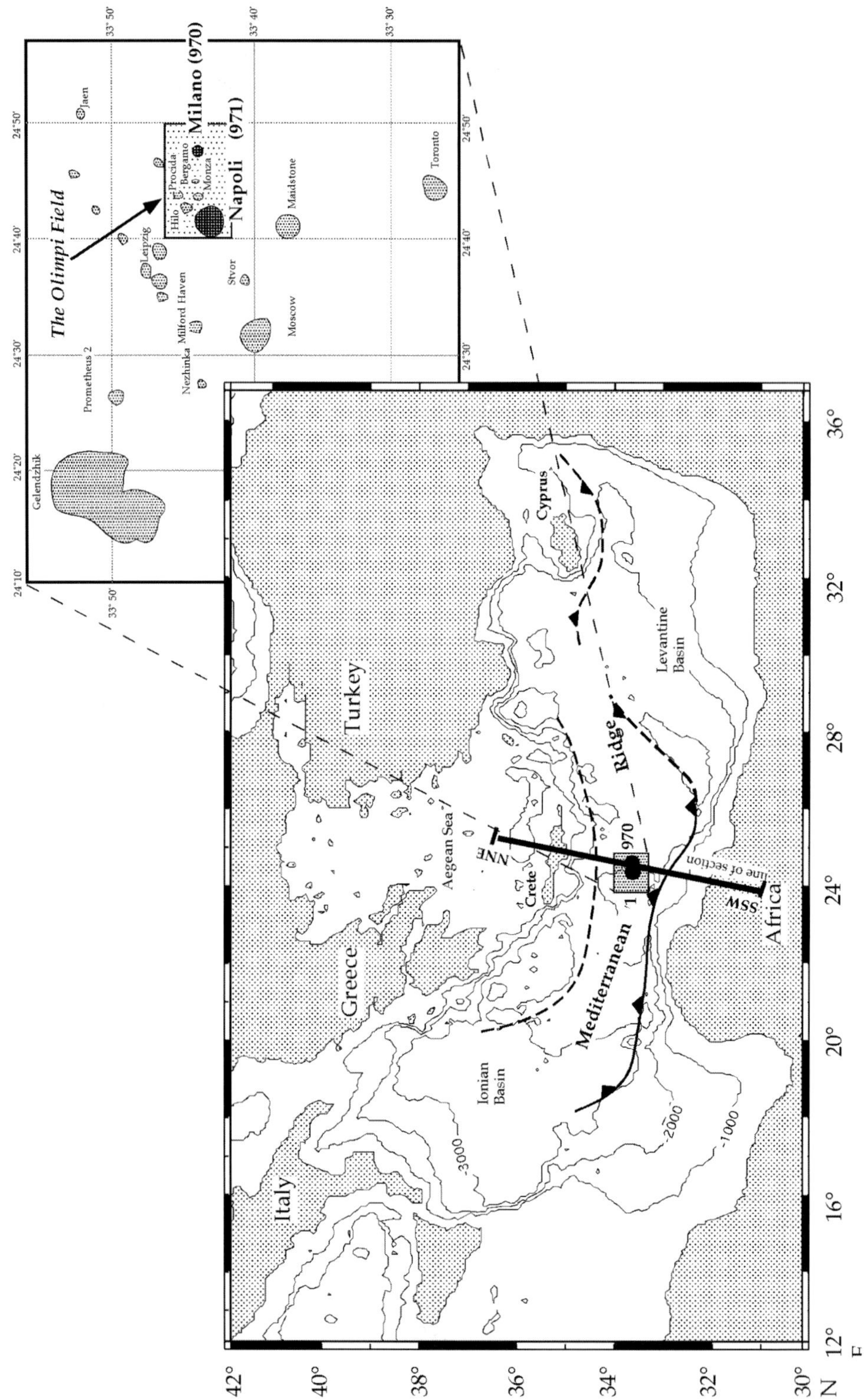
The Olimpi Field
Milano (970)
Napoli (971)
Hilo
Procida
Bergamo
Monza
Jaen
Leipzig
Milford Haven
Stvor
Maidstone
Toronto
Moscow
Nezhinka
Prometheus 2
Gelendzhik
Turkey
Greece
Italy
Africa
Aegean Sea
Crete
Cyprus
Levantine Basin
Ionian Basin
Mediterranean
Ridge
970
NNE
SSW
line of section
-3000
-2000
-1000
42°
40°
38°
36°
34°
32°
30° N
12° E
16°
20°
24°
28°
32°
36°

as they usually contain a certain amount of clasts) were recovered and provided information concerning the anatomy, age and origin of these enigmatic structures (see Robertson & Kopf 1998). The mechanisms of formation of mud volcanoes are strongly dependent on the consolidation and saturation conditions of fine grained sediments at depth. A model of evolution of the state of consolidation of such sediments following a specific stress path, which eventually leads to the formation of mud volcanoes and sedimentary diatremes following gas expansion in the pore spaces, has been proposed by Brown (1990).

In this study, the findings of the ODP drilling campaign (Emeis & Shipboard Scientific Party 1996) together with analyses of facies, clast composition and physical factors such as viscosity, particle size, strength and permeability on the material recovered (Kopf *et al.* 1998) were combined with geophysical data and quantitative approaches from earlier expeditions (Camerlenghi *et al.* 1995). This extensive data compilation was undertaken as a large number of variables is required to calculate the rate of mud efflux and the diameter of the feeder channel of the mud dome from Poiseuille's and Stokes' laws, and to estimate the depth of origin of unconsolidated mud using a quasi-static (rather than a dynamic) approach, which was already published in gas exploration literature (Cherskiy 1961). The potential driving forces, the episodicity and the mechanisms of mud extrusion on the Mediterranean Ridge are discussed on the basis of the above findings.

Geological setting: mud volcanoes near the inner escarpment

Mud domes occur on the Mediterranean Ridge in several zones, the largest of which is the Olimpi field and its surroundings (Fig. 1). In this zone, which is elongated subparallel to the local ENE–WSW tectonic grain of the accretionary complex (Fusi & Kenyon 1996), the majority of the mud structures have a dome-like shape (e.g. Limonov *et al.* 1994; Camerlenghi *et al.* 1995). Initial site surveys during ODP Leg 160 (Emeis & Shipboard Scientific Party 1996) revealed the flat-topped, slightly asymmetrical nature of the volcanoes with depressional features (e.g. a shallow moat as well as inward-dipping reflectors underlying the flanks; see seismic sections in Figs 2 and 3), which have also been documented from mud volcanoes elsewhere (Barber & Brown 1988).

The main features in the collision zone between the African and Eurasian Plates are shown in an interpreted cross-section (Fig. 4) perpendicular to the strike of the Mediterranean Ridge (Fig. 1, NNE–SSW-oriented line). The tectonic setting of the mud volcanoes at the inner ridge of the accretionary complex corresponds to the outcrops of backthrust faults. The wedge consists of Cretaceous (possibly earlier) to Pleistocene sediments, including Miocene evaporites (Chaumillon & Mascle 1995), which are indirectly indicated to be brines (Emeis & Shipboard Scientific Party 1996).

Fig. 1. Map of the Eastern Mediterranean showing the location of the Milano mud volcano (Site 970) and Napoli mud volcano (Site 971). The N–S-oriented line across the accretionary wedge shows the location of Fig. 4. (Note the areas of small mud domes near the outer deformation front, which are shaded.) The inset shows the Olimpi area at the inner deformation front with its mud volcanoes varying in size and shape (after Kopf *et al.* 1998).

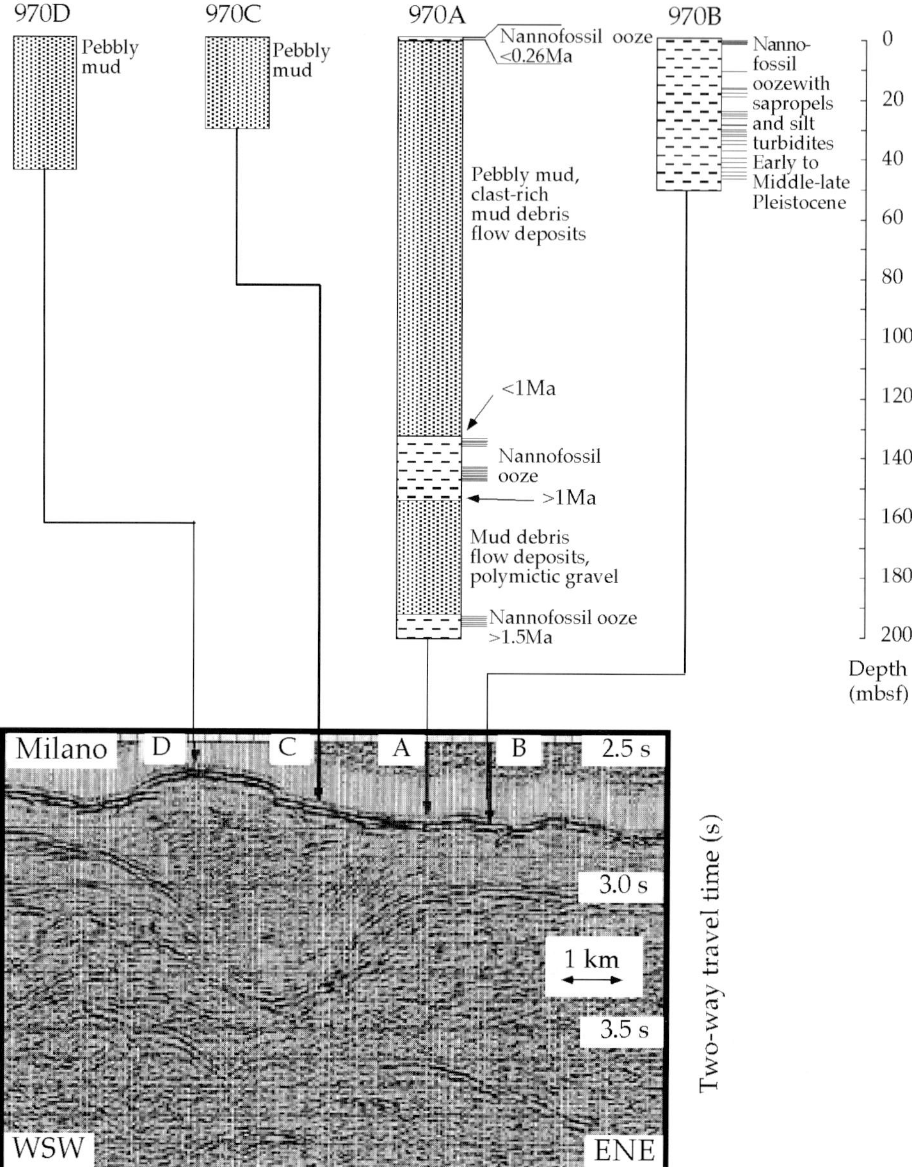

Fig. 2. Summary of the lithostratigraphy of the Milano mud volcano and their location on the migrated time sections. (Note the presence of inward-dipping reflectors beneath the flanks of the mud dome.)

Database

General results from ODP drilling

A brief summary of the findings from two transects of holes across the Milano (Site 970) and Napoli (Site 971) mud volcanoes is given in Figs 2 and 3.

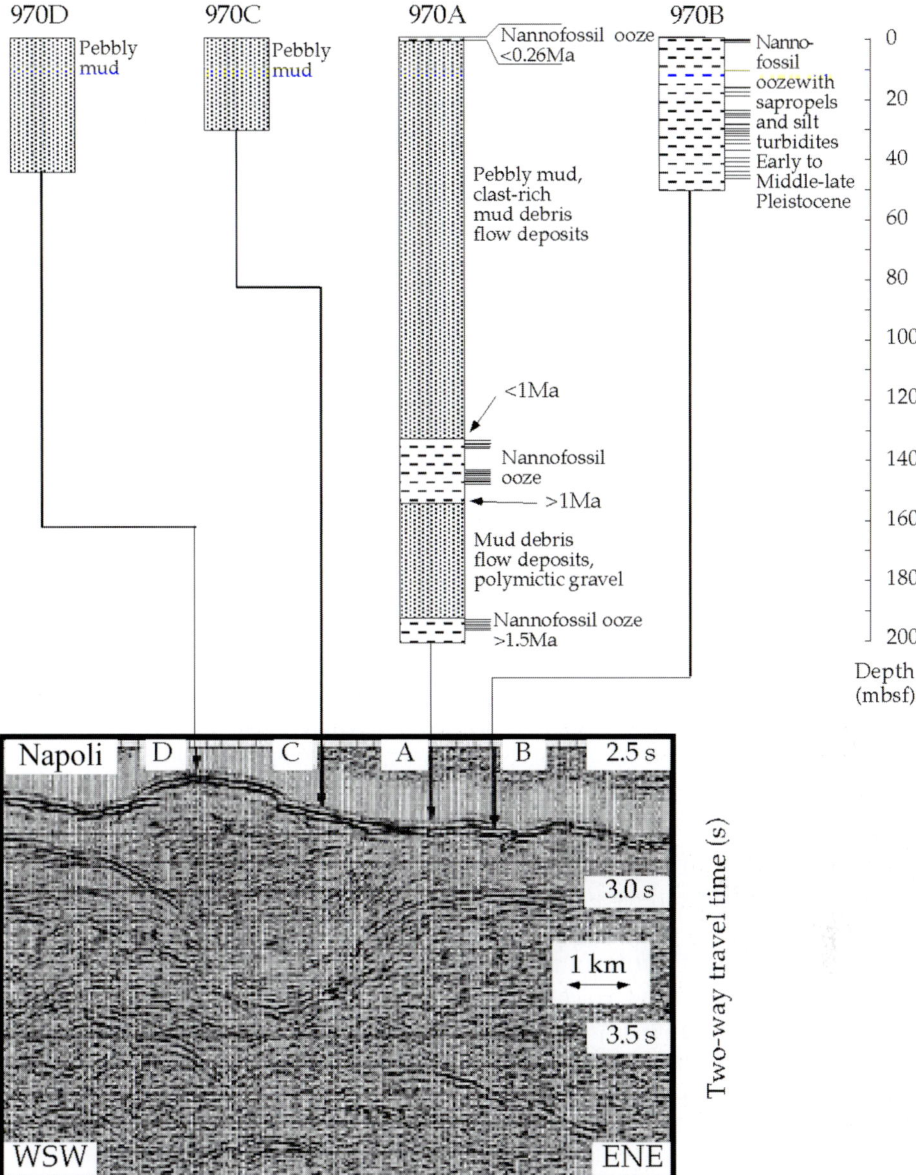

Fig. 3. Summary of the lithostratigraphy of the Napoli mud volcano and location of sediment types on the migrated time sections. (Note the presence of inward-dipping reflectors beneath the flanks of the mud dome as well as a moat encircling the structure.)

Milano mud volcano. The sediment recovered from the four-hole transect from topographic base to apex (Fig. 2) consisted of typical hemipelagites and various types of mud breccia (Emeis & Shipboard Scientific Party 1996; see Table 1). At the top of the successions recovered from the basal holes (970A, 970B), as well as interbedded with the mud breccias, the sediments comprise nannofossil oozes with sapropel beds and thin- to medium-bedded turbidites. Mud breccias consist of polymictic gravels

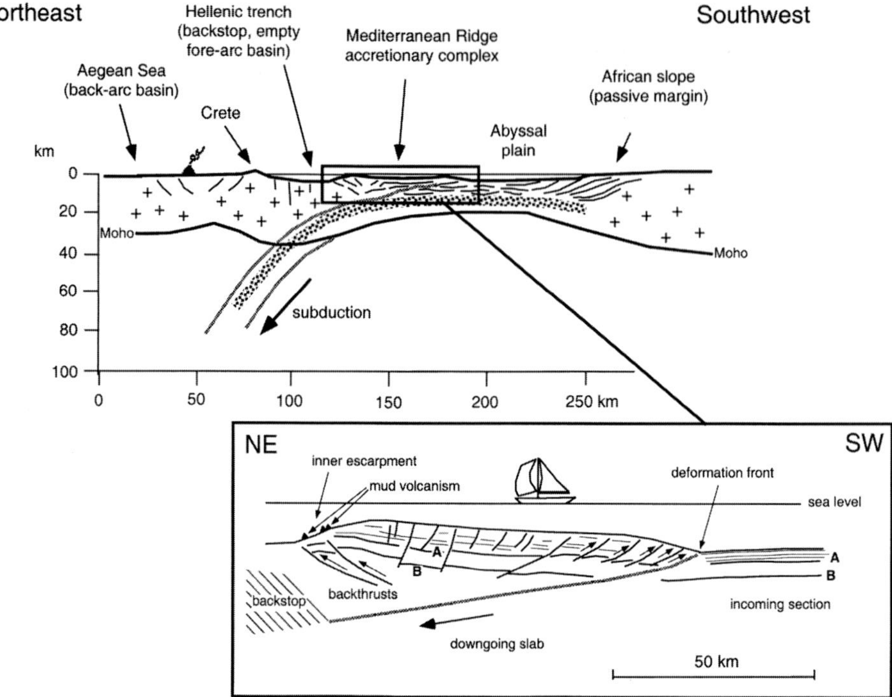

Fig. 4. Interpreted cross-section showing the tectonic setting of the mud volcanoes at the inner escarpment of the Mediterranean Ridge accretionary complex. In the sketch, A is the presumed level of the base of the Plio-Quaternary sediments and B is the base of the Messinian evaporites (after Chaumillon *et al.* 1996). (For location, see Fig. 1.)

(Fig. 5a) at the base (970A), which are overlain by mud debris flow deposits with clast contents between 25 and 65% (Fig. 5b). The uppermost (i.e. youngest) part of mud breccias were pebbly muds (only 10–20% clasts; Fig. 5c), which also build the flank (970C) and crest (970D) of the dome (Fig. 3). Critical findings from the drilling were the duration of mud volcanic activity over at least 1.5 Ma (determined biostratigraphically from the lowermost pelagic sediments at Hole 970A), the layering within the mud breccia as an indication for extrusive outflow rather than diapirism, and the general fining upward trend in the mud breccia (possibly the result of an initial discharge of highly overpressured fluids).

Napoli mud volcano. Five holes were drilled at the Napoli mud dome, again demonstrating interfingering between hemipelagites with largely clast-bearing, matrix-supported mud debris flow deposits. By comparison with the Milano dome, the overall clast abundance is only 15–25% of the total volume. The matrix contains a mixture of Pleistocene, mid-Miocene, Oligocene and Eocene nannofossils. The crestal holes (971D, 971E) recovered dominantly gaseous (e.g. methane, H_2S, higher hydrocarbons) silty clays with scattered clasts of mudstone, siltstone, and angular fragments of coarsely crystalline halite of presumed Messinian age (Figs 3 and 5d). Mud extrusion here began at *c.* 1.5 Ma (Robertson & Shipboard Scientific Party of ODP Leg 160 1996). Layering in mud breccias, although less obvious than at the Milano dome, was observed.

Table 1. *Summary table of the primary observations and bulk and dry densities of the different lithologies drilled at ODP Sites 970 and 971*

'Mud breccia' recovered during Leg 160	Pebbly muds	Mousse-like silty and sandy clays	Matrix supported mud debris flow deposits	Polymictic gravels
Sorting	massive	well sorted	poor to moderate	normal grading common, reverse grading and chaotic beds rare
Fabric	matrix supported	matrix supported	matrix supported	clast supported with rare matrix-supported fabrics
Matrix	silt–sand	silt and sandy clay	dominantly silty clay	?
Grain population	mixed pebble to silt grade	clay with rare sand, silt, or pebbles	three groups: silty clay; cllasts up to 2 cm across; pebbles (> cm)	gravel to clay
Angularity of clasts or pebbles	angular–rounded	subangular–angular	angular–rounded	sub-rounded
Gas-derived texture	slightly gaseous	mousse-like	?	?
Upper contact	Gradational in 970A with overlying hemipelagic sediment	not exposed	not exposed	sharp contact with overlying hemipelagic sediment in Core 970A-18X
Clast size	up to 120 mm	up to 40 mm	up to 140 mm (FMS indicates up to 0.6 m)	up to 50 mm
Bulk density ($kg\,m^{-3}$)	1800–1900	1200–1400	1700–2300	1800–2050

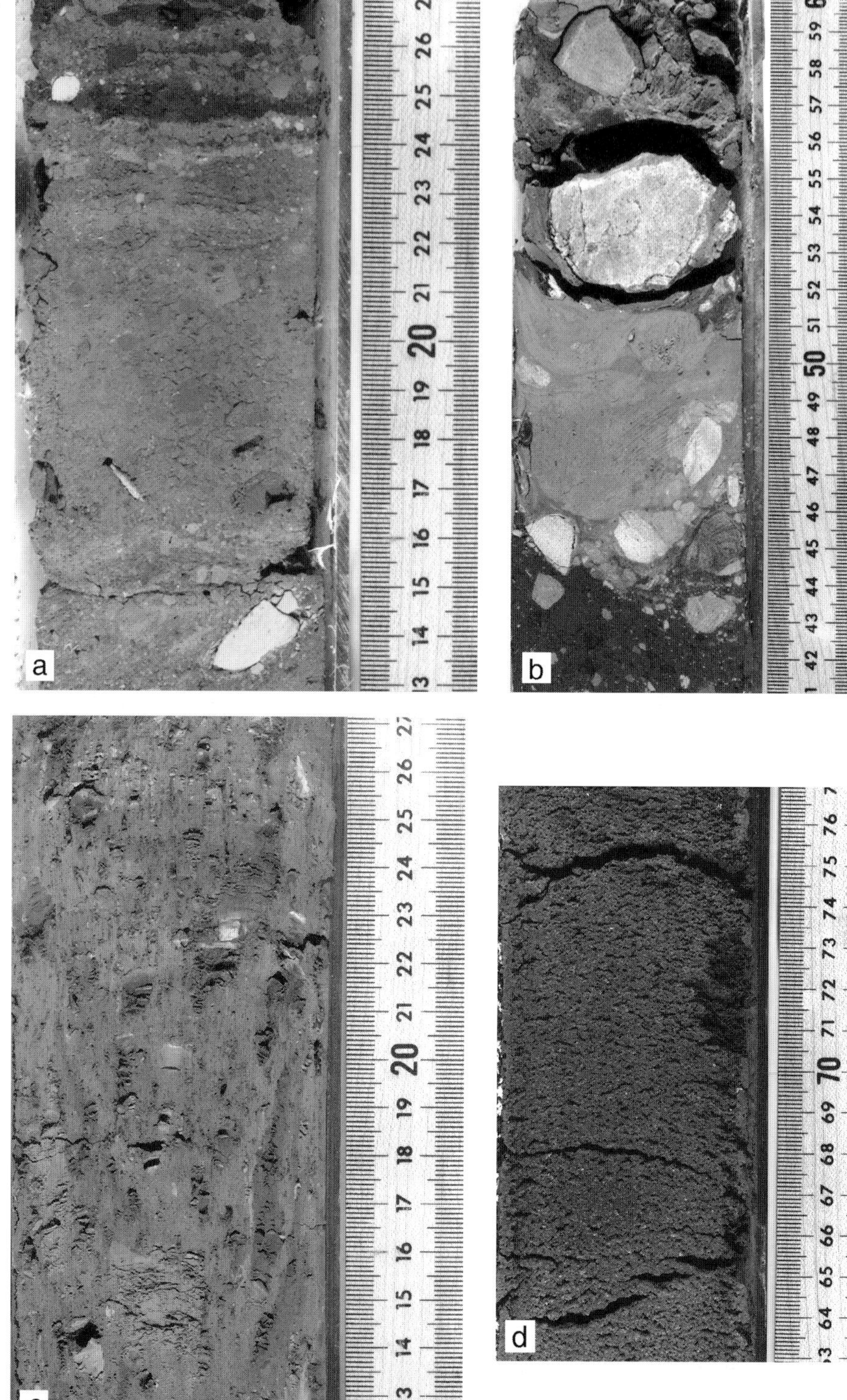
a
b
c
d

Seismic and bathymetric database

Seismic lines. The two seismic lines shown (Figs. 2 and 3) and additional cross-sections perpendicular to them (Emeis & Shipboard Scientific Party 1996) were interpreted to discriminate mud breccia from layered hemipelagic sediments. This has been based on the absence of discrete layering and continuous reflections right beneath the domes. Instead, a characteristic diffuse, seismically opaque pattern is observed when mud occurs (see published sections by, e.g. Limonov *et al.* (1994, 1995), or Camerlenghi *et al.* (1995)). To quantify the cross-sectional area, time–depth conversion was carried out using velocity data from both ODP Leg 160 and previous ODP cruises at convergent margins. The necessity to include data from the accretionary wedges off Southern Chile (Behrmann *et al.* 1992) and Nankai (Taira *et al.* 1991) resulted from the shallow terminal depth of the drillholes on the Mediterranean Ridge. From P-wave velocities determined on cores and with downhole logging devices, a 'best-fit' graph was calculated from the combined dataset (Fig. 6).

The depth-corrected interpretations of all four cross sections are shown in Fig. 7. If compared with previous work on migrated time sections (Camerlenghi *et al.* 1995, fig. 5), it is apparent that the shape of the dome both on the sea floor and as interpreted at depth (i.e. the base tapering off towards the 'feeder channel') is similar in the depth-corrected section. It should be noted also that for each mud volcano a minimum and maximum size interpretation is attempted (see caption of Fig. 7).

Bathymetric data. Three published bathymetric maps of the area were chosen to estimate the area covered by mud extrusions (Limonov *et al.* 1994; Hieke *et al.* 1996). The shape of the domes was defined by the deepest closed contour. This may imply underestimates, as ODP drilling demonstrated that the mud flows occasionally reach the opposite flank of the moat (e.g. Fig. 4, hole 971A). We shall return to this aspect in the discussion.

Fault zone geometries, sizes of mud domes and feeder channels

To estimate mud efflux quantitatively with time, cross-sectional area of the conduit (or width of a planar fault zone) has to be defined, both at depth and closer to the sea floor. Such data are not available for submarine mud domes. Thus, we took the approach to compile published data on the dimensions of faults and fault zones in accretionary systems, estimates from geophysical data, and observations on land, e.g. the geometry of dykes and the sizes of gryphons of active mud volcanoes.

Fault zone geometry. Investigation of fault geometry in accretionary systems is based on the relationship between mud volcanoes and active backthrust faults cross-cutting the wedge (Fig. 8). As a result of clast-rich mud breccia being the typical deposit recovered, and because of the angularity of the majority of the clasts (Flecker & Kopf 1996), we assume that fault zone geometries are crucial to initiate mud volcanism. Once the

Fig. 5. Core photographs of (**a**) polymictic gravel from section 970A-18X-2, 13–27 cm; (**b**) mud debris flow deposits from section 970A-14X-1, 41–60 cm; (**c**) pebbly mud from section 970A-6X-01, 13–27 cm; (**d**) mousse-like silty clay with tiny halite microclasts (light spots) from section 971D-1H-3, 63–77 cm.

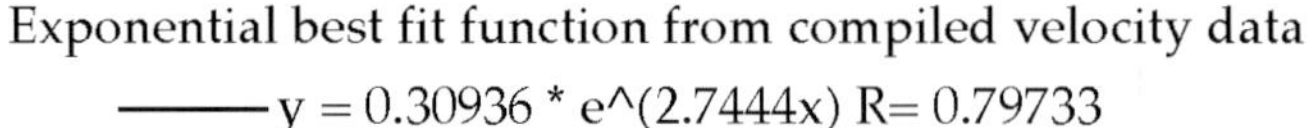

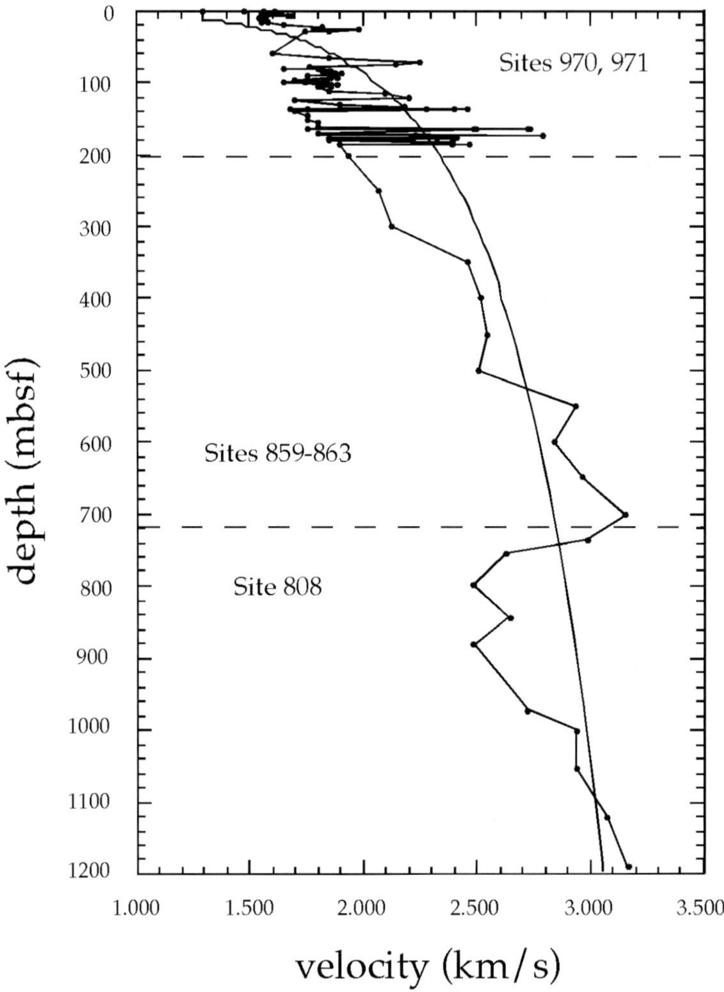

Fig. 6. Calculated 'best-fit' function from P-wave velocities obtained from ODP Legs 160 (Emeis *et al.* 1996), 141 (Behrmann *et al.* 1992) and 131 (Taira *et al.* 1991), which was used for depth conversion of seismic lines across the Milano and Napoli domes.

Fig. 7. Depth-corrected, interpreted cross-sections across the Milano (**a**) and Napoli (**b**) domes. Dark shading indicates the minimum area covered by mud breccia (min.); light shading (max.) indicates the area covered by mud breccia in addition to the minimal area, so that our maximum models result from min. + max.; it should be noted that for the NW–SE cross section over the Napoli dome, no clear distinction between generations of mud flow (i.e. min. and max. interpretation) could be made, so that the depth of 800 mbsf was chosen as a boundary according to the distinction made from the WSW–ENE cross section. (See text.)

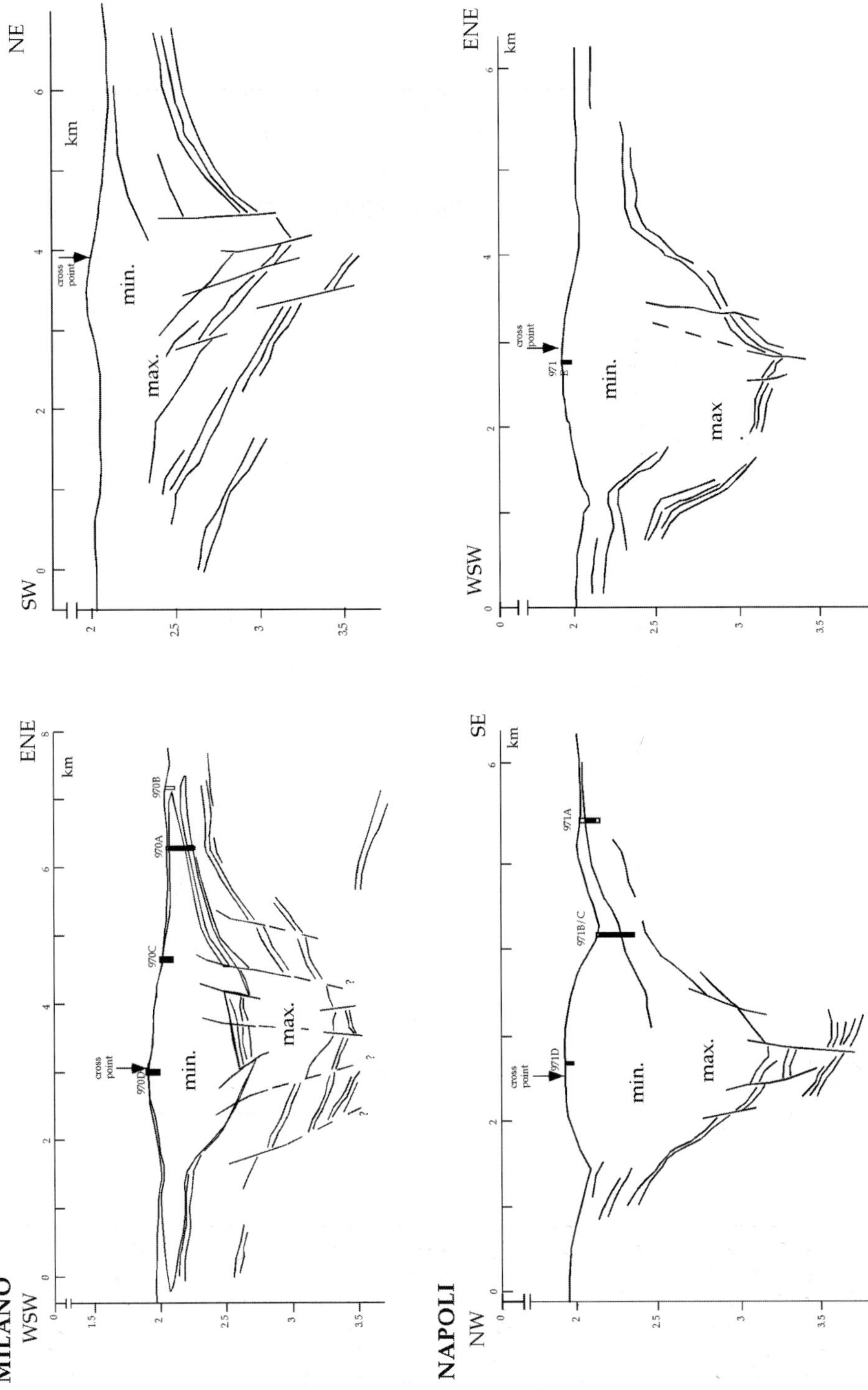
MILANO
WSW
ENE
km
cross point
970D
970C
970A
970B
min.
max.
NAPOLI
NW
SE
km
cross point
971D
971B/C
971A
min.
max.
SW
NE
km
cross point
min.
max.
WSW
ENE
km
cross point
971 E
min.
max

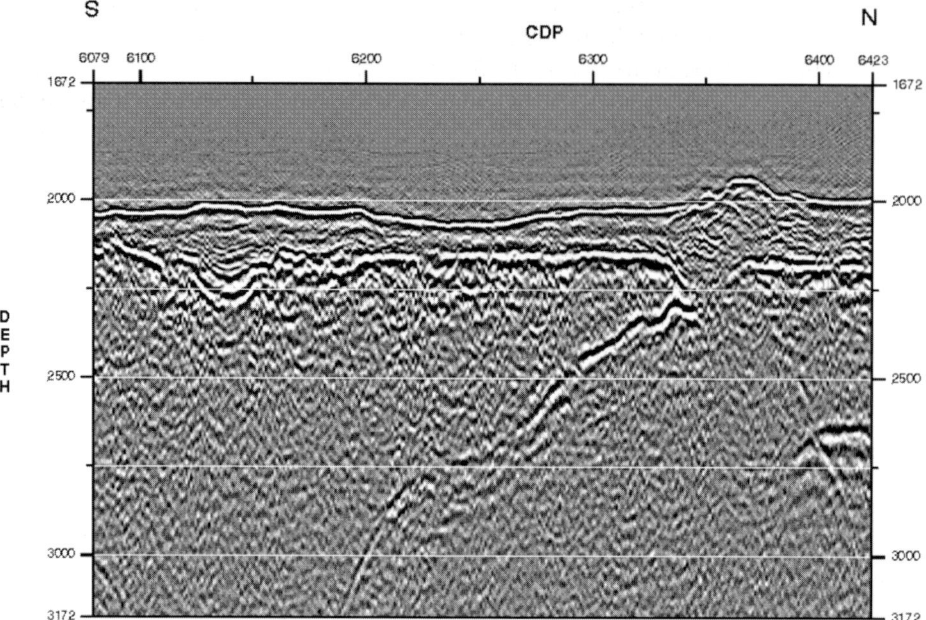

Fig. 8. Pre-stack depth-migrated seismic section across mud volcano 'Lich' from the central part of the Mediterranean Ridge. (Note the strong relationship to the backthrust fault along which mud migrates upwards.) The more or less continuous reflectors at sea-floor level cross-cutting the dome suggest the occurrence of a narrow feeder channel, at least within the limits of the seismic resolution. Modified from Kopf *et al.* (1999).

initial eruption has taken place, the fault breccia is replaced by mud, and the mud keeps the conduit open (be there extrusion or not). Consequently, first of all the geometry of the potential path has to be assessed. The dynamic interplay of stress (onto the fault plane as a result of convergence, and by the mud as a result of gas expansion, etc.) thereafter is very complex, and is not the subject of this study.

As a backthrust to a mud volcano has never been drilled, we have to rely on similar features in other areas. The most extensive and reliable database on structural features is provided by the Initial Reports of the Ocean Drilling Program. As claystones and siltstones from the toe of the Mediterranean Ridge accretionary prism (Sites 972 and 973, Emeis & Shipboard Scientific Party 1996) are very similar to accreted sediments elsewhere, we reviewed structural data from the Barbados Ridge (Mascle *et al.* 1988; Shipley *et al.* 1995), Nankai (Taira *et al.* 1991), Southern Chile (Behrmann *et al.* 1992), Cascadia (Westbrook *et al.* 1994) and Costa Rica (Kimura *et al.* 1997) convergent margins. The most abundant microfaults usually range from 1–5 mm width, but are not thought to penetrate the accreted strata completely. Fault zones (thrusts) were penetrated off Barbados (1 cm–1.5 m at Site 671; Mascle *et al.* 1988), Nankai (frontal thrust zone of 8 m thickness at Site 808; Taira *et al.* 1991); and Cascadia (13 and 50 cm wide at Site 891; Westbrook *et al.* 1994). The basal décollement zone was drilled off Barbados (c. 40 m thick at Site 671, Mascle *et al.* 1988; and 53 m thick at Site 949, Shipley *et al.* 1995) and Costa Rica (well-defined zone of only several centimetres at Site 1043, Kimura *et al.* 1997). In extensional settings, active fault zones have also been penetrated (e.g. in the Woodlark Basin; Taylor *et al.* 1999). Both activity of such faults as well as fluid discharge (and elevated fluid pressures along them) are believed to occur in an analogous manner. Again, splays of the Moresby Seamount detachment fault zone in the Woodlark Basin were found to be a few metres in thickness. As activity along the detachment and the décollements elsewhere, but also along the backthrusts in the Mediterranean Ridge's apex is evidenced, we trust the geometries to be very similar. As it is furthermore obvious that the mud volcanoes directly relate to such faults (see Fig. 8), we are confident that the compiled data provide clear evidence that independent of the lateral tectonic stresses (note that we do not only different convergence rates, but extensional settings as well), the width of a highly permeable fault zone is generally tens of centimetres to several metres. Even if we take our highest findings of some tens of metres off Barbados, the final value is significantly less if we consider that only a fraction of the entire fault zone is available for migration of fluids, or in our case mud. By far the bulk part of, for example, a décollement zone is fractured, well-consolidated mudrock, and pore volume (both primary, but more importantly secondary as a result of brecciation and hydrofracture) is generally less than 10 or 20%. Hence, a fault zone of, for instance, 50 m width equals effectively a slit of only 5–10 m width. Moreover, this is misleading given that the pore space occurs irregularly over the entire interval, so that friction of the individual channels and pores lowers the fluid or mud discharge considerably. This aspect will turn out to be crucial for simplifications of our quantitative estimates below, and we will take it up in the discussion of the results of this study.

Geophysical data. Feeder channels of submarine mud volcanoes were often mentioned in previous work (Limonov *et al.* 1994), but none the less we generally lack precise description of them. Estimates, however, were taken from geophysical data that have a limited resolution. Bright spots and 'opaque' patterns found on seismic reflection profiles, although several kilometres wide, have been interpreted by previous workers as feeder channels of mud volcanoes (Ivanov *et al.* 1996). Such a hypothesis is not very convincing bearing in mind the problems associated with seismic surveys as a consequence of the evaporite-bearing sediments and the rough sea floor known as 'cobblestone topography' (Camerlenghi *et al.* 1992). The crestal areas of mud domes imaged via side-scan sonar and related techniques usually do not allow definition of feeder channels or gryphons.

Observation on land. Our third set of data is taken from descriptions of related phenomena, such as the ascent of magma, as well as overpressured sediments on land, e.g. in clastic dykes and polygonal fault systems in mudrocks (Lonergan *et al.* 1998), or in continental mud volcanoes (e.g. Yassir 1989).

Ascending magmas usually follow feeder channels, fractures and faults. Dependent on magma viscosity, the critical width of a cylindrical or planar feeder structure can be estimated (Petford *et al.* 1994, fig. 2). For magmas, the narrowest dykes observed in the field are only several centimetres wide (Reddy *et al.* 1993), whereas dykes are commonly several metres to tens of metres wide (e.g. Corry 1988). Viscosities of felsic magmas, however, range in the order of 10^5 to 10^{10} Pa s (Petford *et al.* 1994) and thus are higher than those for gaseous mud (see below).

Mud volcanism on land is a worldwide phenomenon. Summaries have been given by Higgins & Saunders (1974) or Yassir (1989). Especially in regions of the former

Soviet Union, abundant mud volcanoes disseminate considerable amounts of unlithified rock, oil and gas (Vereshchagin & Kovtunovich 1970; Jakubov *et al.* 1971). In Azerbaijan, more than 200 mud volcanoes are known and have been described in great detail (Jakubov *et al.* 1971), showing feeder channels and gryphons between 0.5 and *c.* 20 m across. Moreover, radial and concentric fractures of 10–150 cm width have been observed in the crestal area before mud extrusion took place. On Sakhalin, similar mud domes with feeder channels of 2 m in diameter were found (Gorkun & Siryk 1968). On Trinidad, Higgins & Saunders (1974) described vents of *c.* 2–5 m in diameter, situated on the mud volcano crest (i.e. crater area). The Atalante mud volcano at the deformation front of the Barbados Ridge accretionary complex has a wide conduit of *c.* 200 m in diameter (Henry *et al.* 1996).

In general, the diameter and relative heights of the features on land are similar to the geometries known from the ocean floor (although there are exceptions of taller and wider geometry, e.g. Timor). For Azerbaijan, the diameter of the mud volcanoes ranges between 400 m and 6–10 km maximum, with heights of 5–500 m relative to the surrounding topography (Jakubov *et al.* 1971). In Trinidad, mud volcanoes are commonly smaller (100 m–2.5 km across) and rise several metres to several tens of metres above the surrounding topography (Higgins & Saunders 1974). The mud volcanoes described in the Olimpi field on the Mediterranean Ridge lie within this range (Camerlenghi *et al.* 1995, table 2), with basal diameters between 700 m and >3.5 km. Heights vary between 80 and 160 m above sea floor. However, much larger features, such as the Pan di Zucchero or Gelendzhik mud domes (or mud dome domains), were found in the vicinity (Fig. 1). Despite these apparent similarities, basic differences between the submarine domes and those on land have to be considered (see discussion below).

Age constraints

Age constraints are essential for estimates of the rate of mud extrusion. Biostratigraphic study of the sediment drilled at Sites 970 and 971 at Milano and Napoli domes recovered a continuous record of nannofossil zones mirroring regular pelagic sedimentation punctuated by mud debris flows. Owing to reworking of the species from some of these zones into the mud breccia, a division between mud breccia and hemipelagites at Milano was made (Fig. 2, lithologs). From both the upper (5–125 m, ranging from 0.26 Ma to <0.99 Ma) and lower (155–195 m, ranging between >1.5 Ma and 1.75 Ma) mud debris flows at hole 970A (Emeis & Shipboard Scientific Party 1996), sedimentation rates of 160–164 m Ma^{-1} can be estimated conservatively. The duration of active mud extrusion is 0.98 Ma or less. The massive debris flow at hole 971A at Napoli (20–80 mbsf (m below sea floor) with an age range of >0.46 Ma to 1.5 Ma; Fig. 3), however, can result in only *c.* 60 m Ma^{-1} sediment accumulation minimum estimate. Mud volcanism at the Napoli dome therefore lasted 1.04 Ma or less. These accumulation rates are not particularly meaningful, as the main part of a debris flow is usually deposited very quickly, and may be followed by a period of settling and sedimentation of the finer components. Rapid deposition is supported by the results from oedometer tests on mud breccia samples (Kopf *et al.* 1998). The compressibility suggests that thin mud breccia intervals should take only 10^2–10^4 a to consolidate, i.e. consolidation is faster than the geological timescale (for discussion, see Camerlenghi *et al.* (1995), fig. 14).

Gas efflux accompanying mud volcanism

Very little is known about the volume of gas dissipating from mud volcanoes, either at the sea floor or on land. This is especially true for submarine domes because of the difficulty of quantitative measurements. Henry *et al.* (1996), however, estimated a minimum fluid expulsion of 40 000 $m^3 a^{-1}$ for the Atalante mud volcano off the Barbados accretionary complex based on constant heat flow data. Its gas flux was even found to be 100 times higher than this (see Table 3c, below, for exact values), which is surprising bearing in mind the size of the mud pie being only 0.003 km^3. The data suggest a fluid supply from a large area into the mud volcanic system (Henry *et al.* 1996). No other quantitative flux rates from mud domes in marine settings are known to the authors.

On land, most of the data available have been gained in oilfields in the former Soviet Union. Cumulative gas efflux estimated for about 200 mud domes in Azerbaijan is 2.5×10^8 m a^{-1} from eruptions plus an additional 2×10^7 m a^{-1} from emission during phases of inactivity (Jakubov *et al.* 1971). Gas efflux estimated for the Pugachevskiy mud volcano on Sakhalin is 1.12×10^7 m a^{-1} (Vereshchagin & Kovtunovich 1970), similar to that of the Dashgil dome in Azerbaijan (1.46×10^7 m^3 gas per year, volume of 260 km^3 mud breccia; Jakubov *et al.* 1971).

In a simplifying approach because of lack of data on the Mediterranean Ridge, the relationship between gas dissipation per day and mud volume was calculated. These ratios were applied to the volume of mud known from the Olimpi field mud volcanoes (Table 3c). Being aware that both scenarios are somewhat different from the Olimpi field area, we used the results as upper (off the Barbados accretionary prism) and lower (on land in Azerbaijan) limit for the Mediterranean Ridge estimate of the volume gas flux Q. As a minimum, we obtained a daily flux of 40–280 m^3 using the 'on land' conversion factor, and *c.* $(3–7) \times 10^6$ to $(20–45) \times 10^6$ $m^3 day^{-1}$ on the basis of the active Barbados mud pie (detailed results are shown in Table 3c). In the latter case, both the flux estimates from an earlier (*c.* 1800 $m^3 day^{-1}$; Le Pichon *et al.* 1990) and a recent submersible cruise (*c.* 11 000 $m^3 day^{-1}$; Henry *et al.* 1996) to the Atalante mud volcano could be used, accounting for the range of the results (see Table 3c for details). For the Mediterranean Ridge mud volcanoes, a value somewhere between the obtained limits for Azerbaijan and Barbados was assumed to apply to the Olimpi field domes, mainly because of the situation neither at the deformation front nor on land, but 100 km hinterlandwards of the deformation front. This yields a gas flux of *c.* $10^3–10^4$ $m^3 day^{-1}$. This is, if absolute values are taken, well within the range for Barbados (see above) and for, for example, Azerbaijan (where the average flux is 3425 $m^3 day^{-1}$). We will come back to the influence of this rather crude estimate on our calculations below.

Viscosity tests

In addition to physical property measurements on mud breccias (e.g. water content, bulk and grain density, porosity, Vane shear strength) as part of the shipboard routines (Emeis & Shipboard Scientific Party 1996), we tested the viscosity of the mud matrix at different water contents. Dynamic viscosity h of the matrix was determined using a Bohlin CVO eccentric rotating disc viscosimeter (principles outlined by Macosko (1993)). For this purpose, the mud breccia samples were left in 4 mol %

sodium hexametaphosphate solution for 48 h to deflocculate. After separation of the fraction <100 mm (75.24 wt % of the entire sample) by sieving, a set of samples of different water content (24, 40, 54 and 60%) was produced for the viscosity tests. Separation was carried out both as the clasts are assumed to be of secondary origin (i.e. collected during mud uprise), and because of the grain-size limit of the apparatus.

Results

Composition and density

Matrix. Calcite and quartz are the main matrix constituents. Kaolinite, sodium-rich smectite, and minor chlorite and illite are the clay minerals present. Detailed information and a full list of X-ray diffraction results have been provided by Robertson & Kopf (1998).

Wet bulk density and grain density data have been outlined by Emeis & Shipboard Scientific Party (1996). In summary, wet bulk densities are in the range 1800–2000 kg m^{-3} for pebbly mud, 1700–2300 kg m^{-3} for mud debris flow deposits with variable clast content, 1800–2050 kg m^{-3} for the polymictic gravel, and only 1300–1700 kg m^{-3} for the virtually clast-free, mousse-like silty clays on the crest of the Napoli dome (Table 1). We have accepted an average value of 1600 kg m^{-3} for mud (i.e. matrix) and 2100 kg m^{-3} for clast-bearing mud breccia in our calculations. It should be noted that these density measures do not include gas, but do include pore water.

Clasts. The different varieties of mud breccia contain clasts of variable composition, size, shape and abundance (Table 2). A detailed description of the clast lithologies and likely provenance has been given by Robertson & Kopf (1998), and will not be repeated here. However, it is important to note that some of the clasts recovered, namely the most abundant, variably lithified calcareous muds and mudstones, are fragmented by non-systematic, dilatant fractures (probably hydrofractures; see Behrmann 1991). Other lithoclasts, dominantly pelagic carbonate and quartzose litharenites of turbiditic origin and variable provenance (Robertson & Kopf 1998), are generally well lithified and *c.* 5 cm in size. The largest clasts recovered during ODP Leg 160 were *c.* 10 cm across (limited by the drilling devices), but Formation MicroScanner images revealed clasts of up to 50 cm in diameter in the borehole wall (Robertson & Shipboard Scientific Party of ODP Leg 160 1996). The latter measure was used for Stokes calculations (see below).

Grain densities of mud breccia, as determined during ODP Leg 160 (Emeis & Shipboard Scientific Party 1996), were used as values for the clasts in our calculations. Out of the entire range (see Table 2), we accepted 2600 kg m^{-3} as an average value.

Viscosity tests

Viscosity. Viscosity tests were run on a set of samples covering the range of water contents from W_L and W_P(see above). The gap between the rotating discs was held between 0.6 and 0.8 mm for the multiple tests, ensuring it was more than 10 times wider than the largest particle of the sample. Results of viscosity determinations with increased frequency in rotation (i.e. deformation rate in s^{-1}) are shown in Fig.

Table 2. *Summary table of major clast types, sizes, cements, shape, density, colour and possible provenance*

Lithology	Size (mm)	Colour	Grain size, sorting	Shape	Carbonate content	Lithification	Provenance	Comments
Claystone, mudstone	1–60	black to dark greenish grey; dark reddish brown	clay size, homogeneous	angular–subrounded	non-calcareous	semi-consolidated	north	homogeneous, sometimes fissile
Calcilutite	5–185	light greenish grey; dark greenish grey	very fine	subangular–subrounded	calcareous	unconsolidated to semi-consolidated	north	homogeneous; occasionally other mudclasts inside
Pelagic limestones	20–40	middle grey	fine silt to sand, poorly sorted	angular–subangular	calcareous	well lithified	north	foram bearing, rarely laminated and fractured, in part healed with calcite; fossils range from Mid-Cretaceous age to present; oldest unreworked clasts are Burdigalian–Langhian in age
Siltstones	5–100	greenish grey, green, white	well sorted, coase to medium silt	subangular–sub-rounded	calcareous	well to semi-consolidated; rarely lithified	south	occasional lamination on a mm scale
Litharenites	10–50	brownish grey	fine and coarse grained; poorly sorted	subangular	calcareous	semi-lithified to well lithified	north	rare mudstone clasts inside, darkened fractures, dominantly strained, fragmented quartz grains from a potential 'ophiolitic' source; intraclasts of serpentinite, sphene, basalt, radiolarian chert
Sandstones, litharenites	30–140	greenish grey	medium to coarse sand, poorly sorted, occasionally well sorted	rounded–subangular	calcareous	semi-lithified to well lithified	south	quartz of both metamorphic and magmatic sources, often subrounded and generally undeformed; exsolved feldspar, mica, opaques, glaucony, benthic and planktic foraminifers, echinoid spines; lamination discernible
Halite	20–50	white	crystalline	angular	–	solid	–	occasionally mudstone clasts inside
Fibrous calcite	10–40	light grey to white	crystalline	subangular	calcareous	solid	–	fibrous growth

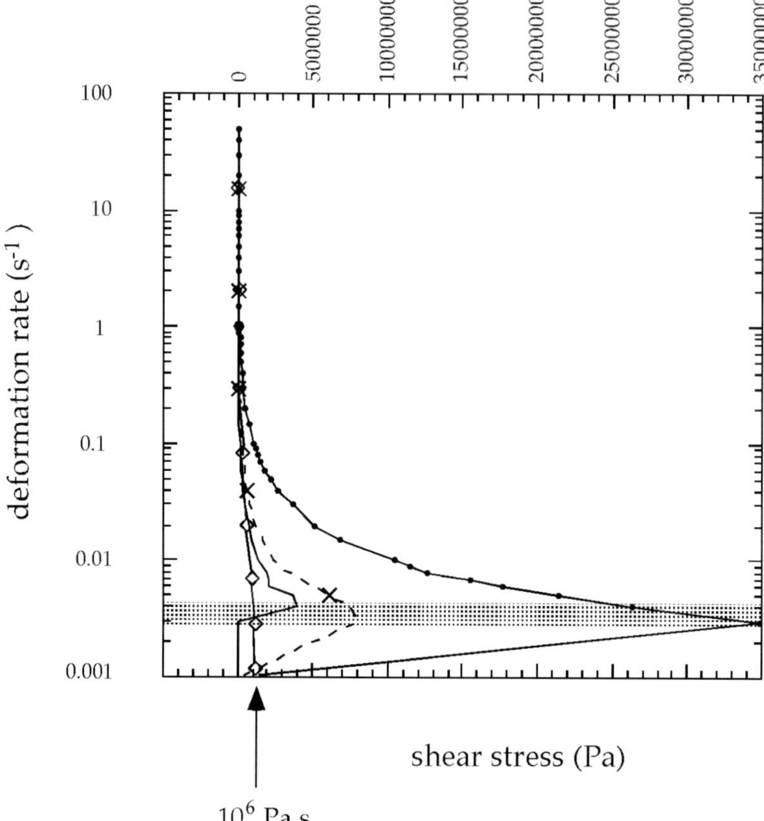

Fig. 9. Viscosity data from mud breccia matrix obtained from frequency sweeps (variable deformation rate v. shear stress). Plateau areas underneath the shaded bar describe the viscosity of the sample. The sample with 60% water content (open lozenges) reaches a plateau at *c.* 10^6 Pa s. The other three samples (54, 40 and 24% from left to right) tend towards higher viscosities, but did not reach a plateau and showed only artefacts beneath the shaded bar. (See text.)

9. Viscosity values determined for frequencies above 10^{-4} Hz have to be treated with caution, especially for the samples with low water contents, as can be seen from the lack of a distinct 'plateau' area of the graph, which usually describes the dynamic viscosity for a sample.

The viscosity determined for the sample with the highest water content (i.e. 60%, which is the liquid limit of the sample; see Fig. 9 and Kopf *et al.* (1998)) was 10^6 Pa s. This value may range up to one order of magnitude higher for the other three samples (24, 40 and 54%), depending on the water content and the number of oscillation tests carried out on the same sample. (Note that the clay-rich samples dry out very rapidly so that the sample with the highest water content automatically provides the most reliable results.) On the other hand, degassing of the same sample immediately after opening the sealed core-liner definitely led to underestimation of the viscosity. For the above reasons we used 10^6 Pa s as an approximation for calculations regarding the flux rate; possible errors resulting from this will be discussed further below.

Viscosity tests (Fig. 9) confirm that the mud is plastic with a shear stress threshold below which it does not deform. Within the range between W_L and W_P, it is, therefore, a non-Newtonian fluid. With increasing water content, the shear stress decreases dramatically, a phenomenon also found while testing other mud volcano samples from Barbados (Henry, pers. comm.). Nevertheless, mud is plastic only for low water contents (i.e. between W_L and W_P), but may behave as a Newtonian fluid at the highest water content tested and above that (as indicated in Fig. 9; see also Lance *et al.* (1998)). According to the water content and density measured on drill-cores, the fact that it was fully degassed when tested, and because of the moussy texture of the undeformed, crestal sediments at the Napoli dome (see Emeis & Shipboard Scientific Party 1996), the mud was treated as a Newtonian fluid in our calculations. Nevertheless, this aspect does not seem to be crucial, as mud movement may well occur under non-Newtonian conditions; it would simply require some modifications of Poiseuille's law (e.g. Turcotte & Schubert 1982).

Theoretical background for quantitative estimates

Our attempt to quantify mud efflux follows three lines: (1) we apply Poiseuille's and Stokes' laws for flow in a rigid, circular pipe to the phenomenon of mud volcanism to estimate both the ascent velocity and the volume flux rate of mud, using the geological and geophysical information compiled above; (2) we calculate the diameter of the feeder channels of the volcanoes, having modified Stokes' law for vertical movement of bodies of different density; (3) we estimate the depth of mud origin on the basis of Cherskiy's equations. Regarding, in particular, the depth estimates under (3), the justification for using a quasi-static approach (rather than a dynamic one) will be discussed in detail in a separate section after having introduced the results of the quasi-static theory.

Poiseuille's law

The discharge rate V of a fluid of viscosity η through a pipe with the radius $d/2$ (where d is the diameter of the pipe) can be modified from Poiseuille's flow law (see equation (6-33) of Turcotte & Schubert (1982)), so that

$$V = \frac{\pi g \Delta p (d/2)^4}{8\eta} \tag{1}$$

where Δp is the pressure gradient over the pipe and g ($= 9.81\ \mathrm{m\,s^{-2}}$) is the acceleration as a result of gravity, giving a volume v. time.

Stokes' law

By contrast, Stokes' law describes the velocity v' of a falling sphere (with radius r) in a fluid of viscosity η (in poise; 10 poise = 1 Pa s), as follows:

$$v' = \frac{2g\Delta\rho(2r)^2}{9\eta} \tag{2}$$

$\Delta\rho$ is the difference between the density of fluid (i.e. mud) relative to the surrounding rock ($\rho_{rock} - \rho_{mud}$). In an extensive numerical study of diapirism, Marsh (1982) modified Stokes' law to

$$v' = \frac{2g\Delta\rho d^2}{9\eta} \quad (3)$$

where d is the diameter of the pipe of the structure (e.g. Marsh 1982). Although there are differences between the ascent of magmas and mud in detail, the principal physics allow allocation of the above relationships (Petford *et al.* 1994). For the Mediterranean Ridge mud volcanism, mud ascent velocities can be calculated on the basis of Poiseuille flow, with η being the relevant viscosity of the mud. Estimates concerning the diameter, however, have to account for the size of the clasts transported within the mud breccia, and are based on Stokes' law.

For volume mud flux estimates (i.e. discharge V, see equation (1)), Stokes' law is modified, as follows:

$$V = \left[\frac{2g\Delta\rho(2r)^2}{9\eta}\right]\pi\left(\frac{1}{2}d\right)^2 \quad (4)$$

where V is the product of the velocity of mud ascent (i.e. height per time interval) multiplied by the cross-sectional area of the feeder channel $\pi(d/2)^2$. Note that the diameter d (of the pipe) is different from $2r$ (of the clast); also, v' (equations (2) and (3)) is velocity (in $m\,s^{-1}$) whereas V (equations (1) and (4)) is a volume flux rate (in $m^3\,s^{-1}$).

For both types of calculations, we require primary input and secondary geological controlling factors, as follows. Density (see above) is put into Stokes' equation as difference ($\Delta\rho$) between mud matrix and accreted wall rock. Wet bulk density values from areas free of clasts were taken as representative for the mud, whereas grain densities were thought to reflect clasts best. Other controlling factors, such as the viscosity of the matrix (η) and diameter of the feeder channel (d), are measured or calculated. Age control is provided by both biostratigraphic dating (see above and Emeis & Shipboard Scientific Party 1996) and indirectly as a result of consolidation behaviour of mud breccia (Camerlenghi *et al.* 1995). Particle size (Flecker & Kopf 1996), Atterberg limits and shear strength, and permeability (Kopf *et al.* 1998), are an important control on viscosity and therefore are crucial for the discussion of the mud extrusion dynamics (see below).

Cherskiy equations

Periodic eruptions of mud volcanoes are considered as a result of constant gas dissipation which may temporarily be hindered or prohibited to allow high fluid pressures to accumulate underneath this (often argillaceous) plug. Two equations (named after the author of the book in which they were published) describe the relationship between the gas pressure at the Earth's surface (i.e. when the shaft of a drillhole is opened; $p_{surface}$) and the pressure of this gas at depth (i.e. where the drillhole cuts the gaseous strata; p_{depth}), and allow calculation of the source depth of the overpressured mud. The theory is based on steady-state gas flux from a primary to a secondary deposit (the argillaceous strata), a pressure build-up and its sudden release by drilling (allowing the mixture to rise), until the drillhole is plugged again and the cycle can start from the beginning (Cherskiy 1961). Gas pressure is assumed as to equal the mud pressure when the source layer is penetrated and blow-up occurs. Applied to

Mediterranean Ridge mud volcanoes, this model involves gas or fluid migration to the mud reservoir, and tectonic activity (e.g. the backthrusts, see Fig. 4) to allow very rapid ascent of the mud–gas mixture. The relationship between gas pressure on the secondary deposit and after eruption of the mud dome has been expressed (Cherskiy 1961) as follows:

$$p_{\mathrm{surface}} = p_{\mathrm{depth}}\,\mathrm{e}^{S} \tag{5}$$

or, in greater detail (if gas composition, gas flux and feeder channel size are known), as

$$p_{\mathrm{surface}} = \sqrt{\left[p_{\mathrm{depth}}^{2}\,\mathrm{e}^{2S} + 1.377\lambda\frac{Q^{2}m_{\mathrm{V}}T}{d^{5}}(\mathrm{e}^{2S}-1)\right]} \tag{6}$$

where m_{V} (coefficient of volume compressibility), T (temperature in K) and λ (coefficient of friction; dimensionless, dependent on d) relate to the gas, d is the diameter of the drillhole (in cm), and Q is the rate of daily gas efflux. S can be calculated from

$$S = 0.03415\frac{h\rho_{\mathrm{gas}}}{Tm_{\mathrm{V}}} \tag{7}$$

where h is the distance between the gaseous horizon and the surface of the hole, and ρ_{gas} is the density of the gas relative to air (in the original text by Cherskiy (1961); in a submarine scenario, it is relative to sea water).

The term p_{depth} can be expressed as a function of the depth (h), as follows:

$$p_{\mathrm{depth}} = h\rho_{\mathrm{mud}}\,g + p_{\mathrm{surface}} \tag{8}$$

where ρ_{mud} is the density of the mud. Solving equation (5) for p_{depth} and combining it with equation (8), we obtain

$$h\rho_{\mathrm{mud}}\,g + p_{\mathrm{surface}} = \sqrt{\left[\frac{p_{\mathrm{surface}}}{\mathrm{e}^{2S}} - 1.377\lambda\frac{Q^{2}m_{\mathrm{V}}T}{d^{5}}(\mathrm{e}^{2S}-1)\right]} \tag{9}$$

Solving this equation using equation (7) allows us to calculate the source depth of the mud (h) as the only unknown variable if we take m_{V} and λ from the literature and use d, T, p_{surface} and Q from our determinations and calculations, as follows:

$$h = \frac{\ln[p_{\mathrm{surface}}^{2}(g\rho_{\mathrm{mud}})^{2}]zT + \ln\left[1.377\lambda\frac{Q^{2}m_{\mathrm{V}}T^{2}}{d^{5}}(g\rho_{\mathrm{mud}})^{2}\right]zT}{(0.1366\rho_{\mathrm{gas}})\left\{\ln\left[1.377\lambda\frac{Q^{2}m_{\mathrm{V}}T^{2}}{d^{5}}(g\rho_{\mathrm{mud}})^{2}\right]0.1366\rho_{\mathrm{gas}}\right\} + \left\{\ln\left[\left(\frac{p_{\mathrm{surface}}}{g\rho_{\mathrm{mud}}}\right)^{2}\right]0.1366\rho_{\mathrm{gas}}\right\}} \tag{10}$$

Quantitative estimates

After having compiled published data of gas dissipation from both submarine and terrestrial mud domes, for simplicity we assume for our calculations that it comprises only methane. An assumption like this is supported as >90 vol. % in any case, and dominantly >98 vol. % of the gas escaping from the domes studied is methane (e.g. Jakubov *et al.* 1971; Yassir 1989; Emeis & Shipboard Scientific Party 1996).

Mud ascent

Obviously (see equations(1)–(3)), there are two variables with a profound effect on mud extrusion rates: the diameter d of the feeder channel, and the viscosity η. We have chosen values of 0.1 and 10 m for d, which reflect more or less the data range we compiled from the literature (see above). None of the intervals of mud extrusive activity lasted longer than 1 Ma (Emeis & Shipboard Scientific Party 1996), and we suspect that they have been significantly shorter (see discussion by Camerlenghi *et al.* (1995) and below). Because of the difficulty of explaining plugging and pressure build-up, however fine the grain size may be, we regard a small feeder channel (or a set of fractures, possibly in the backthrust fault zone) as the most likely geometry. The most reliable value from our viscosity results (i.e. 10^6 Pa s; see above) constrains an enormous range of velocities from 6 m a^{-1} to almost 38 000 km a^{-1} (Table 3a), indicating the profound influence of the width. Although appearing unrealistic as a steady-state assumption over long periods, very rapid upward migration of mud at an initial stage in this order of magnitude is physically reasonable as a result of the density inversion within the prism. The metastable state (i.e. with the overpressured mud being kept at depth without pore fluid being allowed to escape) becomes unstable as soon as way is given to the less dense material of low viscosity, e.g. through a pipe of reasonable width. The effect of varying the controlling factors (namely, viscosity and the diameter of feeder channels as estimated for the Milano and Napoli domes) on mud lift velocity is discussed beneath.

Diameter of feeder channels

We modified Stokes' law to calculate the diameter of the feeder channel in the case of Milano and Napoli mud volcanoes (equation (2)). The volume flux of mud (V) as a minimum and maximum volume (vol_{mud}) estimate per time interval ($t = 1$ Ma; see section on age constraints above) allows us to solve equation (3) to determine the diameter of the conduit, as follows:

$$d = \sqrt{\left(\frac{\pi \mathrm{vol}_{\mathrm{mud}} 9\eta}{tg\Delta\rho r^2}\right)} \tag{11}$$

For the Milano dome, we can calculate the diameter d of the conduit to range between 1.99 and 2.33 m. The diameter of the Napoli feeder channel ranges between 2.73 and 2.97 m. Using Stokes' law, we can show that the conduit of the Milano and Napoli mud domes ranges within small limits, i.e. for the favoured maximum interpretations between 2.33 and 2.97 m in diameter (Table 4a). The average value obtained from the favoured maximum interpretations is 2.65 m. This measure is right within the range used earlier for the ascent velocity calculations (i.e. diameters of 1–10 m) and is in agreement with the compiled geometries of faults and feeder channels (see Discussion). Vice versa, the feeder channel widths obtained were used to estimate mud ascent velocity, as introduced above (for results, see Table 3a). We want to point out here that our estimates of the width of the conduit are based on the assumption of a cylindrical, unplugged geometry. In nature, however, they may be partly blocked (either by clasts, or simply by fragmented host rock, such as a fault breccia) or non-circular (i.e. a slit along a fault plane of a certain extension may be rectangular), and hence may be much wider than our conservative estimate using the pipe flow law.

Table 3. *Results from quantitative estimates*

(a) Ascent velocities of mud from Stokes' law

Diameter of feeder channel (m)	Velocity ($m\,s^{-1}$)	Velocity ($km\,a^{-1}$)
0.1	0.0000218	0.687
0.825	0.0014	44
1	0.00218	68.75

(b) Mud dome areas

(i) Area covered by mud extrusions (weighing)

Publication	Feature	Weight of cut areas (g)	Equivalent area (km^2)
Cita & Camerlenghi (1990)	Milano	0.0044	5.83
	Napoli	0.0088	11.66
Limonov *et al.* (1994)	Milano	0.0038	5
	Napoli	0.0094	12.5
Hieke *et al.* (1996)	Milano	0.003	4
	Napoli	0.0071	9.5

(ii) Area covered by mud extrusions (calculated)

Dome	Calculated mean average area from paperweight method (km^2)	Area calculated from diameter (Camerlenghi *et al.* 1995)
Milano	5.41	7.42
Napoli	12.08	10.18

(iii) Volume mud volcanoes

Dome	Minimum volume (km^3) conservative interpretation	Minimum + maximum volume (km^3) regular interpretation
Milano	3.3892	7.8494
Napoli	9.541	14.5863

(c) Gas efflux rates estimated for Olimpi mud domes

Feature	Mud breccia volume (km^3)	Gas efflux rate ($m^3\,a^{-1}$) and [$m^3\,day^{-1}$]*	Factor for conversion
Dashgil	260	1.5×10^7 [40000]	1733
Milano	5.41 (min.)	1.43×10^5 [39.2]	
	7.42 (max.)	6.21×10^5 [171]	
Napoli	10.18 (min.)	2.7×10^5 [73.9]	
	12.08	10.1×10^6 [277]	

* Average daily flux of 140 m^3/day from Milano and Napoli data.

Duration of mud volcanic activity for Milano and Napoli domes

The volumes of mud extruded at the Milano and Napoli domes were estimated in two steps. The four depth-corrected seismic sections across the Milano and Napoli mud volcanoes (Fig. 7) allowing us to define the area of mud breccia in two cross-cutting planes roughly perpendicular to each other at each dome. Every interpretation involved a minimum (conservative) and maximum estimate of cross-sectional area covered by mud (minimum and maximum areas in Fig. 7). Volume then can be approximated as a series of truncated cones. The volcanoes were then sliced horizontally into truncated cones of 50 m thickness whose upper and lower radii were calculated as the linear average of the four radii at each level. The sum of all truncated cones yielded mud breccia volumes of 3.4 km^3 (minimum) and 7.8 km^3 (maximum) for the Milano dome and 9.5 km^3 (minimum) and 14.6 km^3 (maximum) for the Napoli dome (Table 3b).

The ascent velocities from Poiseuille's and Stokes' laws and the volume of mud breccia from the geophysical data (see above) were then combined to compute the duration of extrusive activity. To equal the volume of the mud dome by a cylinder having the assumed (or estimated) diameter of the feeder channel (see Table 4a), the height H can be calculated as

$$H = \frac{\mathrm{vol}_{\mathrm{mud\ volcano}}}{\pi(\frac{1}{2}d)^2} \tag{12}$$

If H is divided by the ascent velocity, the duration of activity required to explain a dome of the present size can be calculated. For the Milano and Napoli domes, the full range of these estimates is shown in Table 4b. If we use our calculated best estimates for the volume, feeder channel size and ascent velocity, mud volcano activity for the Milano dome lasted between 31 and 60 ka, whereas it would have taken between 12 and 17 ka to build up the Napoli dome (Table 4b). If, on the other hand, wider channels are used the diapiric model (see equation (3) and Marsh (1982)), the ascent velocity increases whereas the duration decreases. A conduit of 10 m width (and its correspondingly increased ascent velocity) already has a profound effect on the results: the duration of activity is only 92 a for the Milano and Napoli domes, i.e. almost three orders of magnitudes smaller (Table 4b). It is, therefore, suggested that mud extrusion may have interrupted the continuous hemipelagic sedimentation on the Mediterranean Ridge for relatively short periods (see Discussion below).

Source or mobilization depth of overpressured mud

The source depth of the mud was calculated using equation (10). To do this, the coefficient of volume compressibility (m_V), the relative density (ρ_{gas}) and coefficient of friction of methane (λ) were compiled from the literature. Depending on temperature and pressure (i.e. depth), methane only occurs as free gas, dissolved in the pore fluid (negligible given the low solubility of methane in pore water, i.e. <2000 ppm for the pT values on the Mediterranean Ridge (Kuo 1996), or is bound as solid gas hydrate (Dickens & Quinby-Hunt 1994). This is particularly important as the Napoli mud volcano lies at the boundary of the hydrate stability field (Fig. 10) e.g. Sloan 1990). The most profound change in thermophysical properties occurs at the transformation from dissolved to gas phase, when friction is reduced to basically zero whereas compressibility increases as a result of volume expansion. On the other hand, the coefficient of compressibility (i.e. the reciprocal of the bulk modulus) at 300 bar and 0°C for methane (0.299 MPa^{-1}) and water (0.364 MPa^{-1}) ranges roughly over the same order of magnitude (e.g. Bartholomé *et al.* 1978; Kaye & Laby 1985), so that we chose 0.3 for our calculations. The coefficient of friction is zero for a gas

Table 4. *Results from feeder channel calculations, duration of mud volcanic activity and depth of mud source*

(a) Feeder channel geometries, calculated on the basis of the modified Stokes' law

Feature	Diameter, min. (m)	Diameter, max. (m)
Milano	1.99	2.33
Napoli	2.73	2.97

(b) Duration of activity of Milano and Napoli domes

Feature	Diameter feeder channel (m)	Ascent velocity ($m\,a^{-1}$)	H_{min} (m)	H_{max} (m)	Duration$_{min}$ (a)	Duration$_{max}$ (a)
Milano	1	60 744	3.465×10^9	3.470×10^9	57 042	57 124
	1.99	59 538	3.465×10^9		58 198	
	2.33	111 895		3.470×10^9		31 011
	10	37 965 185	3.465×10^9	3.470×10^9	91.3	91.4
Napoli	1	60 744	3.480×10^9	3.488×10^9	57 289	57 421
	2.73	210 880	3.480×10^9		16 502	
	2.97	295 401		3.488×10^9		11 808
	10	37 965 185	3.480×10^9	3.488×10^9	91.7	91.9
Average	12.6588	187 227	3.4725×10^9	3.479×10^9	18 547	18 582

(c) Depth estimates for Olimpi mud domes based on Cherskiy (1961)

Feature the gas flux estimate was taken from	Q (m^3/day)	Feeder channel (m)	Density ($kg\,m^{-3}$)	Depth of mud, h (mbsf)
Olimpi*	10^3	2.65	1600	1685
	$\mathbf{10^3}$	**2.65**	**2100**	**1710**
	10^3	2.65	2600	1730
	10^4	2.65	1600	1639
	10^4	2.65	2100	1662
	10^4	2.65	2600	1680
	10^3	1	2100	1660
	10^3	2	2100	1704
	10^3	3	2100	1727
	10^3	10	2100	1808
Dashgil†	40	2.65	2100	1794
	280	2.65	2100	1740
Atalante‡	1800	2.65	2100	1697
	11000	2.65	2100	1660

The favoured set of values as well as the resulting depth are given in bold type.

* Note that the term 'Olimpi' stands for a representative dome from within the Olimpi field; gas flux and feeder channel diameter estimated for Milano and Napoli domes are used.

† Gas flux from Dashgil mud volcano is used (Jakubov *et al.* 1971).

‡ Gas flux from Atalante mud pie is used, see (Le Pichon *et al.* 1990 – upper row) and (Henry *et al.* 1996 – lower row).

and 0.02 for water at temperatures around the freezing point. The latter value was assumed to represent the fluid–gas hydrate state (i.e. a sorbet-type, or slushy consistency rather than gas) in our model. The density of methane relative to sea water was deduced from the relationship between the molecular weights as *c.* 0.889. Obviously,

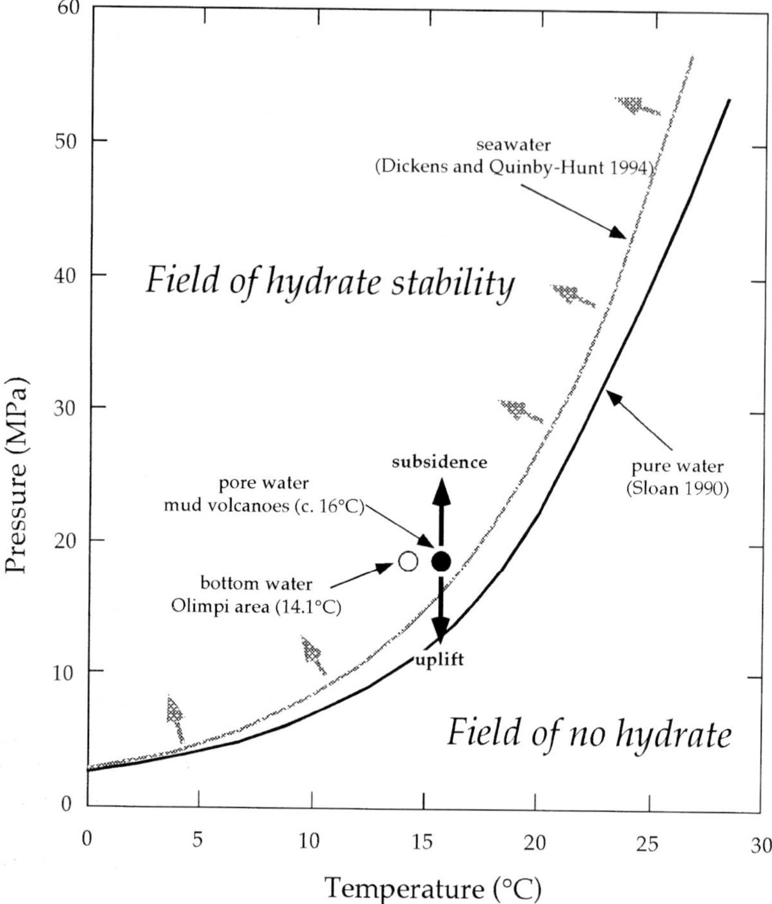

Fig. 10. Phase diagram for natural gas hydrates illustrating the possible position of the Milano and Napoli mud domes during their history. Phase boundaries are taken from the literature as outlined; the grey boundary is extrapolated for sea water. The shift of the boundary for enhanced salinities, comparable with those found at the crest of the Napoli dome, is indicated by grey arrows. Black arrows mark the shift of data points caused by vertical movements of the Mediterranean Ridge or mud dome.

this does not account for either a gradual increase of density with depth nor phase changes. It should be noted that the pT-value for an Olimpi mud dome is not far from the gas hydrate stability field (Fig. 10), and note that only little methane is dissolved (Kuo 1996) relative to the presumed large amounts of both free, gaseous methane as well as a potential amount of methane bound as gas hydrate (De Lange 1997; De Lange & Brumsack 1998)).

The temperature of *c.* 14°C (Fig. 10) or 287 K has been measured during ODP Leg 160 (Emeis & Shipboard Scientific Party 1996); $p_{surface}$ (i.e. the hydrostatic pressure on the sea floor) can be easily calculated for an approximate depth of 2000 m below sea level, which is almost exactly 20 MPa for ocean water (see Fig. 9). As far as the density of the overlying strata goes (termed ρ_{mud} in the above equations), we can assume mud of 1600 kg m^{-3} (i.e. average bulk density of mousse-like mud), 2100 kg m^3 (i.e. average bulk density for clast-bearing mud breccia) or 2600 kg m^3 (i.e. average density of the clasts which are dominantly mudstones from the prism), dependent on whether accreted rock or an already mud filled fracture or conduit overlies the mud reservoir. Both densities were used (see results in Table 4c) The average diameter of the conduit d (2.65 m) was calculated for the Milano and Napoli domes as demonstrated above and, therefore, is burdened with some uncertainties. For the daily gas flux Q, the estimate of 10^3–10^4 m^3 day^{-1} was used.

Using equation (10), the source depth was calculated for the Milano and Napoli mud domes on the basis of a varying set of controlling factors (see Table 4c). Methane is treated as free gas, but with a coefficient of friction of 0.02 (assuming some interaction between the methane and mud). For the most likely average values of 2100 kg m^3 mud breccia density, average conduit width of 2.65 m and a gas flux of 10^3 $m^3 day^{-1}$, the estimated source depth is 1710 mbsf (Table 4c). The effect of variation in density (i.e. of the mousse-like mud or the clasts), gas flux or diameter of the feeder channel is relatively small. For the range of values tested, the depth of the mud reservoir was found to be between 1640 and *c*. 1800 mbsf (for details, see Table 4c). As can be seen from equation (10), neither the very limited variation in temperature (of only a few degrees) or hydrostatic pressure (depending on the bathymetry in the Olimpi field) nor the variation in values as above significantly change the results.

Discussion

For clarity, the discussion is divided into three main parts. In an introductory part, we shed light on the positive and negative aspects of quasi-static versus dynamic approaches to tackle mud extrusion dynamics. This is supported by additional model calculations and will, it is hoped, allow the reader to assess the validity of what has been done in this study. In the second part, our results concerning origin and mobilization depth of the mud are put into the regional tectonic context. In the third part, the results of ascent and its episodicity are discussed within the context of other geological scenarios.

Quasi-static versus dynamic modelling of the source depth of the mud

From a first glance at Cherskiy's parameter-rich, and moreover empirical equations, it may appear to the reader that a simpler relation may capture the basic mechanism of mud extrusion. However, an excursion through thermodynamic modelling suggests that no easy solution to estimate the actual depth of the mud can be provided.

Concerning the flux estimates, we used a dynamic approach (i.e. Poiseuille flow) but tried to keep it as simple as possible. This was merely because the more variables to be considered, the more likely it is that the uncertainties may heavily outweigh the known controlling parameters. To illustrate this, we want to conduct a simple analysis for an isotropic hole filled with lower-density mud relative to the surrounding rock. Much depends on the assumed stiffness (Poisson's ratio) of the host sediment, which enables it to maintain the hole open in that the lateral stress is only a fraction of the vertical stress. Typically, this stress ratio in deep-sea sediments is 0.6–0.7 (trending to higher values in tectonic regimes). Hence, the mud in the hole (or conduit) needs only 0.7 of the density of the host rock for its hydrostatic stress to balance the formation pressure at any depth. For the above example of 2100 kg m^3 for the surrounding rock and 1600 kg m^3 for the mud, the mud denser than 1600 kg m^3 will balance the lateral stress if the Poisson's ratio is 0.7. Even for lower mud densities, only a small excess pressure is required to keep the conduit open. This can result from fluid overpressure at depth, which is transmitted up the column (if the latter is sealed at the top), or will tail off and drive flow if the conduit is open. Adding a constant-with-depth lateral tectonic stress would mean that greater mud densities can be stable without widening the conduit, or lower mud densities require a larger pore fluid pressure to maintain an open hole. Ignoring these complexities (and we have to ignore them, because *in situ* the lateral stress does not seem to affect, for example, the width of fault zones; see the section on fault zone geometry above), we can use Poiseuille's flow (equation (1)) to estimate discharge rates. Similar to what has been determined above (see the section on mud ascent), the flow rate for a mud with a density of $1600\,kg\,m^3$, a viscosity of 10^6 Pa s and a driving force of $1\,MPa\,km^{-1}$ is *c.* $300\,m^3\,day^{-1}$ for a hole of 10 m width, and *c.* $18\,m^3\,day^{-1}$ for a slit of $5\,m \times 20\,m$. This shows that high-viscosity mud does not allow rapid ascent, so that lower viscosities (if water segregates and entrains the mud or mousse-like textures develop) are more likely.

So far, this agrees well with our approach, illustrating neatly that the estimates for both the ascent velocities as well as the conduit diameter may be close to what is physically expected (bearing in mind all the uncertainties, such as clast abundance and size). However, from here we can follow various paths introducing other variables. If the conduit has an elastic wall, then the stress will increase as a linear function of width on either side of an equilibrium width, providing a positive or negative contribution. If the entire wall of the conduit is allowed to be elastic, it may close at the top and remain open and tapering lower down. Episodic eruptions will then be periods when the walls open all the way to the top, and profiles of such a behaviour could be derived for different mud densities, conduit widths or other factors. Given that such assumptions will lead us onto fairly unsafe grounds where the controlling variables (i.e. mud density, conduit width, but also the stress regime as a result of regional tectonics and physical properties) have been and have to be estimated with a considerable amount of uncertainty themselves, we will not include such assumptions. Otherwise, quantitative estimates without well-justified error bars may simply be numbers without great significance. Also, we do not wish to discuss other obvious related phenomena, such as: how the system may change with time, for example, when the mud reaches its liquid limit; whether segregation processes are relevant with time, for example, when previously dewatered, squashed mud aligns along the conduit walls; gas-assisted lift of the mud; and many more.

Summarizing these paragraphs, we draw the reader's attention to our initial aim of quantitatively estimating mud extrusion dynamics with the simplest possible approach. Regarding the mud ascent, the size of the conduits and the episodic activity, we consequently based all calculations on Poiseuille's and Stokes' laws and on recent drilling results (see the various sections above).

Concerning the mud mobilization depth, we had to face an even greater difficulty: there are no dynamic models which on one hand allow us to tackle the subject in a simple way, and on the other hand require very few variables. The latter aspect in particular is crucial in a field such as mud volcanism where very little is known for sure. In short, we meant to avoid a situation where the majority of the factors rely on an estimate rather than knowledge (e.g. the stress field of the system for a more sophisticated mud discharge estimate; see above). Despite the apparent overparametrization of Cherskiy's approach, many of the variables can be easily obtained from the literature, and have an only marginal effects on the resulting depth estimate. In fact, all the crucial factors, which strongly control the results, are more or less well constrained by data (see the section on source or mobilization depth of overpressured mud, above). Moreover, Gorkun & Siryk (1968) have successfully applied the Cherskiy approach to mud volcanoes on Sakhalin, so that we felt confident to test a modified set of equations for the Mediterranean Ridge accretionary complex. As the impact of the various parameters has been discussed above (see also Table 4), we now discuss the geological relevance of the resulting depth estimate.

Source and source depth of mud and their implications

From previous studies in the Eastern Mediterranean Sea (Chaumillon *et al.* 1996) it is evident that the Olimpi mud volcano field (Fig. 1) can be viewed as being representative for an area where the Mediterranean Ridge suffers maximum deformation. Its volcanoes at the inner deformation front seem to be different from those in a belt near the Cyrenaica Peninsula closer to the outer deformation front (i.e. plate boundary), both in terms of size and composition. The Olimpi domes appear to be significantly larger and contain more clasts (according to back-scatter and deep tow sonar images). In both locations, however, the driving force of mud volcanism is believed to relate to regional compression between the converging plates (Camerlenghi *et al.* 1995). This results in dewatering due to tectonic stresses near the front and allows uprise of trapped mud at the inner ridge (see Fig. 1). The age and source depth of the mud are still uncertain, but there are hints in favour of a Late Miocene origin. Both halite clasts recovered from the Napoli crestal holes (hole 971D) as well as pore-water salinities (Emeis & Shipboard Scientific Party 1996) relate the source of the mud to Messinian deposits. Where Messinian oozes have been drilled elsewhere in the Mediterranean (e.g. ODP Leg 160, hole 968 (Emeis & Shipboard Scientific Party 1996), and ODP Leg 10^7, holes 653B and 654A (Kastens *et al.* 1987)), these resemble the matrix of the mud breccias in terms of their dark grey colour and composition, and contrast strongly with the Plio-Quaternary sediments above and Tortonian sediments beneath. Moreover, Messinian mud recovered at hole 653B during ODP Leg 10^7 has noticeably low shear strengths, which coincides with results from shear box tests on the Napoli mud breccia samples (Kopf *et al.* 1998). The brackish fauna found in the matrix which may have been deposited in a palaeoenvironment similar to those deduced for the period during the Messinian drawdown in the Mediterranean (Rouchy & Saint Martin 1992), are consistent with this model. Vitrinite reflectance data on the matrix are compatible with a model where the mud is assumed to have been subducted during Messinian time and then migrated along the décollement for some 4 Ma (on the basis of a convergence rate of 25 mm/a between Africa and Eurasia; Kastens 1991) before it migrated upward (Schulz *et al.* 1997).

The depth from which the mud originated has been the subject of various studies and approaches. Evidence from thermal maturity data on the mud breccia matrix samples from the Napoli dome suggest a mobilization depth between 4900 and 7500 mbsf (Schulz *et al.* 1997), which coincides with the décollement depth of 5300–7000 mbsf predicted by Camerlenghi *et al.* (1995) in a geometric approach. The results of Schulz *et al.* (1997) have to be re-evaluated after having measured thermal maturity of both clasts and matrix from Milano dome mud breccias. Here, evidence is provided for the immaturity of the mud, leading to an estimated depth of *c*. 2000 mbsf (Schulz & Kopf, unpublished data). It should be noted that the same thermal gradients and geometric constraints were applied as by Schulz *et al.* (1997), according to the Lopatin method (Waples 1983). Camerenghi *et al.* (1995), however, suggested two other depth intervals, which were preliminarily related to mud extrusion: the minimum source depth of 400 ± 50 mbsf for the mud underneath what is believed to represent the Messinian evaporites, and a minimum depth of 2200 mbsf for the gas triggering mud uprise underneath. Our calculations indicate a source depth of *c*. 1650–1800 mbsf (Table 4c), lying between the two estimates by Camerlenghi *et al.* (1995). A moderate source depth of the mud, in our opinion, is not in contradiction to a former transport at the décollement level, as the mud initially may have followed the plate boundary, has then risen (whether above a backthrust fault or other zone of weakness), and finally has been extruded at the inner ridge, possibly after temporary incorporation into the wedge (see Fig. 4). However, a very long transport of the material (both mud and clasts) is unlikely, given that the majority of clasts are subangular and that the sorting is poor (Flecker & Kopf 1996). On the other hand, both maturation of the organic matter and rounding of the matrix components may have been minimized by near-lithostatic pore fluid pressures at the décollement level.

As demonstrated in the case of the Barbados (Martin *et al.* 1996) and postulated for the Nankai subduction systems, gas, predominantly methane, occurs in the décollement zone. This implies that high-pressure fluids derived deep within the subduction zone control the variation in mud volcano activity by variable mobilization and pulsed upward migration (Yassir & Bell 1994). The clast-rich mud breccias at the base of the domes drilled suggest an explosive initial stage, with mud debris flows building the bulk material of the cone. Later, only episodic extrusion of small amounts of gas-bearing, clast-free mud takes place (Robertson & Shipboard Scientific Party of ODP Leg 160 1996). Alternatively, evaporite bodies within the accretionary prism may have a similar effect, trapping mud temporarily before access is given to extrusion (e.g. as a result of dissolution or tectonic processes; see Kopf *et al.* (1998)). Arguments for an episodic, explosive extrusion of overpressured mud are fractured mudstone clasts (Robertson & Kopf 1998), poor sorting of the mud breccia (Flecker & Kopf 1996), occurrence of the coarsest mud breccias at the base of the successions recovered by drilling, and very low permeabilities from tests on undisturbed cores (Kopf *et al.* 1998). Involvement of Messinian salt in release of the mud is suggested from a strong seismic reflector on migrated time sections (Emeis 1996), located *c*. 300–400 m beneath the sea floor at the Milano dome, evaporite bodies at comparable depths elsewhere in the Eastern Mediterranean (Robertson & Shipboard Scientific Party of ODP Leg 160 1996), halite clasts within the mud breccia, and previous estimates by Camerlenghi *et al.* (1995; see above). Salt deposits behave more plastically than other accreted strata during the continuing collision of the wedge with the rigid backstop of Cretan continental crust. As a consequence, initially homogeneous slices of salt may have been tectonically severed, thus losing their ability to retain overpressured mud. The high fluid content and low specific gravity of the muds cause them to rise. This implies that a dominantly tectonic driving force of regional compression on the Mediterranean Ridge can be invoked for mud volcanism (Camerlenghi *et al.* 1992; Robertson & Kopf 1998). Pore-water dissolution may also have contributed to the destruction of salt seals. However, although considerable quantities of Messinian evaporites were incorporated into the accretionary prism (e.g. Chaumillon *et al.* 1996), no direct evidence has been provided for their presence some 100 km northward of the deformation front in the Olimpi area.

Episodicity and velocity of mud ascent

A period of extrusive activity of 10^2–10^4 a on the basis on compressibility data of mud breccia has been put forward by Camerlenghi *et al.* (1995). The upper limit of this estimate corresponds to the order of magnitude of our estimates for the duration of mud extrusive activity ((1.2–5.8) $\times 10^4$ a; see Table 4b). This indirectly supports our feeder channel estimates as reasonable, although the ascent velocities found according to Poisseuille flow appear to be very high (Table 3c). On the other hand, the ascent velocities estimated for the mud are similar to those known from magma rise. As shown in various studies, rhyolitic liquids have viscosities between 10^4 and 10^8 Pa s and rise through dykes of 1–5 m width (e.g. McBirney 1984; Bacon 1992). On the other hand, a comparison of mud and magma has a limited validity because of the differences in scale. For magmas coming from mantle depth, both viscosities and ascent velocities are initially very low, but can reach values of several tens (and rarely hundreds) of kilometres per hour as a result of pressure release and gas expansion (Baloga *et al.* 1995). As mentioned above, gas expansion as a consequence of evaporation of methane, which was dissolved in the pore water at depth, as well as from gas hydrate dissociation can play an important role in mechanisms driving mud effusion at shallow depth (Fig. 9). Therefore, we are confident that for short periods of time in the upper part of the accretionary prism, high ascent rates may well be reached on the Mediterranean Ridge. It should be noted, however, that this evidence is indirect (i.e. from pore-water study; see De Lange & Brumsack (1998)), and that no solid gas hydrate has even been found on the Mediterranean Ridge.

Discontinuous gas and fluid supply may be a combination of a steady biogenic and/or thermogenic methane delivery and dissociation of solid gas hydrate or exsolution of methane from the pore fluid. Carbon and oxygen stable isotope signatures on carbonates from mud breccias (both clasts and matrix) indicate neither thermogenic nor biogenic methane having been precipitated in considerable amounts (Kopf, unpublished data). From geochemical data, however, the volume of methane within and underneath the Milano dome has been estimated to be at least 5×10^9 m^3 (De Lange 1997). Our gas flux estimates of a mud volcano of the size of the Milano or Napoli dome range somewhere around 10^3 m^3 methane dissipating per day, and are, although somewhat crude, a conservative estimate (see above). This becomes obvious if flux is regarded as independent of the size of the dome. Some volcanoes on Sakhalin (e.g. Pugachevskiy mud volcano; see Vereshchagin and Kovtunovich 1970) and in Azerbaijan (Dashgil mud volcano; Jakubov *et al.* 1971) on land, expel more than 30 000 m^3 day^{-1} (with an average of 3425 m^3 day^{-1} for the 200 mud domes studied in Azerbaijan; Jakubov *et al.* (1971)). Moreover, the favoured gas flux estimate for the Mediterranean Ridge (10^3 m^3 day^{-1}) is more than one order of magnitude lower than the most recent measure for the submarine Atalante mud volcano off Barbados (Henry *et al.* 1996). The possible impact of gas on the viscosity, which should facilitate mud extrusion (Macosko 1993), as well as on the pressure (Scarfe 1973) could not be taken into account in the calculations, as there is no reliable estimate of how much gas is involved. However, the solubility of methane in pore water is only some 2000 ppm (at 25°C and 20 MPa; e.g. Kuo 1996), so that gas exsolution from the pore fluid has presumably no profound effect. Nevertheless, we want to acknowledge that free gas (as undissolved admixture within the pore water or underneath the mud reservoir) can lead to a considerable increase in the flow rate of both mud and fluid. One likely source for methane is the destruction of gas hydrates, as there are indications for their presence (Emeis & Shipboard Scientific Party 1996). Unfortunately, the possible effect on the density at the phase boundary ($920\,kg\,m^{-3}$ for stoichiometric gas hydrate (i.e. 5.75 H_2O, 1 CH_4) and only $0.715\,kg\,m^{-3}$ for gas; both at 1 atm, 0°C) could not be acknowledged in our simplified estimates.

From our calculations, mud volcano feeder channels were found to be narrower than previously assumed (Ivanov *et al.* 1996). The two mid-sized mud volcanoes of the Olimpi field range between only 2 and 3 m in diameter (assuming they are empty cylinders). Our results also imply that mud flows through a set of fractures or a small conduit rather than a crater of several metres (Higgins & Saunders 1974; Yassir 1989). Given the episodicity of mud extrusion (Figs 2 and 3), a narrow conduit has been proposed as the most likely geometry (Kopf *et al.* 1998). Low permeability and fine grain size of the mud breccia facilitate temporary plugging of the feeder structure to allow pressure build-up for further mud volcanic activity. The estimates are supported by seismic cross-sections across the central areas of several mud volcanoes in the central part of the Mediterranean Ridge. Here, continuous reflectors across the feature (Fig. 10) suggest a limited width of the conduit (possibly below the resolution of the seismic survey), given it is in the centre of these domes. In addition, a narrow feeder channel is suggested from analogue modelling where during tests with mud of different viscosity variable feeder channel geometry led to different shapes of the mud volcano (Lance *et al.* 1998). Wide conduits led to pie-like features (like Atalante, where the channel is estimated to be almost 200 m wide; Henry *et al.* 1996), whereas narrow conduits allow build-up a cone. The mud volcanoes studied are distinctly cone shaped and, therefore, we suggest that they are derived from a narrow conduit.

Volume flux (or mud discharge), as derived from feeder channel geometry, is found to be very high. For the period of extrusive activity, durations sufficient to build up the Milano or Napoli mud volcanoes range between 12 and 58 ka (see Table 4b for details). The feeder channel diameter estimated for the Milano and Napoli domes (average 2.65 m) results in durations of activity of *c.* 18.5 ka. These values do not coincide with those based on biostratigraphic dating of the hemipelagic successions interbedded within the mud debris flows (almost 1 Ma for either the Milano or the Napoli dome; see Emeis & Shipboard Scientific Party 1996), and show this to be a rather blunt tool to set time constraints to the extrusive activity. On the other hand, if there are fragments attached to the conduit wall that may partly block the feature, or if there are clasts sinking against the rising mud, the estimated discharge rates may decrease considerably. However, our calculations suggest phases of apparent inactivity being interrupted by highly episodic mud extrusion.

Summary and conclusions

(1) The Milano and Napoli domes are interpreted as mud volcanoes, and their formation is thought to have begun *c.* 1.5 Ma ago. Early cone build-up by clast-dominated gravel deposits suggests an explosive, diatreme-like initial eruption. Later episodic activity of the volcanoes can be explained by the hydraulic regime of a deeper fluid

source, possibly related to levels as deep as the décollement zone. We assume that gas supply (of predominantly methane) facilitates pressure build-up and mud extrusion. Our gas efflux estimates of *c.* 10^3 m a^{-1} are conservative and may underestimate the actual dissipation.

(2) The feeder channel diameters of the mud volcanoes are calculated to be probably between 2 and 3 m, assuming that they are pipe shaped and unplugged. *In situ*, however, they may be partly blocked or non-circular, and hence may be much wider. A conduit of very limited width is in agreement with seismic cross-sections showing almost continuous reflectors within the domes, and with observations on land. We conclude that during extrusive activity, a narrow conduit or a set of open fractures allows outflow of viscous mud, whereas temporary plugging by the low-permeability mud breccias, or valving of the flexible conduit wall as a result of lateral tectonic stresses, accounts for periods of inactivity. A narrow channel is also supported by the cone-shaped geometry of the volcanoes. From the feeder channel geometry, the ascent velocity (i.e. Poiseuille flow) of overpressured mud (at a constant viscosity of 10^6 Pa s) is estimated to be around 60–300 km a^{-1}, which is considerably higher than estimated for other mud volcano areas.

(3) On the basis of the ascent velocity and feeder channel geometry, durations of activity sufficient to build up the Milano and Napoli mud domes are accepted to be *c.* 12–58 ka, which is a considerably shorter period than that indicated from biostratigraphic dating of mud breccia interbeds. Small variation in feeder channel width, however, profoundly affect the duration of activity, so that these estimates of episodicity have to be treated with caution. Nevertheless, we propose mud volcanism to be a highly episodic phenomenon.

(4) For the mud volcanoes studied, we estimate a source depth of 1700 ± 50 mbsf, i.e. within the accretionary wedge of >5–7 km thickness at the apex of the Mediterranean Ridge. Mobilization from décollement depth, as previously proposed, cannot be supported with our estimates.

The authors thank C. Friedrich and W. Schemionek for assisting with the viscosity measurements in the laboratory of the Material Research Centre at Freiburg University. P. Henry and D. Tanner are thanked for discussion. Numerous suggestions by referees B. Clennell, Y. Mart and J. L. Vigneresse helped to improve the manuscript. Funding for this research was received through DFG Grant Be-1041/10, and BASF AG, Germany.

References

BACON, C. R. 1992. Partially melted granodiorite and related rocks ejected from Crater Lake caldera, Oregon. *Transactions of the Royal Society, Edinburgh, Earth Sciences*, **83**, 27–42.

BALOGA, S., SPUDIS, P. D. & GUEST, J. E. 1995. The dynamics of rapidly emplaced terrestrial lava flows and implications for planetary volcanism. *Journal of Geophysical Research*, **100**(B12), 24509–24519.

BARBER, A. J. & BROWN, K. M. 1988. Mud diapirism: the origin of mélanges in accretionary complexes? *Geology Today*, **4**, 89–94.

BARTHOLOMÉ, E., BICKERT, E., HELLMANN, H., LEY, H., WEIGERT, W. M. & WEISE, E. 1978. *Ullmanns Encyklopädie der technischen Chemie*, **16**, Weinheim: Verlag Chemie.

BEHRMANN, J. H. 1991. Conditions for hydrofracture and the fluid permeability of accretionary wedges. *Earth and Planetary Science Letters*, **107**, 550–558.

——, LEWIS, S. D., MUSGRAVE, R. J. & Shipboard Scientific Party Leg 141, 1992. *Proceedings of the Ocean Drilling Program, Initial Reports, 141*. Ocean Drilling Program, College Station, TX.

Brown, K. M. 1990. The nature and hydrogeologic significance of mud diapirs and diatremes for accretionary systems. *Journal of Geophysical Research*, **95**(B6), 8969–8982.

—— & Westbrook, G. K. 1988. Mud diapirism and subcretion in the Barbados Ridge accretionary complex: the role of fluids in accretionary processes. *Tectonics*, **7**, 613–640.

Camerlenghi, A., Cita, M. B., Della Vedova, B., Fusi, N., Mirabile, L. & Pellis, G. 1995. Geophysical evidence of mud diapirism on the Mediterreanean Ridge accretionary complex. *Marine Geophysical Research*, **17**, 115–141.

——, ——, Hieke, W. & Ricchiuto, T. S. 1992. Geological evidence of mud diapirism on the Mediterranean Ridge accretionary complex. *Earth and Planetary Science Letters*, **109**, 493–504.

Chaumillon, E. & Mascle, J. 1995. Variation laterale des fronts de deformation de la Ride mediterranéenne (Mediterranée orientale). *Bulletin de la Société Géologique de France*, **166**, 463–478.

—, — & Hofmann, H. J. 1996. Deformation of the western Mediterranean Ridge: importance of Messinian evaporitic formations. *Tectonophysics*, **263**, 162–190.

Cherskiy, N. V. 1961. *Konstruktsii gazovykh skvazhin*. Gostoptekhizdat.

Cita, M. B., Ryan, W. F. B. & Paggi, L. 1981. Prometheus mud-breccia: an example of shale diapirism in the Western Mediterranean Ridge. *Annales Géologique des Pays Helléniques*, **3**, 543–570.

Corry, C. E. 1988. *Laccoliths: Mechanisms of Emplacement and Growth*. Geological Society of America, Special Publication, **220**.

De Lange, G. J. 1997. The occurrence of gas hydrates in sediments of Eastern Mediterranean mud domes as inferred from ODP Leg 160 porewater results. *5th TTR Post-Cruise Conference, Amsterdam, Meeting Abstracts*, 11.

—— & Brumsack, H. J. 1998. Pore water indications for the occurrence of gas hydrates in Eastern Mediterranean mud dome structures. *In*: Robertson, A. H. F., Emeis, K.-C., Richter, C. & Scientific Party (eds) *Proceedings of the Ocean Drilling Program, Scientific Results 160*. Ocean Drilling Program, College Station, TX. 569–574.

Dickens, G. R. & Quinby-Hunt, M. S. 1994. Methane hydrate stability in sea water. *Geophysical Research Letters*, **21**, 2115–2118.

Emeis, K.-C. & Shipboard Scientific Party. 1996. *Proceedings of the Ocean Drilling Program, Initial Reports, 160*. Ocean Drilling Program, College Station, TX.

Flecker, R. & Kopf, A. 1996. Clast and grain size analysis of sediment recovered from the Napoli and Milano mud volcanoes, ODP Leg 160 (Eastern Mediterranean). *In*: Emeis, K.-C. & Shipboard Scientific Party, 1996. *Proceedings of the Ocean Drilling Program, Initial Reports, 160*. Ocean Drilling Program, College Station, TX. 529–532.

Fusi, N. & Kenyon, N. H. 1996. Distribution of mud diapirism and other geological structures from long-range sidescan sonar (GLORIA) data in the Eastern Mediterranean Sea. *Marine Geology*, **132**, 21–38.

Gorkun, V. N. & Siryk, I. M. 1968. Calculating depth of deposition and volume of gas expelled during eruptions of mud volcanoes in southern Sakhalin. *International Geology Review*, **10**, 4–12.

Henry, P., Le Pichon, X., Lallemant, S. *et al.* 1996. Fluid flow in and around a mud volcano field seaward of the Barbados accretionary wedge: results from Manon cruise. *Journal of Geophysical Research*, **101**(B9), 20297–20323.

Hieke, W., Werner, F. & Schenke, H.-W. 1996. Geomorphological study of an area with mud diapirs south of Crete (Mediterranean Ridge). *Marine Geology*, **132**, 63–93.

Higgins, G. E. & Saunders, J. B. 1974. Mud volcanoes—their nature and origin. *Verhandlungen der Naturforschenden Gesellschaft Basel*, **84**, 101–152.

Ivanov, M. K., Limonov, A. F. & van Weering, T. C. E. 1996. Comparative characteristics of the Black Sea and Mediterranean Ridge mud volcanoes. *Marine Geology*, **132**, 253–271.

Jakubov, A. A., Ali-Zade, A. A. & Zeinalov M. M. 1971. *Mud Volcanoes of the Azerbaijan SSR*. Baku: Academy of Sciences of the Azerbaijan SSR.

Kastens, K. A. 1991. Rate of outward growth of the Mediterranean Ridge accretionary complex. *Tectonophysics*, **199**, 25-50.

——, Mascle, J. & Shipboard Scientific Party 1987. *Proceedings of the Ocean Drilling Program, Initial Reports, 107*. Ocean Drilling Program, College Station, TX.

KAYE, G.W.C. & LABY, T .H. 1985. *Tables of Physical and Chemical Constants*. Harlow: Longman.

KIMURA G., SILVER E. E., BLUM P, and Shipboard Scientific Party Leg 170 (1997) *Proceedings of the Ocean Drilling Program, Initial Reports, 170*. Ocean Drilling Program, College Station, TX.

KOPF, A., ROBERTSON, A. H. F., CLENNELL, M. B. & FLECKER, R. 1998. Mechanism of mud extrusion on the Mediterranean Ridge. *GeoMarine Letters*, **18**(3), 97–114.

——, ——, SCREATON, E. J. *et al.* 1999. *Backstop hydrogeology of a wide accretionary complex south of Crete, Eastern Mediterranean Sea*. ODP full proposal 555.

KUO, L.-C. 1996. Gas exsolution during fluid migration and its relation to overpressure and petroleum accumulation. *Marine and Petroleum Geology*, **14**(3), 221–229.

LANCE, S., HENRY, P., LE PICHON, X. *et al.* 1998. Submersible study of mud volcanoes seaward of the Barbados accretionary wedge: sedimentology, structure and rheology. *Marine Geology*, **145**, 255–292.

LE PICHON, X., FOUCHER, J. P., BOULEGUE, J. *et al.* 1990. A mud volcano field seaward of the Barbados Accretionary Complex: a submersible survey. *Journal of Geophysical Research*, **95**, 8931–8943.

LIMONOV, A. F., KENYON, N. H., IVANOV, M. K. & WOODSIDE, J. M. (eds) 1995. *Deep-sea depositional systems of the Western Mediterranean and mud volcanism on the Mediterranean Ridge*. UNESCO Report in Marine Sciences, **67**.

——, WOODSIDE J. M. & IVANOV, M. K. (eds) 1994. *Mud volcanism in the Mediterranean and Black Seas and shallow structure of the Eratosthenes Seamount. Initial results of the geological and geophysical investigations during the Third 'Training-through-Research' Cruise of the R/N* Gelendzhik (*June–July, 1993*). UNESCO Report in Marine Sciences, **64**.

LONERGAN, L., CARTWRIGHT, J. & JOLLY, R. 1998. The geometry of polygonal fault systems in Tertiary mudrocks of the North Sea. *Journal of Structural Geology*, **20**, 529–548.

MACOSKO, C. W. 1993. *Rheology—Principles, Measurements, and Applications*. New York: VCH.

MARSH, B. D. 1982. On the mechanics of igneous diapirism. *American Journal of Science*, **282**, 808–855.

MARTIN, B. M., KASTNER, M., HENRY, P., LE PICHON, X. & LALLEMENT, S. 1996. Chemical and isotopic evidence for sources of fluids in a mud volcano field seaward of the Barbados accretionary wedge. *Journal Geophysical Research*, **101**(B9), 20325–20345.

MASCLE, A., MOORE, J. C. & Shipboard Scientific Party Leg 110 1988. *Proceedings of the Ocean Drilling Program, Initial Reports, 110*. Ocean Drilling Program, College Station, TX.

MCBIRNEY, A. R. 1984. *Igneous Petrology*. Freeman, Cooper, San Francisco, CA.

PETFORD, N., LISTER, J. R. & ROSS, R. C. 1994. The ascent of felsic magmas in dykes. *Lithos*, **32**, 161–168.

REDDY, S. M., SEARLE, S. M. & MASSEY J. A. 1993. Structural evolution of the High Himalayan Gneiss sequence, Langtang Valley, Nepal. *In*: TRELOAR, P. J. & SEARLE, M. P. (eds) *Himalayan Tectonics*. Geological Society, London, Special Publications, **74**, 375–389

ROBERTSON, A. H. F. & KOPF, A. 1998. Origin of clasts and matrix within Milano and Napoli mud volcanoes, Mediterranean Ridge accretionary complex. *In*: ROBERTSON, A. H. F. ET AL. (eds) *Proceedings of the Ocean Drilling Program, Scientific Results, 160*. Ocean Drilling Program, College Station, TX. 575–596.

—— & Shipboard Scientific Party of ODP Leg 160 1996. Mud volcanism on the Mediterranean Ridge: Initial results of Ocean Drilling Program, Leg 160. *Geology*, **24**(3), 239–242.

ROUCHY, J.-M. & SAINT MARTIN, J.-P. 1992. Late Miocene events in the Mediterranean as recorded by carbonate–evaporite relations. *Geology*, **20**, 629–632.

SCARFE, C. M. 1973. Viscosity of basic magmas at varying pressures. *Nature*, **241**, 101–102.

SCHULZ, H. M., EMEIS, K.-C. & VOLKMANN, N. 1997. Organic carbon provenance and maturity in the mud breccia from the Napoli mud volcano: indicators of origin and burial depth. *Earth and Planetary Science Letters*, **147**, 141–151.

SHIPLEY, T. H., OGAWA, Y., BLUM, P. & Shipboard Scientific Party Leg 156 1995. *Proceedings of the Ocean Drilling Program, Initial Reports, 156*. Ocean Drilling Program, College Station, TX.

——, STOFFA, P.L. & DEAN, D.F. 1990. Underthrust sediments, fluid migration paths, and mud volcanoes associated with the accretionary wedge off Costa Rica: Middle America Trench. *Journal of Geophysical Research*, **95**(B6), 8743–8752.

SLOAN, E. D. 1990. *Clathrate Hydrates of Natural Gases*. Marcel Decker, New York.

STAFFINI, F., SPEZZAFERRI, S. & AGHIB, F. 1993. Mud diapirs of the Mediterranean Ridge: sedimentological and micropalaeontological study of the mud breccia. *Rivista Italiana di Paleontologia e Stratigrafia*, **99**, 225–254.

TAIRA, A., HILL, I., FIRTH, J. V. & Shipboard Scientific Party Leg 131 1991. *Proceedings of the Ocean Drilling Program, Special Reports, 131*, Ocean Drilling Program, College Station, TX.

TAYLOR, B., HUCHON, P., KLAUS, A. & SHIPBOARD SCIENTIFIC PARTY 1999. *Proceedings of the Ocean Drilling Program, Initial Reports*, **180**. UDP, College Station, TX.

TURCOTTE, D. C. & SCHUBERT, G. 1982. *Geodynamics*. J. Wiley & Sons, New York.

VERESHCHAGIN, V. N. & KOVTUNOVICH, Y. M. (eds) 1970. *Geologiya SSSR: Ostrov Sakhalin*. Moscow. Nedra, **18**.

WAPLES, D. W. 1983. Physical–chemical models for oil generation. AAPG Treatise of Petroleum Geology, Reprint Series, **8**, 27–42.

WESTBROOK, G. K., CARSON, B., MUSGRAVE, R. J. & Shipboard Scientific Party Leg 146 1994. *Proceedings of the Ocean Drilling Program, Initial Reports, 146 (Part 1)*. Ocean Drilling Program, College Station, TX.

YASSIR, N. A. 1989. *Mud volcanoes and the behaviour of overpressured clays and silts*. PhD thesis, University College London.

—— & BELL, J. S. 1994. Relationships between pore pressures, stresses and present-day geodynamics in the Scotian Shelf, Offshore Eastern Canada. *AAPG Bulletin*, **78**, 1863–1880.

Index

Page numbers in *italics* refer to figures and page numbers in **bold** refer to tables